Science
Horizons
Year Book

1993

P. F. COLLIER, INC.

NEW YORK   TORONTO   SYDNEY

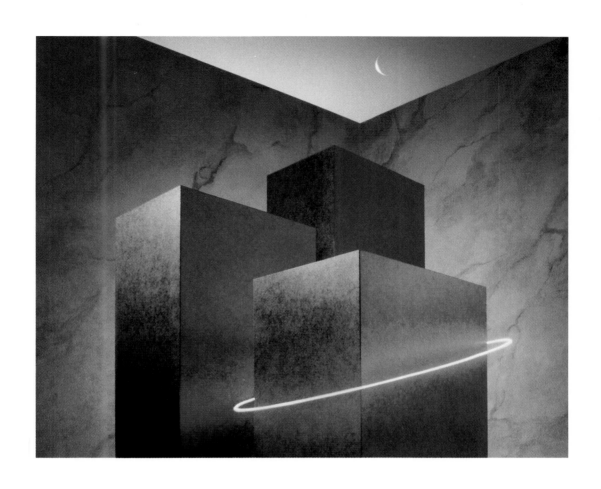

Science
Horizons
Year Book
1993

Published 1993 by P.F. Collier, Inc.

This book is also published under the title Science Annual 1994

Copyright © 1993 by Grolier Incorporated

Library of Congress Catalog Card Number 64–7603

ISBN 0-02-942579-4

Printed in the United States of America

STAFF

Editorial Director
Lawrence T. Lorimer

Executive Editor
Joseph M. Castagno

Director, Annuals
Doris E. Lechner

Editor
Lisa Holland

Art Director
Eric E. Akerman

Photo Researcher
Lisa J. Grize

EDITORIAL

Managing Editor	Jeffrey H. Hacker
Design Director	Nancy A. Hamlen
Production Editor	Sheila Rourk
Editorial Assistant	Karen A. Fairchild
Copy Editors	David M. Buskus Meghan O'Reilly
Proofreaders	Stephan Romanoff Patricia A. Behan
Chief, Photo Research	Ann Eriksen
Manager, Picture Library	Jane H. Carruth
Assistant Art Director	Elizabeth A. Farrington
Assistant Photo Researcher	Linda R. Kubinski
Chief Indexer	Pauline M. Sholtys
Indexer	Linda King
Financial Manager	Marlane L. McLean
Editorial Librarian	Charles Chang
Manager, Electronics	Cyndie L. Cooper
Production Assistant	Carol B. Cox
Staff Assistant	Audrey M. Spragg

MANUFACTURING

Director of Manufacturing
Joseph J. Corlett

Senior Production Manager
Christine L. Matta

Production Manager
Barbara L. Persan

Production Assistants
A. Rhianon Michaud
Jennifer K. Fish

Contributors

JAMES A. BLACKMAN, Professor of pediatrics, University of Virginia, Charlottesville, VA
SIAMESE TWINS

DAVID BJERKLIE, Science writer, *Time* magazine
HIGH-TECH OLYMPIANS

DEBORAH BLUM, Science writer, *Sacramento Bee*
NUCLEAR DETECTIVES

BRUCE BOWER, Behavioral sciences editor, *Science News*
BEHAVIORAL SCIENCES REVIEW

LINDA J. BROWN, Free-lance writer
ENDANGERED SPECIES REVIEW

MARK CALDWELL, Professor of English, Fordham University, New York, NY
THE TRANSPLANTED SELF

ROSALIND CARTWRIGHT, Coauthor, *Crisis Dreaming: Using Your Dreams to Solve Your Problems* (HarperCollins, 1992)
Coauthor, DIRECTING YOUR DREAMS

ANTHONY J. CASTAGNO, Energy consultant; manager, nuclear information, Northeast Utilities, Hartford, CT
ENERGY REVIEW
IN MEMORIAM

THEODORE A. REES CHENEY, Associate professor, Fairfield University, Fairfield, CT
THE SEVEN WONDERS OF THE ANCIENT WORLD

JAMES R. CHILES, Free-lance writer specializing in technology and history
FIGHTING TODAY'S FIRES

NEIL F. COMINS, Professor of Physics, University of Maine, Orono, ME
LIFE ON AN OLDER EARTH

LIZ CRUTCHER, Free-lance writer
RAPID RAILS

DONALD CUNNINGHAM, Free-lance writer based in Washington, DC
TECHNOLOGY REVIEW
TRANSPORTATION REVIEW

GODE DAVIS, Free-lance writer
OCEANOGRAPHY REVIEW
THE PLIGHT OF THE MANATEES

JAMES A. DAVIS, Department of mathematics, University of Richmond, Richmond, VA
MATHEMATICS REVIEW

JERRY DENNIS, Free-lance writer and author of *It's Raining Frogs and Fishes* (HarperCollins, 1992)
PARTY ANIMALS

JOSEPH DeVITO, Free-lance writer
TIME IN A CAPSULE

FREDERICK ENGLE, Geographer, Center for Earth and Planetary Studies, Smithsonian Air and Space Museum
Coauthor, LOOKING AT EARTH

DAVIS S. EPSTEIN, Meteorologist and free-lance writer
THE ONSLAUGHT OF ANDREW
WEATHER REVIEW

BARNABY J. FEDER, Correspondent, *The New York Times*
QUIRKY KEYBOARDS

SANDY FRITZ, Science writer specializing in prehistory
WHO WAS THE ICEMAN?

TERRY GEORGE, Senior editor, *Travel Holiday* magazine
THE GREENING OF HOLLYWOOD

SUSAN GILBERT, Free-lance writer and editor of *The New York Times Good Health Magazine*
THE POWER OF LIGHT

MARIA GUGLIELMINO, Registered dietitian and exercise physiologist
NUTRITION REVIEW

KATHERINE HARAMUNDANIS, Free-lance writer specializing in science and technology; member, American Astronomical Society
ASTRONOMY REVIEW

PETER HARBEN, Consultant on the markets for industrial minerals, chemicals, and certain metals
THE STRATEGY OF STRATEGIC MINERALS

ERIN HYNES, Free-lance writer
BOTANY REVIEW

VINOD K. JAIN, Free-lance writer
CHEMISTRY REVIEW

CHRISTOPHER KING, Managing editor, *Science Watch*, Institute for Scientific Information, Philadelphia, PA
NOBEL PRIZE: PHYSICS AND CHEMISTRY
NOBEL PRIZE: PHYSIOLOGY OR MEDICINE
THE SUBATOMIC MENAGERIE
TROUBLE IN CYBERSPACE

GENE KNAUER, Writer and partner at Knauer, Gehrig
THE RETURN OF THE GEODESIC DOME

FRED W. KOONTZ, Curator of mammals, New York Zoological Society
BEAM ME UP SCOTTY!

LYNNE LAMBERG, Coauthor, *Crisis Dreaming: Using Your Dreams to Solve Your Problems* (HarperCollins, 1992)
Coauthor, DIRECTING YOUR DREAMS

ROBERT LANGRETH, Associate editor, *Popular Science* magazine
BIONIC BOTANY

LOUIS LEVINE, Department of Biology, City College of New York, New York, NY
GENETICS REVIEW

MARK LEWYN, Washington correspondent, *Business Week* magazine
TEACHING A COMPUTER TO TELL A "G" FROM A "C"

MICHAEL LIPSKE, Free-lance writer specializing in natural history
PLANTS THAT EAT MEAT

BETH LIVERMORE, Free-lance writer specializing in health and the environment
ENVIRONMENT REVIEW

THERESE A. LLOYD, Managing editor, *Oilfield Review*
PHYSICS REVIEW

LINDA MARSA, Contributor, *Omni* magazine
SCIENTIFIC FRAUD

DENNIS L. MAMMANA, Resident astronomer, Reuben H. Fleet Space Theater and Science Center, San Diego, CA
RETURN TO MARS
SPACE SCIENCE REVIEW
WHEN HEAVEN FALLS TO EARTH

THOMAS H. MAUGH II, Science writer, *Los Angeles Times*
THE HUMAN GENOME PROJECT

ELIZABETH McGOWAN, Free-lance writer
THE SAGA OF THE DEAD SEA SCROLLS

MARTIN M. McLAUGHLIN, Free-lance consultant; former vice president for education, Overseas Development Council
FOOD AND POPULATION REVIEW

RICHARD MONASTERSKY, Earth sciences editor, *Science News* magazine
GEOLOGY REVIEW
PALEONTOLOGY REVIEW
SEISMOLOGY REVIEW
VOLCANOLOGY REVIEW

GEORGE NOBBE, Free-lance wildlife writer
GOING INTO ORBIT

DENNIS NORMILE, Far East correspondent, *Popular Science* magazine
SUPERCONDUCTIVITY GOES TO SEA

DAVID A. PENDLEBURY, Editor, *Science Watch,* Institute for Scientific Information, Philadelphia, PA
FEMTOCHEMISTRY: CHEMISTRY AS IT HAPPENS

DEVERA PINE, Free-lance science writer and editor
WHAT HAPPENED AT THE EARTH SUMMIT?

ABIGAIL W. POLEK, Free-lance writer
COMPUTER REVIEW

MICHAEL ROGERS, Novelist and senior writer for *Newsweek* magazine
THE ECOLOGY OF WAVES

CATHY SEARS, Contributing editor, *American Health* magazine
JUNGLE POTIONS

LINDA SHINER, Senior editor, *Air & Space/Smithsonian*
WELCOME TO SPACE CAMP

NEIL SPRINGER, Free-lance writer specializing in health and science
AGRICULTURE REVIEW
PUBLIC HEALTH REVIEW
RESTORING MOUNT RUSHMORE

JANE STEVENS, Free-lance science writer
FAMILIAR STRANGERS

DOUG STEWART, Free-lance writer
NO HONKING MATTER

PRISCILLA STRAIN, Program manager, Center for Earth and Planetary Studies, Smithsonian Air and Space Museum
Coauthor, LOOKING AT EARTH

LEONARD RAY TEEL, Free-lance writer and instructor at Georgia State University, Atlanta, GA
A WEATHER EYE ON THE WEATHER CHANNEL

JENNY TESAR, Free-lance science and medical writer; author, *Scientific Crime Investigation; Global Warming*
HEALTH AND DISEASE REVIEW

JIM URBAN, Correspondent, Associated Press
LSD MAKES A COMEBACK

PETER S. WELLS, Professor of anthropology, University of Minnesota, Minneapolis, MN
ANTHROPOLOGY REVIEW
ARCHAEOLOGY REVIEW

JOHN H. WHITE, JR., Senior historian emeritus, National Museum of American History, Washington, DC
HORSE POWER

DELTA WILLIS, Free-lance writer and author of the upcoming book *The Sand Dollar and the Slide Rule*
DESIGN BY NATURE

RICHARD WOLKOMIR, Free-lance writer and coauthor of *Junkyard Bandicoots and Other Tales of the World of Endangered Species* (John Wiley, 1992)
AMERICAN SIGN LANGUAGE
THE CONTEMPTIBLE COCKROACH
THE MAGIC OF TOPIARY

CARL ZIMMER, Associate editor, *Discover* magazine
THE BODY ELECTRIC

Contents

Features

TECHNOLOGY

REVIEWS

ASTRONOMY and SPACE SCIENCE

CONTENTS

RETURN TO MARS

by Dennis L. Mammana

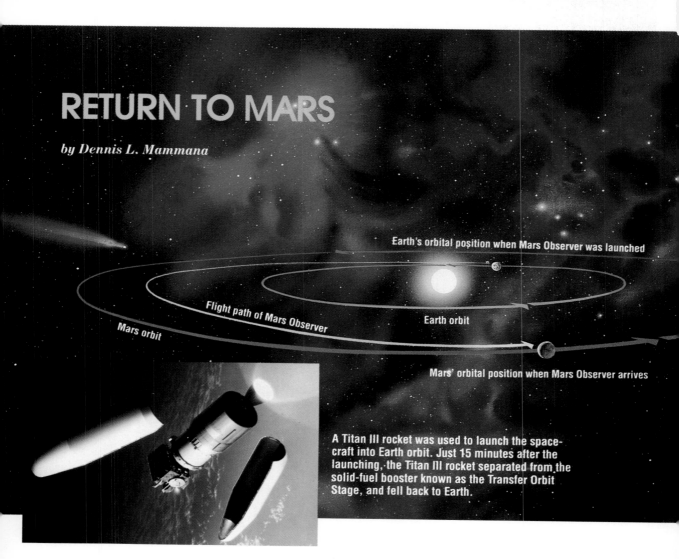

Earth's orbital position when Mars Observer was launched

Flight path of Mars Observer

Mars orbit

Earth orbit

Mars' orbital position when Mars Observer arrives

A Titan III rocket was used to launch the space-craft into Earth orbit. Just 15 minutes after the launching, the Titan III rocket separated from the solid-fuel booster known as the Transfer Orbit Stage, and fell back to Earth.

It has been 17 years since an American robot probe last visited Mars, the Red Planet. But on September 25, 1992, the Mars Observer spacecraft set sail on an 11-month, 450-million-mile (725-million-kilometer) voyage to map the Red Planet, explore its geological and climatological history, and search for evidence that life might once have existed there. The mission has opened an entirely new era of space exploration—one that will pave the way for human exploration of Mars.

The Spacecraft

The Mars Observer, or MO, as its engineers affectionately know it, is an entirely new generation of planetary explorer. Built for the National Aeronautics and Space Administration (NASA) by General Electric's Astro-Space Division in Princeton, New Jersey, this 5,700-pound (2,585-kilogram) craft was constructed economically using existing designs for Earth-orbiting satellites.

The craft's main body is shaped like a large box measuring about 3.25 feet (1.1 meters) high, 7.0 feet (2.2 meters) wide, and 5.0 feet (1.6 meters) deep.

One large solar array, consisting of six 6.0- by 7.2- by 0.3-foot (183- by 219- by 9.1-centimeter) solar panels, collects sunlight and converts it to electricity to operate the spacecraft. From the main body, three booms extend outward and hold the scientific instruments that will probe Mars.

The Transfer Orbit Stage ("TOS") performed the calculations needed to align the spacecraft for its interplanetary mission. Once it reached a speed of nearly 26,000 miles per hour, TOS separated from the Mars Observer, and sent the craft hurtling toward Mars.

The Journey

After being delayed for only 38 minutes, the Titan III rocket lifted off from Launch Complex 40 at Cape Canaveral Air Force Station in Florida, carrying with it the MO.

Fifteen minutes later the Titan III rocket separated from the solid-fuel booster known as the Transfer Orbit Stage (TOS), and fell back to Earth. TOS could now perform the necessary calculations to align the spacecraft for its interplanetary cruise. Once speeding along at 25,575 miles (41,150 kilometers) per hour, TOS separated from the Mars Observer, and sent the craft hurtling outward toward Mars.

During launch the spacecraft's main communication antenna, instrument booms, and solar array were folded up for protection. But once safely in interplanetary space, the craft partially unfurled this equipment.

During this long "cruise" phase, scientists tested the craft's equipment and made trajectory corrections to guide the spacecraft accurately and safely toward its target.

After 11 months of travel, the Mars "orbit insertion" phase began. The craft's speed was slowed by onboard rockets so that the gravity of Mars could lasso the craft into polar orbit. The date: August 24, 1993.

Over the next four months, onboard rocket thrusters will gradually move the MO into a nearly circular orbit 235 miles (378 kilometers) above the Martian surface. Once in final orbit, MO will be low enough to per-

MARS: INSIDE AND OUT

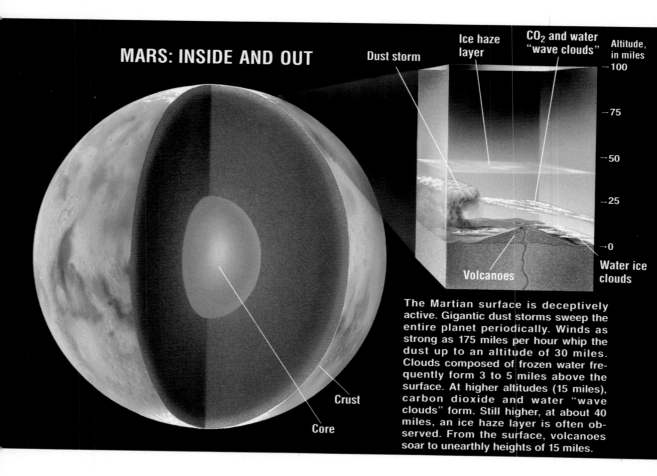

Dust storm

Ice haze layer

CO_2 and water "wave clouds"

Altitude, in miles
- 100
- 75
- 50
- 25
- 0

Volcanoes

Crust

Core

Water ice clouds

The Martian surface is deceptively active. Gigantic dust storms sweep the entire planet periodically. Winds as strong as 175 miles per hour whip the dust up to an altitude of 30 miles. Clouds composed of frozen water frequently form 3 to 5 miles above the surface. At higher altitudes (15 miles), carbon dioxide and water "wave clouds" form. Still higher, at about 40 miles, an ice haze layer is often observed. From the surface, volcanoes soar to unearthly heights of 15 miles.

form a detailed study of Mars, but high enough to prevent the planet's thin atmosphere from slowing the spacecraft and causing it to fall.

Once safely in orbit, MO can fully deploy its solar array and instrument booms. Its main communication antenna—a 4.75-foot (1.45-meter) diameter parabolic antenna—will be raised on its 20-foot (6-meter) boom and turned toward Earth. The craft then will power its instruments to begin conducting the mission experiments.

The Mission Begins

MO's final orbit takes it over the Martian equator at the same local time during each orbit. In other words a hypothetical Martian looking toward the sky might see the spacecraft pass over at 2:00 P.M. and 2:00 A.M. every day and every night. This kind of orbit is necessary so that MO's observations of the sunlight falling on the Martian surface are the same every day, and so that daily atmospheric variations are distinguishable from seasonal ones.

Finally, on January 13, 1994, the mission's mapping cycle will begin. For one full Martian year—687 Earth days—MO will observe and photograph the Martian surface, supplying waiting scientists on Earth with mounds of data daily.

In its near-circular mapping orbit, the Mars Observer will rotate once every orbit to keep its instruments aimed toward the planet. This will allow instruments to view the planet continuously and uniformly during the entire Martian year.

The rotation and orientation of the spacecraft will be controlled by horizon sensors, a star sensor, gyroscopes, and reaction wheels, much the same as on Earth-orbiting satellites. The horizon sensors, adapted from a terrestrial design, continuously locate the horizon and provide control signals to the spacecraft. The star sensor, which will be used for attitude control during the 11-

month cruise, will perform as a backup to the horizon sensors during the all-important mapping orbit.

Once during each 118-minute orbit, the spacecraft will enter the shadow of Mars and rely on battery power for about 40 minutes. The battery is charged by the craft's large solar panel, which generates enormous power every time the craft enters sunlight.

Onboard microprocessors and solid-state memories control the spacecraft and its instruments, and store scientific and engineering data on tape recorders for daily playback to Earth. Whenever Earth is in view, data can be sent back in real time.

Previous Missions to Mars

The Mars Observer is not the first spacecraft from Earth to visit and study the Red Planet. In the mid-1960s, the Mariner spacecraft flew by the planet and sent us the historic first pictures of the cratered Martian surface. Then, in the mid-1970s, the Viking landers looked for signs of life at two landing sites, while its orbiters made the first global maps of the planet. These missions illuminated a mysterious and fascinating world—strikingly familiar, yet completely foreign.

Mars is a strange world of pink skies, towering mesas, fields of boulders, rolling sand dunes, and vast extinct volcanoes— some so high and broad that they easily dwarf those on Earth. A huge canyon, more than 3,000 miles (4,800 kilometers) long, 30 miles (50 kilometers) wide, and 5 miles (8 kilometers) deep, if placed on North America, would stretch from San Francisco to New York City.

The Martian atmosphere, composed mostly of carbon dioxide gas, is quite thin; much like the Earth's atmosphere at an altitude of 20 miles (32 kilometers). Yet, as thin as it is, it can cause regional dust storms to sweep across the planet and obliterate it from our view. Temperatures on Mars are so low that during the winter its atmosphere freezes and turns to snow, while during summer the thin Martian air can become as warm as an October afternoon in New England (50° F, or 10° C).

The presence of dried-up river valleys with large, sinuous channels suggests that catastrophic floods occurred in Mars's distant past—even though there's hardly a drop of water there now. What little water does exist may lie entirely frozen in its northern polar ice cap, and beneath the Martian surface as permafrost.

Solving the Mysteries of Mars

By comparing Mars with the Earth, scientists hope to understand how and why these two planetary neighbors evolved differently. Mars Observer will try to answer some of the questions and mysteries revealed by its robotic predecessors.

For example, what causes the frequent, planet-wide dust storms? How does the surface composition vary across the planet? Are Martian volcanoes still active? Does Mars have a magnetic field or a molten core? Where did the water go that created the planet-wide network of riverbeds?

And what are the mysterious face-shaped plateaus and other odd landforms that resemble the pyramids and sphinx of Egypt? Some scientists have suggested that these are monuments built by ancient civilizations on Mars, but most believe they are natural rock formations. Mars Observer will take detailed photographs to find out.

Mission of Mars Observer

Now that it's in orbit, Mars Observer will perform a global, wide-ranging study of the Martian atmosphere and surface, and perform a scientific inventory of the entire planet. Scientists hope to gain enough information about Mars's dynamic weather cycles to influence our understanding of climates, past and present, on Earth.

MO's primary objectives are to identify and map the surface elements and minerals, measure its surface topography and features, define in global terms the gravitational field, determine the distribution, abundance, sources, and destinations of volatile materials (such as carbon dioxide and water) and dust over an entire seasonal cycle, and explore the structure and circulation of the atmosphere.

"Mars Observer will examine Mars much like Earth satellites now map our weather and resources," says Dr. Wesley

PREVIOUS MISSIONS TO MARS

MISSION	COUNTRY	LAUNCH DATE	ARRIVAL AT MARS	REMARKS
Sputnik 22	U.S.S.R.	10/24/62	—	Blew up during ascent.
Mars 1	U.S.S.R.	11/1/62	—	Lost contact due to faulty antenna.
Sputnik 24	U.S.S.R.	11/4/62	—	Disintegrated in space.
Mariner 3	U.S.	11/5/64	—	Contact lost after launching.
Mariner 4	U.S.	11/28/64	7/14/65	Mars flyby; provided first close-up images of Mars.
Zond 2	U.S.S.R.	11/30/64	—	Passed by Mars; returned no data.
Mariner 6	U.S.	2/24/69	7/31/69	Mars flyby; provided high-resolution photos of planet.
Mariner 7	U.S.	3/27/69	8/5/69	Mars flyby; provided high resolution photos of planet's southern hemisphere.
Mariner 8	U.S.	5/8/71	—	Malfunctioned after launch.
Cosmos 419	U.S.S.R.	5/10/71	—	Became stranded in Earth's orbit.
Mars 2	U.S.S.R.	5/19/71	11/27/71	Landing capsule made unsuccessful attempt to soft land; orbiter continued to transmit data.
Mars 3	U.S.S.R.	5/28/71	12/2/71	Successful soft landing; camera transmitted for only 20 seconds.
Mariner 9	U.S.	5/30/71	11/13/71	Orbited Mars; photographed planet's surface and its moons.
Mars 4	U.S.S.R.	7/21/73	2/10/74	Never achieved Mars orbit.
Mars 5	U.S.S.R.	7/25/73	2/12/74	Achieved Mars orbit, but operated only a few days.
Mars 6	U.S.S.R.	8/5/73	3/12/74	Transmission stopped when landing rockets were fired.
Mars 7	U.S.S.R.	8/9/73	3/9/74	Descent module malfunctioned, bypassing planet.
Viking 1	U.S.	8/20/75	6/19/76	Landed 7/20/76; transmitted thousands of photos.
Viking 2	U.S.	9/9/75	8/7/76	Landed 9/3/76; discovered water frost; provided weather reports.
Phobos 1	U.S.S.R.	7/7/88	1/89	Disabled by ground control error.
Phobos 2	U.S.S.R.	7/12/88	1/89	Achieved Mars orbit; studied Martian surface, atmosphere, and magnetic field.

Huntress, director of NASA's Solar System Exploration Division. "It will give us a vast amount of geological and atmospheric information covering a full Martian year. At last we will know what Mars is actually like in all seasons, from the ground up, pole to pole."

David Evans, project manager, NASA's Jet Propulsion Laboratory (JPL), agrees. "It will tell us far more about Mars than we've learned from any previous missions to date. We want to put together a global portrait of Mars as it exists today, and, with that information, we can begin to understand the history of Mars.

"The first humans to set foot on that planet will certainly use Mars Observer maps and rely on its geologic and climatic data," he says, looking to the future.

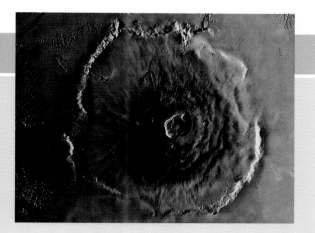

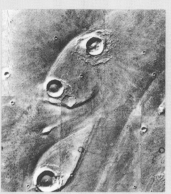

Among the prominent features on the Martian surface is Olympus Mons (above), one of the largest volcanoes in the solar system. Formations that resemble streambeds (left) have led astronomers to think that water once flowed on Mars. The Viking 2 lander took dramatic photos of the planet's rocky surface (below).

Science Operations

Carrying out all these experiments is an awesome responsibility, undertaken by dozens of researchers from universities and scientific institutions in the United States, Russia, France, Germany, and Great Britain.

Fortunately, participating scientists need not move from their homes to the project's headquarters at the Jet Propulsion Lab-oratory in Pasadena, California, to work with the spacecraft or to help plan its activities. Instead, they will communicate and access MO data through a computer workstation in their own offices or homes.

More than 60 such workstations will be connected to the project database at JPL. This database, with about 30 gigabytes of on-line storage, will be electronically available to the science-instrument investigators through NASCOM data links.

Tools On Board

To accomplish its ambitious mission, Mars Observer will use an arsenal of seven powerful instruments to examine the planet from top to bottom, and provide researchers with daily global maps of Mars. These will work together to probe the planet across most of the electromagnetic spectrum.

The *gamma-ray spectrometer* will measure the intensity of gamma rays that are emitted by the natural radioactivity of such elements as uranium, iron, and silicon in the ground below the Martian surface, or by the interaction of cosmic rays with the atmosphere and surface.

Through these measurements, scientists can determine the chemical composition of the surface, element by element, with a resolution of only a few hundred miles. They can also measure the presence of volatile chemicals such as the water and carbon dioxide locked up as permafrost in the surface, and the varying thickness of the polar ice caps.

The *thermal emission spectrometer (TES)* will do prospecting from orbit by measuring the infrared emission coming from the planet. When viewing the surface beneath the spacecraft, the spectrometer has six fields of view, each covering an area 1.9 by 1.9 miles (3.0 by 3.0 kilometers).

The spectrometer will help researchers learn the chemical composition of surface rocks and ice, and map their distribution across the Martian surface. It will investigate the advance and retreat of the polar ice caps, and the distribution of atmospheric dust and clouds over four Martian seasons.

The *laser altimeter (MOLA)* will help construct a more accurate picture of the planet's topography. By shooting short pulses of laser light at the planet and timing the return of their reflections, MOLA can determine the height of surface features below to a precision of several yards.

With these data, scientists hope to gain a better understanding of the relationship among the Martian gravity field, the surface topography, and the forces responsible for shaping the large-scale features of the planet's crust.

The Mars Observer was constructed economically using existing designs for Earth-orbiting communications and weather satellites. The craft is expected to remain in operation for at least four years.

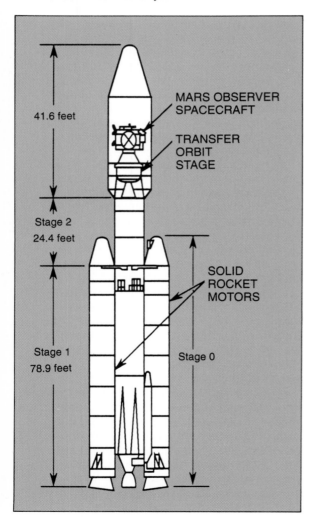

41.6 feet

MARS OBSERVER SPACECRAFT

TRANSFER ORBIT STAGE

Stage 2
24.4 feet

SOLID ROCKET MOTORS

Stage 1
78.9 feet

Stage 0

The *Mars Observer camera* will photograph the Martian surface (only on the day side) with the highest resolution ever accomplished by a civilian spacecraft. Instead of capturing an image all at once, the camera gathers the scene below by incremental widths, using the spacecraft's orbital motion to sweep out the picture.

The low-resolution system will use two wide-angle cameras to capture global views of the Martian atmosphere and surface to provide scientists with a daily "weather map" of the Red Planet. Moderate resolution images will monitor changes in the surface and atmosphere over hours, days, weeks, months, and years.

The high-resolution camera system is powerful enough to pick out details as small as 5 feet (1.5 meters) across—small enough to spot the Viking landers abandoned on the Martian surface years ago. These highest-resolution images will cover patches of ground just 1 mile (1.6 kilometers) square, and thus only a small percentage of the planet's surface area. This mode will be used selectively because of the high data volume required for each image.

The *magnetometer* and a related sensor called an *electron reflectometer* will search the planet for a magnetic field. To date, Mars is the only planet in the solar system (besides Pluto) for which a magnetic field has not yet been detected. The Mars Observer will also scan the surface material for remnants of a magnetic field that may have existed in the distant past.

The *pressure modulator infrared radiometer* (PMIRR) will measure infrared radiation coming from the tenuous Martian atmosphere to produce global models of that atmosphere. Through its measurements, scientists can learn how temperature and water vapor in the atmosphere vary around the planet and with the seasons, and how dust particles are transported and deposited throughout the Martian year.

The spacecraft's *radio system,* designed to communicate with scientists on Earth, will be used in a seventh experiment.

Each time the spacecraft passes behind the planet or reappears on the opposite side, its radio beam will briefly pass through the

Data from this mission may help pave the way for human exploration—and perhaps colonization—of Mars.

Martian atmosphere on its way to Earth. By studying how the waves are distorted by the atmosphere, scientists can learn more about the upper atmospheric density and temperature than with any other MO instrument. And, since tiny variations in the Martian gravity field will alter the frequency of the radio signals as the Mars Observer flies overhead, scientists hope to study these changes to create a global gravity map.

Amassing the Data

At MO's maximum distance from Earth, its radio signals will require about 20 minutes to make the journey. These weak signals, carrying important data, will be captured by NASA's three Deep Space Communications Complexes, located in Goldstone, California; Madrid, Spain; and Canberra, Australia. They will provide continuous communication links with the craft while in Mars orbit.

During its 687-day mapping mission, the Mars Observer will send to Earth about 120 megabytes of data every day, for a total of about 80 to 90 gigabytes (about 600 billion bits) of information. It will soak up the equivalent of a dozen CD-ROMs each week, and will provide waiting researchers more scientific information than has been gathered by any previous planetary missions (except the current Magellan mission to the planet Venus) to date.

The Future of Mars Observer

The Mars Observer is designed to last for three years. However, if the craft, its fuel, and its instruments hold out, the mission may be extended. Scientists are now planning to use MO to help the Russian "Mars 94" spacecraft carry out its mission beginning in September of 1995.

MO will act as a radio relay for the Russian craft, which will deploy several small instrumented penetrators and landers on the Martian surface, storing their data and forwarding it to Earth.

If still operating three years later (1998), the NASA craft could repeat its relay role with a follow-up Russian mission. This mission would release a balloon into the Martian atmosphere, and possibly deploy land stations or rover vehicles to move around the surface under their own power.

In its way, Mars Observer will help pave the way for the first humans to one day explore this fascinating and mysterious world.

LIFE ON AN OLDER EARTH

by Neil F. Comins

Our view of the universe is flavored by time. The Earth's surface, eclipses, and the stars in the sky appear the way they do only because we have captured them at this unique time in the evolution of the universe. What would the universe look like if evolutionary processes on Earth had taken twice as long to produce creatures who were aware of their own existence—say, 9 billion years instead of 4.5 billion years?

How Things Happened

Let's first consider a few of the myriad changes that have occurred in our solar system. Five billion years ago, it didn't exist. Billowing clouds of gas and dust thrown into space by exploding stars collided where the Sun and planets would someday reside. There were no clues to the profound changes yet to come.

Then, about 4.7 billion years ago, the clouds of gas and dust were driven together. The gravity of the resulting cloud forced it to collapse in on itself. In this cloud the pieces of the solar system began condensing into existence, an event that occurred about 4.6 billion years ago. The massive Sun formed in the central, thickest part of the cloud. The remaining debris formed four solid inner planets (Mercury, Venus, Earth, and Mars), and four massive, gaseous outer planets (Jupiter, Saturn, Uranus, and Neptune), satellites of these planets, and various icy and rocky smaller bodies that we now call comets and asteroids. It is likely that Pluto didn't orbit then where it does now, but came to reside there after a collision with another icy dwarf planet.

The coalescence of space debris into planet Earth took less than 50 million years. After that, innumerable space rocks and ice chunks bombarded the young Earth for almost 700 million years, increasing its mass and providing water and organic materials. This cosmic organic matter may have started life on Earth. A collision with a Mars-sized object ejected material from Earth's crust and mantle, material that then came together to form the Moon.

The original atmosphere on Earth was lost to space as collisions expelled some gases while other, lighter gases, such as hydrogen, escaped from Earth's gravity. Subsequent volcanic activity released large amounts of carbon dioxide, producing a second atmosphere. This atmosphere, much like the one that still exists on Venus, was much denser than Earth's present atmosphere. The young oceans absorbed some of the carbon dioxide; rocks absorbed still more carbon dioxide to form carbonate rocks like limestone. Atmospheric pressure decreased as a result, but the remaining air still contained primarily carbon dioxide. Around 2 billion years ago, primitive plants began to evolve and started to convert the carbon dioxide into oxygen and carbon during photosynthesis, returning the oxygen to the air.

By about 1 billion years ago, our present nitrogen-oxygen atmosphere was in place. Shortly after, single-celled animals called protozoa started evolving into multicellular organisms. Creatures developed specialized organs 600 million years ago, and primitive mammals evolved about 200 million years ago, shortly before the reign of the dinosaurs. One of the first human species to walk upright was *Australopithecus afarensis,* the oldest example of which is "Lucy," who lived about 4 million years ago. The early hominids evolved into *Homo erectus* almost 2 million years ago, and the first member of our species, *Homo sapiens,* appeared only 250,000 years ago. Our modern ancestors, *Homo sapiens sapiens,* appeared about 35,000 years ago.

Delaying Evolution

The sequence of events that enabled Earth to support humans at this time in our planet's history was by no means inevitable. For instance, Earth need not have cooled off enough to support a solid surface after only 700 million years. Earth was initially molten; indeed, radioactive elements and the pressure of the planet's matter pushing inward have kept much of the interior molten even to this day. The outer layer, or crust, lost heat into space, eventually cooling into a semisolid, or plastic, layer. If the fraction of radioactive elements were higher, it could have taken hundreds of millions of years more for the surface to stabilize and for solidification to occur.

Astronomers theorize that our universe was created by a titanic explosion, known as the Big Bang, which ejected matter in all directions. Earth was created by the coalescence of space debris, and its continual bombardment by cosmic organic matter may have provided the seeds for life.

Also, the conversion of our atmosphere from primarily carbon dioxide to oxygen and nitrogen demanded not only that the carbon dioxide be removed in sufficient quantities, but that the compounds that combine with free oxygen be saturated—that is, that they contain as much oxygen as they could and still be able to chemically bond. This latter point is by no means trivial; oxygen is extremely reactive. Even without combustion, oxygen combines with many compounds such as carbon monoxide in the atmosphere and iron in rocks. Until all such reactants were saturated with oxygen, there was not sufficient free oxygen for animals to breathe. Slight changes in surface or atmospheric chemistry could have delayed a stable oxygen-and-nitrogen atmosphere by hundreds of millions of years or even longer.

Along the same vein, some astronomers conjecture that the original carbon dioxide atmosphere of Earth may have been many times denser than the secondary atmosphere plants started attacking. After all, the present atmosphere of Venus is 95 times thicker than ours and contains primarily carbon dioxide. If the Moon formed from a collision, the energy in that impact would have had sufficient power to eject most of Earth's early atmosphere into space. Without that loss of carbon dioxide, the thick atmosphere would trap heat from the Sun's rays and be hot enough to slow absorption of carbon dioxide by the rocks and oceans. It easily could have taken an extra billion years to change into an oxygen-and-nitrogen atmosphere. Also, plants living in a thicker carbon dioxide atmosphere may have generated oxygen more slowly. They may consider oxygen to be toxic, as anaerobic bacteria do, so the production of oxygen might have required the evolution of different plant forms to cope with the changing atmosphere.

Collisions by comets or asteroids may have caused more extinctions of living things on Earth. The mass extinction of the dinosaurs and half of everything else alive on the planet 65 million years ago is one example of this. If an extinction had occurred at a critical juncture in evolution, such as when multicellular organisms were first forming or when internal organs began differentiating, then later evolutionary steps could have been delayed by millions of years. Perhaps this did happen, delaying our evolution by a hundred million years.

Finally, if the primate line leading to humans had faltered and no other creature had developed a complex enough brain, the cycle of evolution may have had to repeat itself, perhaps several times, before self-aware beings appeared. A mass extinction in recent history could have even wiped out advanced life. It is feasible, then, that intelligent beings might not have appeared until billions of years later than they—we—did. Let's imagine that it took 9 billion years, instead of only 4.5 billion, until humans evolved.

A Different Earth

Would these later "humans" be like us? The evolutionary process that led to humans is unique, and the likelihood is small that the sentient, or self-aware, creatures who evolved billions of years after we did would look like us. But no matter. Recent advances in the theory of neurological development suggest that eventually a species would evolve that would be self-aware and curious about its surroundings. But the "New Earth" they inhabit would look very different from the one we enjoy today.

Collisions with innumerable space rocks and volcanic activity (left) changed Earth's initial atmosphere until it could support primitive plant life and, billions of years later, dinosaurs. Further collisions with asteroids (below) may have caused dinosaur extinction.

The continents and oceans would be different. The crust of Earth is composed of granitic continental plates somewhat like closely packed icebergs. The plates have been moving relative to each other through recorded geologic history. Geologists believe that 350 million years ago a massive continent called Gondwana lay in the Southern Hemisphere. This land mass drifted north, connecting with other continents about 260 million years ago to form one huge, primeval supercontinent called Pangaea, which is Greek for "all land." Then about 100 million years ago, the plates slowly moved apart from Pangaea's core, which now forms Africa, dispersing into the continents of today.

The continents are still drifting, and, 4.5 billion years from now, their distribution will be completely different. No pattern of land that we recognize today will remain intact. The Mediterranean Sea will be gone, and the Atlantic may be as wide as the Pacific is today, while the Pacific will probably be much narrower. The islands of the Pacific, from Japan south to Australia, may be a single connected unit. New islands will grow over hot spots in the mantle, just as the Hawaiian Islands do today. The mountains in the Andes and Himalayas will have tried to rise higher, only to come tumbling down under the force of their own weight. The world map of New Earth would be unrecognizable to today's humans.

Another major change is the length of one day. It will be almost twice as long as our day. The Moon will be about 1.5 times farther away, appearing correspondingly smaller in the sky, and take 1.8 times as long to travel through the sky. The longer day and more distant Moon are intimately related through the ocean tides created by the Moon's gravity. As tidal waters rub against the solid Earth, the resulting friction slows Earth's rotation and currently causes the day to grow 0.002 second longer each century. The energy lost as Earth spins down must be retained in the Earth-Moon system, so the Moon must speed up in its orbit around Earth. This increase in speed currently forces the Moon to move farther away from Earth at a rate of 1.5 inches (4 centimeters) per year.

Since the Moon appears smaller in the sky, future humans will never see a total eclipse of the Sun. The best they will see is an annular eclipse, with a broad ring of the Sun showing around the edges of the Moon. Total lunar eclipses will still occur because Earth's shadow will remain larger than the size of the Moon.

Changes in the Night Sky

Currently Polaris (also called the North Star), although about 680 light-years away, appears almost exactly over Earth's north rotational pole. But in all likelihood, humans of the future won't see a pole star for two reasons. First, Earth's rotational axis precesses, or wobbles, by 23.5 degrees over a period of about 25,800 years, making the pole wander around the sky. We are fortunate that in recent centuries there has been a bright star to mark the North Pole. Second, many bright stars that come to lie near the pole may no longer be visible, having used up their nuclear fuels during the intervening 4.5 billion years. Polaris itself will explode as it nears the end of its life cycle, as nuclear reactions run away deep in its interior. Perhaps new bright stars will form that will take turns as future polar-guide stars as Earth's rotational axis precesses, much as Polaris and Thuban in Draco have in the past.

Many of the stars that we see now will no longer exist, having exploded as either supernovas or planetary nebulas. These include most of the brightest stars—Sirius in Canis Major, Vega in Lyra, Capella in Auriga, Rigel and Betelgeuse in Orion, Antares in Scorpius, Regulus in Leo, and Deneb in Cygnus, to name a few. Constellations such as the Big Dipper or Orion will all be different, too, since even those stars that still exist in 4.5 billion years will have moved relative to each other.

A Brighter Sun

The Sun will still have nearly the same surface temperature and yellowish hue that it has now. But in 4.5 billion years, the Sun will appear about twice as bright because it will be about 60 percent bigger. The extra 4.5 billion years will have begun to take their toll on the Sun's nuclear-fuel supply. The

Sun is a gaseous ball composed of 2,000 trillion trillion tons of matter. This gas contains primarily hydrogen and helium, with traces of other elements. For every million atoms of hydrogen, there are about 85,000 helium atoms and only about 1,000 of any other kind of atom. Pressure from all that mass crushing down on the center of the Sun is high enough that hydrogen atoms fuse together—or "burn"—in the core to form helium. This simultaneously creates new energy, which keeps the Sun from collapsing further and provides the energy that allows it to shine.

The Sun has fused hydrogen into helium throughout its present lifetime of 4.5 billion years, using up less than half of the available hydrogen in its core. In another 4.5 billion years, 80 or 90 percent of the available hydrogen in the core will have been converted into helium. Serious questions about the fusion rate in the Sun still remain, but future humans will face a Sun running low on hydrogen.

When this fuel runs out, the gas temperature and pressure will drop, and the interior of the Sun will collapse under the weight of the surrounding mass. The pressure in the collapsing gas will build up sufficiently for a shell of hydrogen to start burning around the core, now helium. This fusion will provide an outward force on the outermost layers of the Sun, pushing them farther out than they are now. The surface of the Sun will expand outward until it reaches the orbit of Venus.

Eventually this hydrogen outside the core will run out. The core of the Sun will continue to contract, trying to replace the heat no longer generated by hydrogen burning. When the internal temperatures reach 100 million degrees, the helium generated by hydrogen burning will itself start to burn. This will happen quickly, forming a carbon-rich core. Around this burned-out core, helium burning will start. Around the helium-burning layer, a shell of hydrogen also will start to burn. The vast energy released by both of these shells will push the Sun's outer layers out until they reach the orbit of Jupiter. Earth will actually be enveloped inside the Sun. The temperature on the surface of this

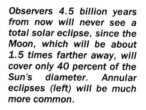

Observers 4.5 billion years from now will never see a total solar eclipse, since the Moon, which will be about 1.5 times farther away, will cover only 40 percent of the Sun's diameter. Annular eclipses (left) will be much more common.

solar-engulfed Earth around 6.5 billion years from now will be around 535,000° F (30,000° C), and everything organic will be burned to a crisp.

Intelligent life 5 billion or 6 billion years from now will face the pressure to leave Earth—and, indeed, our solar system. They will need to colonize planets that orbit around younger—and therefore more stable—stars in order to live. It is likely that humans in the near future will develop technology that will permit them to move off Earth in search of mineralogical and economic gain, whereas the future beings of our speculation will need to move off to another solar system to save their species. The aging Sun will give future life a focus and a goal—survival.

WELCOME TO SPACE CAMP

by Linda Shiner

A small, specialized team at the U.S. Space and Rocket Center in Huntsville, Alabama, is busy with a project that captures the essence of space exploration: braving the unknown, preparing for the unpredictable, and laying the groundwork for communication with the alien world. For eight days at a time, these people house, instruct, and chaperon groups of 50 to 100 human teenagers.

The U.S. Space and Rocket Center is a National Aeronautics and Space Administration (NASA) visitor center, a space theme park, and home of U.S. Space Camp. Space Camp is really several camps in one: a five-day "introduction to space sciences and exploration" for little kids (grades 4 through 6); a five-day academy of intensive astronaut and mission training for medium-size kids (grades 7 through 9); and an eight-day academy of lectures, training, and career counseling for big kids (grades 10 through 12). They "train" in a 70,000-square-foot (6,500-square-meter) warehouse, where various simulators, buoyancy tanks, and other contraptions teach the campers about microgravity and spaceflight while providing them the degree of dizziness youngsters always seem to be seeking. There are also programs for adults and teachers, but, for the most part, the campers are kids.

All campers in the eight-day version, "U.S. Space Academy Level II," cap off their experience with a 12-hour simulated shuttle mission, a practicum I had the chance to observe one day last winter, when the participants included 14 Australian students, eight kids from the U.S., and one South African.

Even though they have been up since about six that morning, and have started that early for the past four days of lectures, films, museum tours, and training, the 23 teenagers on the mission are cheerily cooperative at 10:00 A.M. as the mission begins. Commander Jennifer Farone, a senior from Eastlake, Ohio, sits in the left seat of the shuttle cockpit, calling out instructions from her mission handbook to the pilot on her right. At each command, Tony Itelli, a sophomore from Florence, Alabama, flips the appropriate number of switches and responds with a crisp "Check." Simulated blue sky and clouds drift serenely past the shuttle windows. Between checks, Itelli, exhibiting a proprietary concern for his ship, points to my tape recorder. "What's that for?" he asks. When informed that I am working on an article about the academy, he gives me a solemn thumbs-up.

Seated (occasionally) behind the commander and pilot, mission specialist Warren Watson from Perth is giving fellow mission specialist Jason Fewings an impressive imitation of what would happen to their bodies should the shuttle fall on its side, fail to achieve orbit, or otherwise crash. Watson's contortions, strikingly similar to those of cartoon figure Roger Rabbit, continue through a series of commander instructions and pilot checks until another mission specialist, Helen Hernandez, calls shrilly to counselor Jeff Wheater from the mid-deck, "Who's driving? Is Jeff driving?" Then she adds, somewhat alarmingly, "Jeff, please don't make us crash."

Using a microgravity chair to simulate a moonwalk (above) is just one of the "out-of-this-world" activities students enjoy at U.S. Space Camp.

Watson, now composed, turns to me and says, "We have great confidence in one another."

Reading from her manual, Farone speaks confidently into a headset that connects her to Mission Control. "This is the commander. The *Enterprise* is configured to launch."

Splat Avoidance Vehicle

Meanwhile, some 50 feet (15 meters) away, aboard a full-scale mock-up of the lab module of Space Station Freedom, another small crew of cheerful teenagers is thriving. During the previous four days, they have received instruction in solar and plasma physics, space biology, astrophysics, remote sensing, optics, and computers; now they are conducting experiments designed to reinforce the lectures.

Lined with lockers and lab equipment, the module is sparsely furnished with an exercise bicycle and a chair in which one can spin. These are in constant use.

As Dire Straits plays softly on a tape deck, station commander Carrie Gorman, a 16-year-old licensed pilot from Massachusetts, and her second-in-command, Australian Michael Watts, monitor computer screens that report the vital statistics of their supposedly fragile (actually plywood) labora-

tory. Brad Walker, who will attend the University of Western Australia in the fall, and his friend Joshua Robins take time out from operating a small robotic arm to show me the splat avoidance vehicle (SAV) Walker has constructed.

At the beginning of the week, each camper was given an egg, a supply of flimsy cardboard, balloons, soda straws, rubber bands, nylon stockings, and Elmer's glue, and was told to construct a "planetary probe that would protect its payload on landing." The SAVs are dropped from a height of 45 feet (13.7 meters) on the last day of camp and evaluated. A broken egg disqualifies a contestant, and the winner is the vehicle with the lowest weight and smallest surface area. Walker, who wants to be an engineer, has built a cardboard sphere with straws radiating outward, porcupinelike. Although his career plans aren't focused on space, at least not yet, he readily admits, however,

In the Multi-Axis Trainer (facing page, left), kids learn to orient themselves inside a spinning spacecraft. Campers also have the opportunity to control a simulated space-shuttle flight (facing page right) and gain a sense of what it's like to walk in space (left). Above, a cutaway of the Hubble Space Telescope illustrates to campers how the instrument uses advanced technology to send images back to Earth.

that "if I were offered a job in space, I wouldn't turn it down." Perhaps with a youthful eye on the economy, he hastens to add "I'd take any kind of engineering job."

"If you want a job in space, you'd almost have to come to America," adds Robins. "There aren't many opportunities in Australia. It's like they say, 'If you invent something, leave the country.'"

"We've got a leak," announces Watts. A digital display has shown the station temperature gradually dropping, so the campers, prodded by a counselor who is also aboard (and who has programmed the temperature drop from her computer), decide to embark on an EVA—extravehicular activity. Australian Scott Boyle and Bianca Jordaan of South Africa suit up. They spend two and a half minutes standing at one end of the module in the area designated "airlock," then go through the plywood double doors into the hostile environment of . . . the warehouse. Five minutes later they reenter, having found a piece of masking tape marked "LEAK" affixed to the outside.

Creating Anomalies

Glitches like leaks, frozen controls, dead batteries, meteor showers, and other astronomical phenomena are zestfully devised by the camp counselors, who otherwise would spend the mission simply hanging around, waiting to answer questions for six hours at a time. (The counselors, most of them in their early twenties, switch shifts halfway through the session.) Rob Mackintosh sits in front of a computer terminal in a dark room—Mission Control. "Here's where we torment all the kids from," Mackintosh says somewhat devilishly, tapping what he calls "an anomaly" into the computer. The anomalies show up on the terminal screens in the shuttle cockpit and station module. In return, two closed-circuit television screens show the activity in the simulators to the team in Mission Control, where nine campers are keeping track of the shuttle and station systems as well as the health of the crews. Campers always overcome the anomalies, partly because if one turns out to be too thorny, the campers' counselors give them clues.

AVIATION CHALLENGE

A Mobility Trainer gives a future astronaut an idea of how it might feel to work in a space station.

Flight director Gavin Shakespeare, an ebullient redhead from Perth, commands the control-room specialists from a kneeling position in his swivel chair. CAPCOM, as the campers call him (unless they forget and call him Shakes), will occasionally spin himself vigorously as the group discusses an anomaly, presumably to clear his head. Since the mission is an exercise in make-believe (and since it is 12 hours long), I try to get into the spirit of things by pretending that what I am observing is real—a rather unsettling experience. What if NASA's real flight director were to kneel in his chair and give himself a spin now and then? What if, while astronauts searched for a leak in the ventilation system of a shuttle, ground controllers were to shout to them, "Wa-a-a-r-rm. You're warm. No— cold, cold, cold! No. Back the other way. Warmer. Warmer. HOT! HOT!" What if the astronauts were to use the pockets of their blue flight suits as a stash for bags of Skittles, SweeTarts, and candy bars? What if the real INCO, the instrumentation and communications officer, were to come to work wearing a Walkman, and have to be elbowed several times by the CAPCOM to give his input on a systems anomaly? CNN might carry entire missions live.

During the second half of the mission, after the crews of the shuttle and station have eaten lunch and exchanged places, Warren Watson falls victim to a medical anomaly: hypothermia. Fellow campers follow standard procedure, wrapping him in a blanket, and mission specialist Allison Goeden from Yankton, South Dakota, generously shares her own body heat by sitting on Watson's lap. From Mission Control a message appears on the station terminal: "Tell Allison that a big kiss will warm him up." An air of expectation descends on the laboratory. It is dispelled by Jason Fewings, who leans past Goeden and gives Watson a big smack on the cheek, sending Watson into more Roger Rabbit-like paroxysms, and Goeden scrambling to her feet with a piercing shriek.

Mission Accomplished

By the end of the mission, the shuttle crew has docked with the space station, two truss structures have been built during shuttle EVAs, a satellite has been repaired (twice), several leaks have been patched during station EVAs (and one crew member, who forgot his tether, has been rescued by fellow astro-campers), dozens of packages of Skittles have been consumed, many medical anomalies have been faced down, addresses have been exchanged, at least two campers have fallen asleep on the job, and the *Enterprise* has returned safely to Earth. As I am leaving the training center, I run into Tony Itelli and congratulate him on completing the mission. Itelli nods his head once and sticks up his thumb.

WHEN HEAVEN
FALLS TO EARTH

by Dennis L. Mammana

Stand outdoors on any clear, dark night and gaze into the star-filled sky. Suddenly the heavens are split by a brilliant flash of light. And now, only moments later, it's gone—vanished somewhere overhead, never to be seen again.

Few things are as exciting to watch as the fiery spectacle known as a shooting star, or meteor. On any clear evening, we can see several meteors streak across our sky each hour. What are these phenomena that surprise, excite, and even frighten?

Sky watchers of old wondered, too. The first meteor observations were recorded as far back as 1809 B.C. And, though they couldn't

A shooting star is actually a speck of cosmic dust that, as it enters our atmosphere, burns up in a magnificent burst of light called a meteor.

MAJOR ANNUAL METEOR SHOWERS

SHOWER	USUAL PEAK DATE[1]	DURATION (days)	USUAL HOURLY RATE[2]	SPEED[3]	PARENT BODY
Quadrantids	Jan. 3	0.4	80	67	Unknown
Lyrids	Apr. 22	1	15	77	Comet Thatcher
Eta Aquarids	May 4	6	35–60	106	Comet Halley
Arietid[4]	June 7	12	100	63	Asteroid Icarus (?)
Zeta Perseids[4]	June 9	8	80	47	Unknown
Beta Taurids[4]	June 29	5	20–40	48	Comet Encke
S. Delta Aquarids	July 29	8	30	66	Unknown
N. Delta Aquarids	Aug. 12	8	20	66	Unknown
Perseids	Aug. 12	8	75–100	66	Comet Swift-Tuttle
Orionids	Oct. 21	2	25–30	106	Comet Halley
S. Taurids	Nov. 3	15	15	47	Comet Encke
N. Taurids	Nov. 5	30	15	48	Comet Encke
Leonids	Nov. 16	2	10–20	116	Comet Tempel-Tuttle
Geminids	Dec. 13	3	60–90	58	Asteroid Phaethon (?)
Ursids	Dec. 22	1	5–20	55	Comet Tuttle

[1]Date of actual maximum may vary by one or two days.　　[2]Hourly rate varies from year to year.　　[3]Miles per second.　　[4]Daytime shower.

prove it, the ancients believed that shooting stars were not really stars, but phenomena that occurred high in the atmosphere of Earth. In fact, the very word "meteor" comes from the Greek *meteoros,* which means "something high in the air."

Not until 1798 were the ancients' ideas proven correct. In that year, two students at Göttingen University, Johann F. Benzenberg and Heinrich W. Brandes, observed 22 meteors simultaneously from a few miles apart, and measured their positions against the stars. After comparing their data, the young astronomers calculated that the meteors they saw disappeared from view while still high above the Earth—from 6 to 133 miles (10 to 214 kilometers). What they were, however, no one knew for sure.

Shooting "Stars"

Today we know that "shooting stars" are not stars at all, but specks of cosmic dust—often as tiny as a grain of sand—that float in space. During the Earth's annual voyage around the Sun, we encounter tons of these *meteoroids* each day. When one falls into our upper atmosphere at speeds of between 7 and 45 miles (12 and 72 kilometers) per second, friction with the air heats it to nearly 5,000° F (2,760° C). Within only a fraction of a second, the particle vaporizes, and

Amazingly, meteor showers gave rise to very few superstitions. In fact, most ancient people believed that they resulted from some sort of atmospheric phenomena—a theory proved in 1798 (below).

leaves in its wake a glowing train of hot air. If such a particle falls at night, we may see its fiery death as a magnificent burst of light among the stars: a *meteor.*

On a clear, dark night, a good observer can spot as many as 10 random, or sporadic, meteors every hour. But at certain times of the year, more may be visible, all seeming to come from the same point in the sky. These meteor showers are today known to be remnants of the most famous of cosmic vagabonds, the comets.

Comets are dirty snowballs that spend most of their time in the icy depths of our outer solar system. Each time a comet rounds the Sun, however, some of its ice turns to gas, releasing into space dust that had been trapped within. During each orbit, the comet deposits more and more dust, creating along its path a meteor stream. The older the comet—that is, the more times it has orbited the Sun—the wider its dusty litter will become.

When the Earth crosses one of these cometary orbits, it sweeps up huge clumps of debris, and provides us with a meteor shower. During such events, dozens of meteors may shoot from the sky each hour, all appearing to come from one place in the sky. That point, known as the *radiant,* is an optical illusion caused by meteors falling toward us along parallel paths.

Meteor showers are named for the constellation in which they seem to originate— the point at which the shower's radiant lies. This practice began in 1866 with Giovanni Schiaparelli, the Italian astronomer best known for his descriptions of the "canals" of Mars. In a letter, Schiaparelli discussed "the falling stars of August 10, stars which from now on I will call 'Perseids' for short, after the constellation from which they appear to spread out to us. . . ."

The Perseid shower, which now reaches its maximum around mid-August each year, is perhaps the oldest and most famous of all meteor showers. It was first recorded on July 17, A.D. 36 by ancient Chinese sky watchers who wrote that "more than 100 meteors flew thither in the morning."

Today the Perseids regularly produce between 60 and 80 visible meteors each hour. But their numbers change from year to year, depending on how the meteoroids are distributed in space. For example, if the Earth collides with a particularly thick clump of particles, the show can be spectacular. Such was the case during the Leonid shower of November 17, 1966. For a short while that evening, stargazers watched as the heavens opened and more than 2,000 meteors fell every minute.

How to Watch

The number of meteors we see also depends on the time of night we observe. More are visible during the wee hours before dawn. The reason for this peculiarity lies in the motions of our Earth.

In the hours following sunset, our hemisphere faces the direction opposite our orbital motion—that is, we are on the trailing side of the Earth and are looking out the Earth's "rear window." Only those meteoroids that catch up to us at a speed of at least 18 miles (29 kilometers) per second can fall into our atmosphere. With clear weather, we might see two to six sporadic meteors each hour during early evening.

But after midnight, we are on the leading side, facing the direction of the Earth's orbital motion. Now when we gaze into the sky, we peer out our planet's "front window" and can see all those particles being swept into the atmosphere "head-on." The nearer to dawn it becomes, the more sporadic meteors we can expect to see—perhaps as many as 14 per hour. This is also the best time to watch meteor showers.

Meteors come in all brightnesses, sizes, and colors, and there is no way to predict what might appear on any given night. If our Earth encounters a much larger meteoroid than is common—say, the size of a marble —it might appear much brighter than any of the stars or planets in the sky. We would see its demise as a brilliant smoking or flaming *fireball.* More than 50,000 fireballs occur in our atmosphere each year, though most are not seen, since they occur in daylight or over unpopulated areas or open ocean. If a fireball seems to pop or even explode, it is then called a *bolide.* Only about one out of every 10 fireballs behaves in this way.

VARIETY OF METEORITES

METEORITE TYPES	FINDS %	FALLS %
Chondrites (stones)	43.7	84.3
Achondrites (stones)	1.0	8.7
Stony Irons	5.2	1.3
Irons	50.1	5.7

To enjoy meteors or meteor showers, absolutely no equipment is needed. Just a reclining lawn chair or sleeping bag under a clear, dark sky, and our eyes. A blanket and a mug of hot coffee or soup might also help, especially if it is cold.

Not only are meteor events fun and exciting to watch, they also offer the opportunity for backyard astronomers to contribute to science. Through patience, well-thought-out observing procedures, and lots of time, even novices can perform meaningful research that can help scientists learn the ages of comets, how they behave from one orbit to the next, and how their dust particles are distributed through space.

Perhaps the simplest work one can do is to count the meteors you see over a period of an hour. Write down the time each meteor appears, its brightness, duration, and color. If any leave a train, describe their appearance and duration. Occasionally binoculars come in handy, since some meteor trains can be watched for several minutes.

Recording the paths of meteors helps to differentiate between random meteors and those of a swarm, since shower meteors originate at the shower's radiant, while sporadic meteors move in random directions across the sky. We can sketch their paths or, for a more accurate and permanent record, take photographs.

Another way to detect meteors—even in the daytime—is with a simple radio. As a meteor falls, it drags long, thin streams of electrically charged air behind it. This air reflects radio waves being beamed outward by distant radio stations. To "listen" to a meteor shower, tune an FM radio to a station slightly beyond its normal range—perhaps one 50 to 100 miles (80 to 160 kilometers) away. (Call a radio station nearby and ask what stations may lie at that distance.) When a meteor falls, the normally silent station will suddenly burst forth loud and clear for a second or two.

Bigger Pieces

While only a few meteors may appear to our eyes (or radios), perhaps 90 million occur each day. Many fall during daylight hours, or in parts of the world that we cannot see. And not all of them burn out in the air. If a piece of space rock is large enough, it might survive its plunge through the atmosphere and crash to Earth. If it does, it then becomes a meteorite.

Meteorites are not molten when they hit. Their fall is so brief that only their surfaces are heated, and this often produces a black glassy crust that covers the meteorite.

Our planet sweeps up more than 220,000 tons (200 million kilograms) of meteoric material each year. Most range from microscopic particles less than a millionth of an ounce to rocks of several pounds. Only one-tenth of these ever survive their fiery plunge and hit the Earth. Most of those—about three-quarters—hit the oceans. But some 5,500 tons (5 million kilograms) hit land each year.

With all of this material falling from the sky, it seems likely that people and objects should be pelted from above constantly. But consider that the land area of our planet is very large—some 57,506,457 square miles (148,941,040 square kilometers). That means that each square mile (2.6 square kilometers) of land encounters less than half an ounce (1/10 of a kilogram) each year.

Nevertheless, big chunks do fall from time to time. For example, 15 miles (24 kilometers) west of Winslow, Arizona, lies an immense geological feature known as Meteor Crater, once described by the Swedish scientist Svante Arrhenius as "the most interesting place on Earth."

It is believed that, about 22,000 years ago during the last Ice Age, a 300,000-ton meteorite plowed at about 30,000 miles (48,000 kilometers) per hour into what is now northern Arizona. When the dust set-

LOCATION OF MAJOR METEOR CRATERS

NAME	LOCATION	DIAMETER IN FEET	NAME	LOCATION	DIAMETER IN FEET
Al Umchaimin	Iraq	10,500	Merewether	Labrador	500
Amak	Aleutian Islands	200	Meteor Crater	Arizona	4,000
Aouelloul	Sahara Desert	825	Mount Doreen	Australia	2,000
Baghdad	Iraq	650	Murgab	Tajikistan	250
Boxhole	Australia	500	New Quebec	Quebec	11,000
Brent	Ontario	12,000	Nordlinger Ries	Germany	82,500
Campo del Cielo	Argentina	200	Odessa	Texas	500
Chubb	Ungava, Canada	11,000	Pretoria Saltpan	South Africa	3,000
Dalgaranga	Australia	250	Serpent Mound	Ohio	21,000
Deep Bay	Saskatchewan	45,000	Sierra Madera	Texas	6,500
Duckwater	Nevada	250	Sikhote-Alin	Siberia	100
Flynn Creek	Tennessee	10,000	Steinheim	Germany	8,250
Haviland	Kansas	60	Talemzane	Algeria	6,000
Henbury	Australia	650	Tenoumer	Western Sahara Desert	6,000
Holleford	Ontario	8,000			
Kaalijarv	Estonia	300	Vredefort	South Africa	130,000
Kentland Dome	Indiana	3,000	Wells Creek	Tennessee	16,000
Kofels	Austria	13,000	Wolf Creek	Western Australia	3,000
Lake Bosumtwi	Ghana	33,000			
Manicouagan Reservoir	Quebec	200,000			

About 25,000 years ago, a gigantic meteorite impact formed the immense Meteor Crater in Arizona.

Only very rarely are meteors visible during the day. One exception occurred on August 10, 1972, when what has been termed a fireball streaked across the sky above Grand Teton National Park.

tled, a huge crater remained—4,000 feet (1,219 meters) across and 500 feet (152 meters) deep, with a rim rising 150 feet (46 meters) above the desert floor. Meteor Crater is but one of many dozens of such impact craters around the world—craters from which we can learn much about the violent impacts of the distant past.

Fortunately, most of the really big pieces of wandering space rock have already been swept up by the planets and their moons. Nevertheless, larger rocks can fall. And when these rocks survive their trip through the atmosphere, they can do damage.

It has been estimated that 5,800 meteorites weighing at least 4 ounces (100 grams) apiece fall each year somewhere on the total land area of our planet. Statistically, this means that, each year, 16 buildings could be damaged by meteorites weighing at least 20 ounces (600 grams). And, every nine years, a meteorite could actually strike a person somewhere in the world. In the United States and Canada, that number translates to one person being hit every 180 years!

To date, no humans have been killed by falling meteorites, although several animals have met their end in this rather distinguished way.

If a meteorite is seen to hit the ground, it is called a fall. Reports of falls come from as far back as ancient Greece and Rome. The ancients held these thunderstones, as they might have called them, in awe, and believed them to be thunderbolts or weapons hurled by angry gods.

It is said that the Roman emperor Elagabalus, upon beginning his four-year reign in A.D. 218 entered Rome in triumph and carried with him a black stone, probably a meteorite, from Emesa, Syria. It was thought to represent the Sun God, and Elagabalus insisted that it be worshiped publicly by his subjects.

Even recently, people have reported remarkably close encounters with falling rocks. On November 30, 1954, Annie Hodges of Sylacauga, Alabama, was napping on her couch, when an 8-pound (3.6-kilogram) meteorite ripped through the roof of her home, bounced off a large console radio, and hit her in the arm and leg.

On November 8, 1982, a 5-inch (13-centimeter)-diameter, 6-pound (2.7-kilogram) meteorite slammed through the roof of a home in Wethersfield, Connecticut, bounced around, and came to rest beneath the dining room table. What made this fall even more unusual was that this house was the second in Wethersfield to be hit in an odds-defying 11 years!

Another recent report came on August 31, 1991, from the small town of Noblesville, Indiana. At about 7:00 P.M. that night, 13-year-old Brodie Spaulding was standing in his yard talking to his 9-year-old friend Brian Kinzie, when they heard a low-pitched whistle followed by a thud. When they looked around the yard, they saw a small black stone sitting in a crater about 3.5 inches (9 centimeters) wide and 1.5 inches (4 centimeters) deep. It was a meteorite, and it had missed hitting the boys by only 11.5 feet (3.5 meters).

A Variety of Samples

If a meteorite is not seen to fall, but is discovered accidentally sometime afterward, it is called a find. Meteorites are often named for the places where they fall, and so have names like Allende (Mexico) and Ahniguito (Greenland).

To date, thousands of meteorites have been recovered, with about six falls and 10 finds added to the list each year. Once the meteorites are recovered, scientists analyze them to determine their makeup and, possibly, their origins. Three main classes of meteorites now exist: *stones* (also known as *aerolites*), *stony irons* (*siderolites* or *lithosiderites*), and *irons* (*siderites*).

Stony meteorites are further divided into two important categories: the *chondrites,* which have inside them small spherical structures known as *chondrules,* and the *achondrites,* which do not.

Stony-iron meteorites contain free metal and stony material in roughly equal amounts. *Pallasites* consist of olivine grains enclosed in metal, while *mesosiderites* are mixtures of metal and silicates.

Iron meteorites consist almost entirely of iron and nickel. While more than 40 different minerals have been identified in these, they are categorized by the amount of nickel they contain. The largest iron meteorite ever found is the 60-ton (54,480-kilogram) Hoba meteorite.

Meteorites are found everywhere on Earth, but one of the best meteorite hunting grounds is along the vast ice sheets of Antarctica, where dark rocks appear in stark contrast to the white snow and ice. Since

The young lady at right holds the dubious distinction of being one of the few Americans who can undeniably trace their automobile damage to a meteorite impact. On October 9, 1992, a 27-pound meteorite fell out of the sky and onto her 1980 Chevy Malibu while it stood parked in the driveway of her Peekskill, New York, home.

1969 many have been found there. Some have lain preserved within the ice for thousands, or even millions, of years.

Whence They Came

Meteorites are important to study, since they represent the only samples of other worlds (except the Moon) that we can test in a laboratory. Which worlds they come from, however, is the focus of much research.

Most appear to be pieces of asteroids —the tiny bodies that orbit the Sun mostly between the orbits of Mars and Jupiter. One small group of meteorites—the *carbonaceous chondrites*—seem to be parts of cometary nuclei that have broken loose and fallen to Earth. Some appear similar in many ways to the lunar rocks collected and returned by astronauts. And still another group may come, not from comets, asteroids, or

Some people have taken up the hobby of meteorite collecting. Robert Haag (below), who travels the world in search of new samples, stands in front of the second-largest meteorite known to exist.

the Moon, but from an even more unlikely place—the planet Mars.

This group of eight unique meteorites are known as the SNCs (after the fall sites named Shergotty, in India; Nahkla, in Egypt; and Chassigny, in France). Their composition of noble gases and nitrogen is similar to that measured for the Martian atmosphere by the Viking spacecraft in 1976. This, coupled with the texture of the meteorites, suggests that they were formed on or in a planet with a strong gravitational field, presumably by extreme heat. But how these rocks could have escaped the Martian gravitational field remains a matter of some debate among astronomers. We won't know the answer, of course, until rock samples can be returned from the Red Planet itself.

Entering the Space Age

As with most other sciences, the study of meteors and meteorites has entered the space age. Ever since 1962, when the *Explorer 16* satellite first studied the penetration of meteoroids into our atmosphere, scientists have used Earth-orbiting satellites and planetary probes to study the number, sizes, and paths of meteoroids throughout our solar system.

Today a number of new meteoroid experiments are being conducted in space. The Galileo spacecraft, on its way toward Jupiter, will collect data on meteoroids in the Jovian vicinity. Ulysses, now being flung high above and below the Sun, will, for the first time, measure the number of meteoroids outside the plane of our solar system. Both carry onboard meteoroid detectors designed to sense meteoroids as small as 0.0000000000000000035 ounce.

In the late 1990s, the Cosmic Dust Collection Facility (CDCF) will fly on board the Freedom Space Station. Its 120-square-yard (100-square-meter) sensing area will measure meteoroid trajectories, and capture and return meteoroids for detailed analysis in terrestrial laboratories.

With all these remarkable new tools, scientists will soon be able to trace back meteoroids' origins to the individual bodies from which they came, and learn about worlds we have never visited.

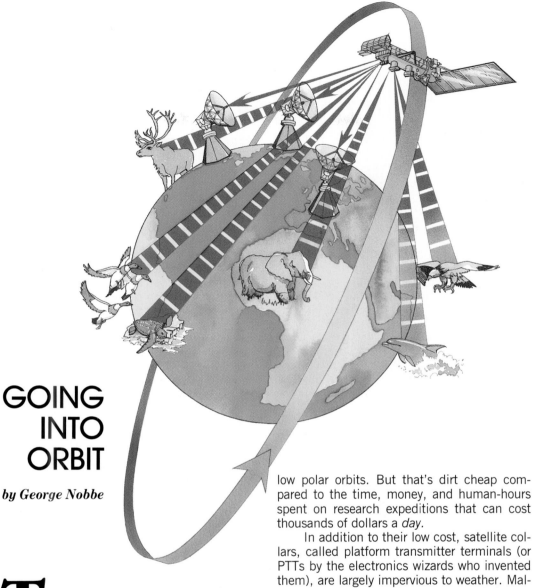

GOING
INTO
ORBIT

by George Nobbe

The newest high-tech weapon in the wildlife biologist's arsenal is the satellite tracking collar. With this radiotelemetry device, scientists can monitor from half a world away the movements of muskoxen on the Arctic tundra, narwhals off the coast of Greenland, migrating sea turtles, and polar bears ranging across international boundaries in their single-minded search for food. Depending on the species being studied, the initial outlay can be as much as $4,000 a collar, plus another $1,500 or so a year to process the information that the gadgetry relays up to Service Argos monitoring systems aboard NOAA (National Oceanic and Atmospheric Administration) weather satellites in

low polar orbits. But that's dirt cheap compared to the time, money, and human-hours spent on research expeditions that can cost thousands of dollars a *day*.

In addition to their low cost, satellite collars, called platform transmitter terminals (or PTTs by the electronics wizards who invented them), are largely impervious to weather. Malfunctions in these battery-powered instruments do occur, and transmissions will occasionally cease for usually inexplicable reasons, but, in general, the devices work well in remote parts of the world that only the most intrepid of well-financed scientists ever dreamed of reaching.

Blubber-borne Transmitters
Compared with more-traditional VHF-radio-telemetry methods, the range of the PTTs is virtually limitless. The signals must reach only as far as the nearest orbiting weather satellite, so researchers no longer have to use ships, planes, and land vehicles to stay within range of their quarries. The PTTs'

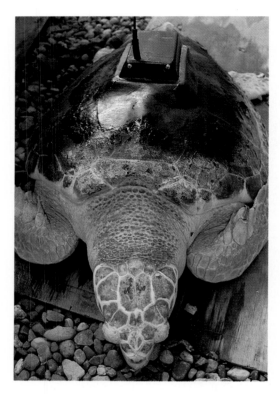

The backpack sported by the loggerhead turtle above carries equipment that allows scientists to use satellites to track the creature's migration routes.

most serious limitation is that the signals will not travel through water. Thus, the satellite system has not altogether replaced VHF-radio transmission as a tracking tool—many scientists, in fact, use both. But in the past seven years, it has become an almost essential one. It is easy to understand why.

Bruce R. Mate, a marine biologist at Oregon State University's Hatfield Marine Science Center in Newport, and a pioneer in the satellite tracking of whales in the Northern Hemisphere, perhaps best describes the value of the technique when he says, "The Arctic can be an unforgiving place, but with satel-

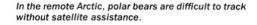

In the remote Arctic, polar bears are difficult to track without satellite assistance.

lites, I can sit here at my computer and get information from any free-ranging whale anywhere on Earth."

Mate studies false killer whales off Hawaii, bottlenose dolphins in the Gulf of Mexico and along the Florida coast, and right whales in Canada's Bay of Fundy. He clamps the transmitters onto the thick blubber of the animals' backs, and the signals are picked up when the creatures surface.

Tagging Tales

Wildlife researchers don't have to understand the Doppler effect or any of the other complex physics involved in this space-age technology, which continues to grow in popularity. Consider the following to be by no means a complete list of recent and ongoing wildlife projects using satellites and PTTs:

• Five musk-oxen in the treeless coastal plain of the Arctic National Wildlife Refuge were tracked by Patricia Reynolds, a biologist with the U.S. Fish and Wildlife Service (FWS), during the winter of 1991—92 to learn more about their distribution, movement, and feeding habits. If the refuge is opened to oil and gas exploration, such data could prove vital to the animals' well-being.

• Wolves had always been studied from fixed-wing aircraft until Warren Ballard of the Alaska Department of Fish and Game began using satellite radio collars in 1988. He is trying to track large adult wolves in that state's Kobuk Valley National Park to determine if they are migratory.

• Thirty caribou and seven musk-oxen were collared in the Yukon Territory during the winter of 1990—91 by the Alaska Fish and Wildlife Research Center in Anchorage in cooperation with the Canadian Wildlife Service. And the center's biologist Tom McCabe is using PTTs to follow moose in Alaska's Yukon Delta Refuge.

• Brown bears have proven less cooperative, according to McCabe. Six grizzlies in the Arctic coastal plain didn't seem to

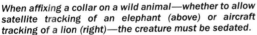
When affixing a collar on a wild animal—whether to allow satellite tracking of an elephant (above) or aircraft tracking of a lion (right)—the creature must be sedated.

mind the 3-pound (1.36-kilogram) transmitters when they were first collared in 1988. By the fall of 1990, however, none of the devices were transmitting. Perhaps the bears had pulled them off when they used steel fencing in the area as a scratching post. McCabe was studying the bears' denning habits and trying to determine if the bruins deliberately sought out caribou during the calving season. Such behavior appears to be more common in the foothills of the coastal plain.

• About 30 female polar bears in the Beaufort, East Siberian, and Chukchi seas have been collared by Steven C. Amstrup and Jerry Garner. The two FWS scientists work with sows because no one has yet designed a collar that will stay on the 45-inch (115-centimeter) necks of the male polar bears. "It's like trying to put a collar on a traffic cone," says Amstrup.

A tragic accident in polar-bear research underscored the value of satellite tracking. Two bear biologists, John Bevins and George Menkens, were killed in October 1990 when their small aircraft crashed in the Beaufort Sea north of Barrow, Alaska. Their colleagues believe that using PTTs in areas where the weather is unpredictable and flying conditions are often hazardous could take some of the risk out of future bear research.

• Dozens of free-ranging manatees, or sea cows, have been tagged in Florida by Jim Reid and other scientists with the U.S. Fish and Wildlife Service's Sirenia Project. They hope to learn more about the animals' movement patterns and ecological requirements. These big marine herbivores are in constant danger of being struck by boats as they feed and travel in rivers and shallow coastal waters. Because radio signals won't travel

Beam Me Up, Scottie

Imagine sitting at a computer in New York City and being able to pinpoint the location of an elephant deep in the forests of Cameroon. It's not a farfetched scenario.

Radio signals from an African forest elephant recently outfitted with a special collar are being transmitted via satellite to a receiving station in France, which in turn transmits the location data to a laboratory in Maryland. Technicians in Maryland then relay the information on to a new laboratory for biotelemetry studies, opened in late 1991 by the New York Zoological Society (NYZS) in its Animal Health Center at the Bronx Zoo.

Locating the lab at the zoo means that society researchers can test and improve expensive and sensitive electronic devices before they are tried in the field. The zoo's Mammal and Animal Health departments are engaged in a joint effort with James Buddy Powell, a field biologist with the society's Wildlife Conservation International. Powell is studying the endangered forest elephant in Korup

The population of Asian elephants at the Bronx Zoo was used to test the accuracy of the high-tech biotelemetry equipment. The results far exceeded expectations.

National Park in Cameroon, West Africa. Because of the difficulty of working in dense forest habitat, this species' habits are not as well known as those of its close cousin, the savanna elephant.

By combining traditional VHF-radio transmitters with a newer generation of satellite transmitters, Powell hopes to increase his odds of following individual animals. He plans to track the

through salt water, the Florida team designed their transmitter to be housed in a floating cylinder with an external whip antenna, all tethered to a belt around the manatee's tail. This unique tag design has enabled manatee tracking to be among the most successful of Service Argos's satellite-telemetry programs. Seven years' worth of data along both coasts of Florida have pinpointed important manatee habitats and documented movements and migrations of these lumbering giants, which can extend from Miami to southern Georgia.

Recently several manatees were tagged in Puerto Rico with PTTs, marking the first satellite tracking of these sirens outside Florida. And a cooperative effort with researchers at James Cook University in Townsville,

Australia, has produced a similar tag for dugongs, Indo-Pacific cousins of manatees.
• John Musick of the Virginia Institute of Marine Science (VIMS) in Gloucester Point has been tracking loggerheads and leatherbacks in and out of the turtle-rich waters of Chesapeake Bay. He says the transmitters can be bolted through the carapaces of hard-shelled sea turtles, but getting something that will stay on the softer shell of the leatherback is extremely difficult.

The research concentrates on the migratory and diving behavior of sea turtles. "We'd like to find out where in the water column they feed, travel, rest," says Musick.

Two sea turtles released late in 1989 rode the coastal currents as far south as Key Largo and Cape Canaveral in search of elephants initially from the ground with the VHF unit, but when he loses an elephant's signal (after all, an elephant can travel through the jungle much faster than a human), Buddy faxes a message to the zoo laboratory, via ham radio and telephone, and the biotelemetry studies team in New York accesses the satellite company's main computer in Maryland to determine the elephant's last satellite-determined location. That information is then faxed back to Powell in Cameroon, so he can resume tracking.

An important element in this project is the zoo's population of Indian elephants. Rather than ship the high-tech equipment to western Africa, transport it to Korup, and then find out it isn't working properly, the society's researchers first tested it out on zoo elephants. Not only can the staff make sure the equipment is working properly, but they can determine if it is "elephant-proof" (which it is), and if they can pinpoint the true geographic location of each animal. On average, locations have been found to be accurate to within about 1,650 feet (500 meters), which may

sound like a lot, but knowing a wild elephant is within three-tenths of a mile of a certain point is a giant step forward.

In April 1992, William Karesh, the NYZS field veterinarian, successfully radio-collared an adult African elephant in Zäire's Garamba National Park. This opportunity to field-test the equipment will also provide information to further refine the satellite-tracking system.

NYZS staff are working on several other wildlife radiotelemetry projects in the new lab. A plastic egg filled with sophisticated sensory equipment is being used to study incubation in white-naped cranes: temperature and humidity within the nest, and even how often the parents turn the eggs. Transmitters are being used to detect ovulation in gaur (a species of Asian wild cattle) and thermoregulation in monitor lizards.

The New York Zoological Society's biotelemetry lab has opened up a whole new world of research possibilities. And it provides both field scientists and zoo curators new insights into the habits of rare and endangered wildlife.

Fred W. Koontz

warmer water and horseshoe crabs. One turtle was back in Chesapeake Bay by June 1990; the other stayed in Key Largo.
• Elephant seals off Macquarie Island in the southern Tasman Sea are the targets of an ongoing study by the Australian National Antarctic Research Expedition. The Japanese have tracked the migration of bottlenose dolphins riding the Koroshio Current off the Kii Peninsula. And the French have monitored the incredible marathon flights of the wandering albatross in the southwestern Indian Ocean.

An International Effort
Designing workable transmitters for this Noah's ark of research animals has become international in scope as well. The leader in

the field is Telonics Inc. of Mesa, Arizona. The Japanese have two companies, Toyocom and Nippon Telegraph and Telephone. Other specialists include Mariner Radar in England and Wildlife Computers, a small Seattle firm run by Roger Hill.

Miniaturization is the key to success in transmitter design, says Hill, who agrees that chasing animals with ships and planes is enormously expensive and inefficient when compared to the more efficient manner by which satellites track wildlife.

Finding out where these animals go used to suffice, but now researchers want to know what they're doing, too, says Hill. He has yet to come up with a sensor that will tell his customers when their subjects are eating. But don't bet against it.

LOOKING AT EARTH

*by Priscilla Strain
and Frederick Engle*

A satellite captured this image of fire-engulfed Yellowstone National Park in 1988. Seventy years earlier, airplanes like the Curtiss Jenny (above) were used to get overviews.

Sometimes the only way to understand what's happening in the world is to rise above it. In ancient times, travelers would hike up hills, and soldiers would climb towers, to better observe the land and prepare for obstacles or dangers. Today aircraft and spacecraft look down on Earth and provide new perspectives on changing populations and ecosystems, as well as on the beauty and symmetry of landscapes.

The use of instruments to gather data from above dates back to the late 1850s, when photographers took their early cameras up in balloons. The first to achieve success was the French photographer and publisher Gaspard Felix Toumachon (also known as Nadar), who produced a crude aerial photo of the outskirts of Paris in 1858. Kites and even pigeons were also used in early attempts at remote sensing.

The airplane provided a far more effective platform for looking down on Earth. One early model to be used this way was the open-cockpit de Havilland DH-4 biplane, which flew as both a bomber and a photo-reconnaissance vehicle during World War I. After the war the DH-4 continued to provide aerial photographs, flying reconnaissance missions for the U.S. Geological Survey (USGS).

In later years, aerial photography found more and more uses. After a severe earthquake, for instance, rescuers would use photos taken from airplanes and helicopters to identify the sites in most desperate need of assistance. And in 1962, when underground coal fires broke out around Centralia, Pennsylvania, aerial heat-detecting imaging systems revealed those areas of the ground that were warmer than normal, enabling workers to target their firefighting efforts.

Aerial photography has also served the public in less obvious ways. In 1979, when workers constructing a recreational park on Neville Island, Pennsylvania, fell ill, public-health officials suspected that some toxic substance was to blame. By studying aerial photos of the island from as far back as 1938, they discovered pools and trenches where hazardous solid and liquid wastes had been dumped for years.

Views from Space

Astronauts on the Gemini missions of the 1960s took the first photographs from space used to study Earth's large geological structures. Those successes paved the way for other space-photography missions, and they also helped inspire another kind of imaging effort, the Landsat (for "land satellite") program. Since 1972 five Landsat spacecraft

The yellow specks above indicate actively blazing sites, while the orange areas represent already burnt areas. A space shuttle observed sediment flowing into the Mozambique Channel in Madagascar (below) due to soil erosion.

have been sent into orbit around Earth, equipped with imaging systems that collect data in different wavelengths of the electromagnetic spectrum. Today Landsats survey soil for erosion and moisture, monitor droughts, map floodplains, explore for oil and minerals, observe the circulation patterns of oceans, track wildlife, and measure pollution.

Each Landsat uses an oscillating mirror that scans Earth's surface from west to east. During these scans, rows of detectors measure the intensity of energy reflected or emitted from the surface, and encode this information as a series of 0's and 1's (binary signals) to be transmitted back to Earth. This process produces what is known as digital imagery.

Data from both visible and thermal wavelengths show the intense heat (in red, above) created by the Chernobyl reactor explosion in 1986. Landsat imagery was superimposed on topographic data to produce this view (right) of Japan's Mount Fuji.

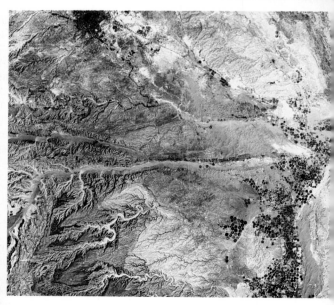

Photos of Madagascar expose the massive soil erosion caused by intensive cultivation (left). Landsat images (above) highlight the dry stream beds that snake across Saudi Arabia's arid Najd plateau.

Digital images are different from conventional photographs. For one thing a digital image's colors are "false"—computer scientists assign the data from each wavelength a primary color (red, green, or blue), then overlay the results to produce a composite image. Though the colors do not necessarily correspond to those the human eye would see, these pictures are nonetheless informative. Red, for example, usually represents near-infrared, and information gathered in these wavelengths indicates the health and vigor of crops and other plant life, with bright red representing healthy vegetation, and lighter red and pink showing vegetation under stress.

Satellites like those of the Landsat system are classified as passive remote-sensing devices because they measure energy naturally reflected or radiated from Earth. Radar is an active sensing system: it provides its own signal, bouncing it off a target, such as an aircraft or a mountain, and recording the returning signal. Because radar systems do not rely on sunlight and are not hampered by cloud cover, they can operate day or night and in any weather. This makes them especially useful in tropical areas, where clouds nearly always block the view of passive sensors. Radar has also proved useful in arid regions to locate ancient waterways now dry and covered by sand. Last fall, archaeologists using radar and other satellite imagery discovered old caravan trails that led to the remains of a long-lost city in Oman—possibly the fabled trading center of Ubar, which flourished more than 2,000 years ago.

Over the past two centuries, humankind has ventured progressively farther into the heavens, pursuing ever more challenging adventures. Today, at the end of the 20th century, more than 200 men and women have flown into space, a score having traveled as far out as the Moon. Yet for many who have participated in these exciting voyages, the best part of all was looking homeward.

Astronomy

Photos of Mars have revealed a jagged surface gash that scientists believe was caused by a huge landslide.

THE SOLAR SYSTEM

New X-ray images of the Sun confirm that clusters of sunspots coincide with X-ray flares in the corona. A single small flare can trigger a chain reaction of larger flares. Other solar observations show that the well-known solar cycle of 11 years for sunspots may be only a component of a longer cycle of 19 years that begins with increased magnetic activity at high solar latitudes.

Images from the Magellan probe show that strong winds on Venus change the planet's topography much as winds move dunes on Earth. Venus's surface also has tectonic scars similar to the trenches on Earth that surround the Pacific basin.

Observations of the asteroid Apollo (3103) 1982BB show that it contains enstatite, a type of iron-free, magnesium-rich silica found in certain meteorites. Researchers believe that those meteorites are fragments of the asteroid, which appears to have come from the asteroid belt between Mars and Jupiter.

The asteroid Gaspra, observed closely in 1992, is the most irregularly shaped object yet known in the solar system; it also has a magnetic field.

A second large asteroid-like object is 1992AD, a 125-mile (200-kilometer)-diameter object in an elliptical path between the orbits of Saturn and Neptune. Astronomers do not know where this strange reddish-colored object came from, but expect to learn from it more about the early solar system.

Comets may be more like frozen mudballs than dirty snowballs, according to new findings, and closer in composition to Pluto and Triton (Neptune's largest moon) than to other solar-system objects. Scientists further believe that short-period comets may come from a region between Neptune and Uranus.

Infrared observations established that Pluto has more nitrogen in its atmosphere than carbon monoxide, a finding that may cause scientists to revise their model of the early solar system.

OUR GALAXY—THE MILKY WAY

Two or possibly three planetlike masses orbiting the pulsar PSR 1257+12 provide evidence that the pulsar wobbles; astronomers suggest that orbiting planets may cause the wobbles. Each body has a mass about three times that of Earth. However, many questions remain about whether pulsars can have planets.

In February 1992, Nova Cygni blazed into view for three days. This was the first nova visible to the naked eye since 1975.

A search for brown dwarfs detected six candidates with the infrared brightness characteristic of these elusive objects. The brown dwarfs are in the comparatively young (about 600 million-year-old) Hyades star cluster, 150 light-years from Earth. Scientists think that brown dwarfs may be the "missing link" between stars and planets.

The Hubble Space Telescope (HST) has found a star at the center of NGC 2440 with a temperature of about 360,000° F (200,000° C)—the hottest star yet discovered. The star is a white dwarf in its final evolutionary phase before burnout.

New observations of the star Beta Pictoris have allowed astronomers to distinguish its bright central core from an outer disk. The disk may be gas and dust in the process of forming a solar system.

Astronomers examining a dark cloud called B355 observed that it is collapsing, probably forming a star. This is the first time actual star formation has been seen.

THE UNIVERSE

The HST observed a gaseous ring around Supernova 1987A, a ring 1.37 light-years in diameter. The supernova is now emitting radio signals that probably originate from an extended clump of gas surrounding the original explosion site, between the gas ring and the extinct star. The blast wave from the supernova is moving outward at roughly 18,600 miles (30,000 kilometers) per second.

Astronomers have discovered about 500 previously unseen low-surface-brightness galaxies. The 24 studied thus far have concentrations of oxygen, neon, sulfur, and nitrogen ranging from $1/10$ to $1/50$ of those in the Milky Way. Because all galaxies start out with hydrogen and helium as their sole chemical constituents, and generate heavier elements only through the birth and death of stars, these low concentrations indicate that the dim galaxies produced their first glimmers of starlight only recently, in the past few hundred thousand years. These galaxies may represent an important component of the "dark matter" long sought to explain the evolution of the universe.

In a search for dark matter, scientists have studied 300 distant quasars. They look for distortions along the line of sight, called gravitational lensing, where light rays of a single image change into several distinct images. So far the most-distant object shown to have undergone such distortion is quasar 1208+101, 12 billion light-years from Earth. The ROSAT satellite also found about 15 "supersoft" stars that emit very-low-energy X rays outside our galaxy. These stars represent a new type of white dwarf.

The galaxy Arp 220 contains huge, youthful star clusters that probably formed after two spiral galaxies collided to form the galaxy. These newly discovered star clusters account for about half the galaxy's brightness and are about 20 million years old. The clusters are doomed to be torn apart within the next few hundred million years by internal tidal forces.

The Hubble Space Telescope has also found an x-shaped spot in the heart of the spiral galaxy M52. The spot may be a black hole. Astronomers found a probable black hole at the core of the galaxy NGC 3115; the black hole has a mass a billion times that of the Sun, and is likely a dead quasar.

Observations from the Cosmic Background Explorer (COBE) verify models of the early universe, showing lumps and bumps that later became galaxies and star clusters.

The brightest object ever observed, IRAS F10214+4724, may be a primeval galaxy in which giant gas clouds are still collapsing to form stars, and where starbirth has only recently begun.

The galaxy Makarian 421, about 40 million light-years from Earth, has a central, quasarlike object emitting high-energy gamma rays, detected for the first time from outside the galaxy. The Earth-orbiting Compton Gamma Ray Observatory (GRO) made the observations.

Extraordinary color photographs of galaxies and other celestial objects help to show the true color evolution of galaxies. The photographs reveal the blue light of hot young stars; brown dust stripes; old, cool, pale-yellow stars in the cores of old galaxies; the red glow of the exterior nebulosity of a planetary nebula; and dark globules of dense dust clouds. Special photographic methods have also revealed massive but nearly invisible objects such as the faint galaxy Malin I.

The HST took the sharpest existing image of gravitational lensing in a search for dark matter. The observations of remote galactic cluster AC 114 are of unprecedented sharpness. With improved accuracy, astronomers can better determine the concentration and mass of the material acting as a lens.

Katherine Haramundanis

Space Science

EXPLORING THE UNIVERSE FROM AFAR

Orbiting telescopes delivered new and important results about the evolution of our universe and of our place in it.

Despite its optical flaws, the Hubble Space Telescope (HST) conducted long-term observations of several planets and moons, supplied strong evidence that massive black holes inhabit the hearts of two nearby galaxies, and helped scientists determine more accurately the expansion rate of our universe.

The orbiting Compton Observatory (CO) teamed with the Roentgen Satellite (ROSAT) to learn more about Geminga, a gamma-ray pulsar. And it detected, for the first time, high-energy gamma rays from active galaxies possibly powered by black holes.

The Cosmic Background Explorer (COBE) settled a quarter-century debate when it discovered variations in the temperature of the cosmic background radiation, the afterglow of the Big Bang. Cosmologists believe that these tiny fluctuations might explain how galaxies and galaxy clusters originated.

On Columbus Day the National Aeronautics and Space Adminstration (NASA) began using ground-based radio telescopes to conduct its High Resolution Microwave Survey (HRMS), a comprehensive, decade-long search for intelligent radio signals from among the stars.

ROBOT PROBES

During 1992 a number of probes were active in exploring the universe, the solar system, and the Earth-Sun environment.

On September 25, Mars Observer (MO) began its long-awaited journey to the Red Planet. Three months after its arrival on August 24, 1993, the satellite will begin its full-Martian-year examination of the planet's weather and geology. (See also the article beginning on page 14.)

Ulysses flew by Jupiter on February 8, 1992, gaining momentum to break free of the plane of our solar system and enter a polar orbit around the Sun. Once there, Ulysses will study the Sun's polar magnetic field and solar wind.

Having mapped 98 percent of the Venusian surface with radar, the Magellan spacecraft lowered its orbit of Venus on September 14, and began another 243-day cycle—this to obtain a global map of Venus's gravity field.

On December 8 the Galileo spacecraft whipped by the Earth a second time, receiving one last gravity assist for its journey to Jupiter. With Galileo on course, scientists continued to try to release its stuck antenna. Once in Jovian

Mae C. Jemison, a physician and chemical engineer, became the first black woman in space when she flew as a crew member aboard the Endeavour in September 1992.

orbit in late 1995, Galileo will launch a probe into the planet's atmosphere, and will study Jupiter's four major moons up close.

Giotto, of the European Space Agency (ESA), most famous for its 1986 rendezvous with the nucleus of Halley's comet, flew within 123 miles (200 kilometers) of Comet Grigg-Skjellerup in July 1992, establishing a new record for the closest visit yet to the core of a comet.

On July 2, 1992, the Solar Anomalous and Magnetospheric Particle Explorer (SAMPEX), the first in a new series of Small Explorer missions, entered orbit and began using the Earth as a giant magnetic shield so that its four instruments can learn the origin of charged particles from space.

A joint U.S./Japanese project, Geotail, the first of five international satellites to help scientists better understand the interaction of the Sun, the Earth's magnetic field, and the Van Allen radiation belts, was launched on July 24, 1992.

SPACE-SHUTTLE PROGRAM
During 1992 eight shuttle missions were scheduled and flown, demonstrating the program's maturity and reliability while reducing operational costs by 9 percent over the previous year.

Highlighting the missions was *Endeavour*'s maiden voyage in May 1992, in which its crew rescued a wayward satellite and set three new spacewalking records.

In September 1992, Dr. Mae C. Jemison became the first African-American female astronaut to fly in space. Also flying during the year were Dr. Claude Nicollier, the first Swiss astronaut, and Drs. Franco Malerba and Mamoru Mohri, the first Italian and Japanese payload specialists, respectively. (For more details, see the chart on pages 54-55.)

SPACE STATION *FREEDOM*
Despite heated debate in both houses of Congress over the station's future, funding in the amount of $2.1 billion was appropriated, $150 million less than the amount requested. More than 400 pieces of developmental hardware now exist, and tests of the station's construction have begun.

INTERNATIONAL COOPERATION
In perhaps the most internationally cooperative year in spaceflight history, U.S. President George Bush and Russian President Boris Yeltsin signed a joint space-research agreement in June 1992. One month later, NASA joined with the Russian firm NPO Energia to study how the Russian Soyuz-TM vehicle could be used as an interim Assured Crew Return Vehicle. In October, NASA and the Russian Space Agency agreed to fly a Russian cosmonaut on a U.S. space shuttle, and a U.S. astronaut on the Russian *Mir* space station. Plans also called for a rendezvous and docking of the shuttle and *Mir*.

Scientists from NASA, the European Space Agency (ESA), the Canadian Space Agency (CSA), the French National Center for Space Studies (CNES), the German Space Agency (DARA), and the National Space Development Agency of Japan (HASDA) cooperated in the International Microgravity Laboratory-1 (IML-1). This carried 29 life-sciences experiments on the longest shuttle mission to date. In September, Spacelab-J, the Japanese space laboratory, carried seven life-sciences experiments into space.

Dennis L. Mammana

Funding for the space station Freedom continued in 1993, but not at the rate hoped for by NASA officials.

UNITED STATES MANNED SPACE FLIGHTS – 1992

MISSION	LAUNCH/LANDING	ORBITER
STS-42	Jan. 22/Jan. 30	*Discovery*
STS-45	Mar. 24/Apr. 2	*Atlantis*

◄ Intelsat VI, a wayward satellite, was successfully captured by astronauts during the space shuttle Endeavour's maiden flight.

STS-49	May 7/May 16	*Endeavour*
STS-50	June 25/July 9	*Columbia*
STS-46	July 31/Aug. 8	*Atlantis*

◄ Astronaut Kenneth D. Bowersox prepares to use exercise equipment aboard Columbia during the shuttle's two-week mission.

STS-47	Sept. 12/Sept. 20	*Endeavour*
STS-52	Oct. 22/Nov. 1	*Columbia*

◄ The Laser Geodynamics Satellite, shown during deployment from the shuttle Columbia, will help scientists make earthquake predictions.

STS-53	Dec. 2/Dec. 9	*Discovery*

UNITED STATES MANNED SPACE FLIGHTS – 1992

PRIMARY PAYLOAD	REMARKS
International Microgravity Laboratory: 54 experiments from 16 countries to study how cells, plants, insects, and humans adapt to microgravity.	• Stayed aloft an additional day because crew had consumed less water and electricity than expected.
Atmospheric Laboratory for Applications and Science (ATLAS-01): Study of the long-term variability in the energy radiated by the Sun and the variability in the solar spectrum.	• One-day liftoff delay due to fuel leaks.
Intelsat VI: Rescue, repair, and redeployment of satellite that had been stranded in a useless orbit since 1990.	• Maiden voyage of orbiter built to replace *Challenger*. • Two failed attempts to catch wayward satellite via robot arm were followed by a successful three-person snag of Intelsat VI. A new booster motor was attached, and the satellite was successfully redeployed. • First use of a drag chute by a landing shuttle.
U.S. Microgravity Laboratory: 31 experiments to study crystals for medical research, such as HIV-related protein crystals with potential for researching AIDS drugs; physiological effects of long-duration spaceflight.	• Record-setting two-week mission.
Tethered Satellite System (TSS): Deployment and recapture of 1,139-pound tethered satellite. Goal: to release satellite 12 miles out to study effects of electrically conducting tether in Earth's magnetic field, and then reel it in.	• Stuck electrical cable delayed release of satellite. • Releasing mechanism stalled, snarling tether. • Four attempts over two days to unreel satellite failed; satellite never went farther than 845 feet from shuttle.
Spacelab J: 43 experiments in materials, crystal growth, and life sciences, including frog-embryology study designed to answer questions about gravity's effect on animal development; bone formation in 30 fertilized chicken eggs.	• All but nine experiments devoted to Japanese-sponsored research. • Mission experiments required 24-hour surveillance by crew members.
Laser Geodynamics Satellite 2 (LAGEOS-2): Deployment of 2-foot satellite with mirrored surface to reflect laser beams as an aid to study movements of Earth's crust and to help formulate earthquake predictions. Launched into 3,600-mile-high orbit that it will occupy for 8 million years.	• LAGEOS-2 contains two messages: a binary formula for numbers 1 through 10; and maps showing Earth's orbit and position of continents as they were 268 million years ago; at launch time; and as predicted for 8.5 million years in future.
DOD-1: Classified satellite deployed by Department of Defense.	• Unexpected maneuver to escape collision from space debris.

EARTH and the ENVIRONMENT

CONTENTS

WHAT HAPPENED AT THE EARTH SUMMIT?

by Devera Pine

It could have been the be-all and end-all of the environmental movement: A 12-day international conference bringing together both world leaders and concerned citizens in a once-and-for-always attempt to devise solutions to the world's ecological woes. Indeed, the United Nations (U.N.) Conference on Environment and Development, held June 3 to 14, 1992, in Rio de Janeiro, Brazil, had a full plate of environmental issues on its agenda, from greenhouse gases and climate change to deforestation and biological diversity. And the conference certainly drew world attention, bringing together more than 8,000 delegates, 1,400 representatives of nongovernmental organizations, some 9,000 journalists, 200,000 local residents, as well as leaders and celebrities ranging from then-President George Bush to the Dalai Lama of Tibet. Described at one point as a "cross between a Cecil B. DeMille and a Robert Altman movie," the conference included exhibitions,

lectures, and, naturally, demonstrations. Not surprisingly, the summit made the front page of *The New York Times* six times in its 12 days; it was described as the largest gathering of world leaders in history.

But despite these outward signs of success, by most accounts the conference was less than effective. At best, the summit was a hopeful starting point for global cooperation on environmental issues. At worst, it was a dismal failure.

In the Making
The germ of the idea for the summit actually began in 1987, when the World Commission on Environment and Development presented a report to the General Assembly of the

What kind of planet will our children inherit? Will there be room to roam, air to breathe, and food to eat? These are just some of the questions discussed at the Earth Summit meeting held in Rio de Janeiro, Brazil, in 1992.

United Nations recommending that the U.N. hold a conference on protecting the environment and promoting sustainable development. This call for a conference came almost 15 years after the first such worldwide meeting on the environment—the U.N. Conference on the Human Environment, held in Stockholm, Sweden, in 1972. Among its achievements, the Stockholm conference created the U.N. Environment Programme (UNEP), called for global cooperation in reducing pollution of the oceans, and established a monitoring network. Many experts also believe that the Stockholm conference opened the eyes of both citizens and governments to the need to protect the environment. In recommending the Rio conference, experts hoped to achieve even more.

In December 1989, the U.N. General Assembly passed Resolution 44/228, establishing the United Nations Conference on Environment and Development (UNCED), commonly known as the "Earth Summit." The stated purpose of the conference was to "elaborate strategies and measures to halt and reverse the effects of environmental degradation in the context of increased national and international efforts to promote sustainable

and environmentally sound development in all countries." Organizers hoped to reconcile the seemingly disparate issues of development and environment; they hoped to plot a course of economic growth and development that followed an environmentally friendly path.

Behind Closed Doors

Negotiations for the conference began with a series of PrepComs—meetings of the UNCED Preparatory Committee, held in Nairobi, Kenya; Geneva, Switzerland; and New York City over the course of two years. At these meetings, government delegates negotiated several draft agreements that would

be finalized and then signed at the summit in Rio. One unique aspect of the Earth Summit was the degree of participation by citizens' groups—purely civilian organizations not affiliated with any government agency. "There was a decision early on that there would be direct access by citizens' groups," says Scott A. Hajost, international counsel for the Environmental Defense Fund, former principle oceans and environmental lawyer for the State Department, and former deputy director of international affairs at the Environmental Protection Agency (EPA). These groups advised the PrepComs right from the start, giving input, for example, on research papers, studies, and reports.

Unfortunately, right from the start, political concerns played a role in determining which environmental issues would and would not be discussed. For example, both the governments of OPEC countries and the United States quashed any attempt to examine the use and environmental impact of fossil fuels. France, which derives most of its energy from nuclear-power plants, headed off any discussion on the use of atomic energy. The United States also kept the disposal of radioactive waste at sea off the agenda, apparently because the U.S. wanted to retain the option of disposing of nuclear submarines at sea, according to Hajost. Other political considerations prevented important issues, such as the threat of militarism to the environment and development, from ever reaching Rio.

In the end, PrepCom delegates managed to negotiate at least some of the text of documents. These included the Rio Declaration and Agenda 21. Any issues not resolved in the PrepComs would be finalized at Rio before the leaders of the world's governments arrived to sign the agreements.

For the United States, however, there remained several hurdles: The first was con-

ERENCE ON
VELOPMENT

June 1992

More than 100 world leaders met to confront the planet's most pressing ills in what is considered the largest gathering for international cooperation in history.

vincing President Bush to attend the meeting. As late as three months before the summit, the president had not publicly stated that he planned to attend. To pressure him to participate, environmental groups set up 800 numbers for people who wanted to send an "Earth Telegram" to the president; they also ran commercials. In one ad shown in movie theaters, actor James Earl Jones presented a cogent argument on why President Bush should attend the conference. In the end, given the pressures of an election year and the fact that all the leaders of the major industrialized nations were attending, President Bush eventually did go to Rio. The other hurdles for the United States came in negotiating its way through the remaining issues.

On Paper
In total, the Earth Summit produced five major documents: the Rio Declaration, Agenda 21, treaties on climate change and biodiversity, and a statement of forest principles. On the whole, most environmentalists consider these documents disappointing. "Greenpeace took the position that on bal-

ance, it was regrettably an historic failure," says Clifton Curtis, Greenpeace International's chief lobbyist for the Earth Summit. "For Greenpeace, each fell far short of the mark." Again, political agendas—very often the U.S. agenda—played a major role in weakening each of these documents, according to environmentalists.

The Rio Declaration
The Rio Declaration on Environment and Development is a set of 27 principles intended to guide nations in environmental protection and development. Although the declaration is not legally binding, experts note that other U.N. declarations—such as the one on human rights—have gained almost the force of law over the years and carry moral weight.

The statement recognizes that economic development must be "sustainable" —in other words, economic progress should work with, not against, the environment. The principles of this document call for the eradication of poverty as an indispensable requirement for sustainable development, appropriate access to information concerning the environment, and the opportunity to participate in decision-making processes.

In addition, the Rio Declaration recognizes the role of women in environmental management; it calls for polluters to bear the cost of pollution (known as the "polluter pays principle"); and it calls upon states to recognize and duly support the identity, culture, and interests of indigenous peoples. The "precautionary principle" of the declaration says that a "lack of full scientific certainty shall not be used as a reason for postponing cost-effective measures to prevent environmental degradation." According to the Citizens Network for Sustainable Development, the Reagan and Bush administrations used this argument to block environmental legislation in the U.S.

Negotiation of the Rio Declaration was not without its conflicts. At one point, for instance, it was rumored that U.S. delegates wanted to rewrite portions of the document, including the statement about the right of individual nations to development. The U.S. argued that in the past, these kinds of statements were used by some governments as a cover for human-rights abuses. In addition, the U.S. objected to a principle that called for the protection of the environmental resources of people under oppression, domination, and occupation.

Because reopening negotiations on the Rio Declaration so late in the negotiating process might have destroyed the entire document, the U.S. ultimately agreed to accept the declaration as long as all references to people under occupation were removed from Agenda 21. In addition, the U.S. issued a statement outlining its objections to other principles in the declaration.

Agenda 21
Agenda 21 is a hefty, 800-page, 40-chapter document that serves as a plan of action for implementing the environmental issues discussed at the summit. Like the Rio Declaration, Agenda 21 is not legally binding.

Agenda 21 is divided into four sections: social and economic dimensions (combating poverty, changing consumption patterns); conservation and management of resources for development (protection of the atmosphere, combating desertification and drought); strengthening the role of major groups (recognizing the role of indigenous, global action for women toward sustainable and equitable development); and means of implementation (financial resources and mechanisms, international institutional arrangements).

The negotiations of many of the points included in Agenda 21 were difficult. Even the document's preamble, for instance, brought forth objections from Saudi Arabia and Kuwait, whose delegates wanted the phrase "environmentally sound energy systems" to read "environmentally safe and sound." The delegations said the change was needed to ensure that the document did not favor nuclear power. According to the *Washington Post,* however, energy analysts suggested that the delegates wanted to use concerns over the safety of nuclear power to "discourage its growth as a competitive threat to oil."

Another point of contention was chapter 4 on changing consumption patterns, especially in the developed countries. "There is language that recognizes overconsumption, but the whole question of consumption of fossil fuels in the North was not adequately addressed," says Hajost.

Money Talks
One of the most fundamental problems with Agenda 21 is its failure to address the issue of financial assistance for implementing environmental programs. At the conference, developing countries of the Southern Hemisphere contended that if the North wants to ensure that they industrialize with an eye to the environment, the North would have to pledge more money to the cause.

In fact, the U.N. estimated it would take $125 billion a year in new aid to fund Agenda 21. To help meet that goal, Agenda 21 calls for increasing the percentage of aid that industrialized countries give to developing nations—from 0.4 percent of their gross national products to 0.7 percent.

At the conference, developing nations at first wanted this increase in aid to be achieved by the year 2000. But as time ran out on the negotiations, they instead agreed to stating that the aid be supplied "as soon as possible." In addition, the document states that industrialized nations should "reaffirm" their commitment to the new level of assistance. As a result, countries that never "affirmed" the aid in the first place—the United States among them—were not really agreeing to contributing more money.

Another goal of the conference was to increase the amount of money provided to the International Development Agency of the World Bank, to restructure the Global Environmental Facility, and to establish an "Earth increment" fund. Many environmentalists decry the lack of funds for environmental projects: "The World Bank spends 1 percent of energy lending on energy efficiency and renewable sources of energy," says Liz Barratt-Brown, senior attorney for the Natural Resources Defense Council (NRDC). "There is clearly international financing that should be rechanneled to sustainable use."

Agenda 21 did not completely resolve the financing issues, though its goal is that the Global Environmental Facility will double its financial contribution in 1993. In addition, several countries announced that they will increase aid to sustainable development: Japan committed to increasing aid to $7 billion over the next five years, the European Community (EC) committed to $4 billion, and the United States said it would contribute $175 million per year. Plus, the United Nations Development Programme will increase funding for environmental projects to $2 billion over the next five years. Overall, however, these pledges amounted to only $6 billion to $7 billion a year; the summit organizers had hoped for $10 billion.

On a more positive note, Agenda 21 did establish a U.N Commission on Sustainable Development. This commission, with members from 53 nations, is charged with ensuring the effective follow-up of the summit.

Hope for the Environment

In addition to Agenda 21 and the Rio Declaration, the Earth Summit included the negotiation and signing of two new treaties—one on biodiversity, the other on climate change. Unlike the other documents of the Earth Summit, both of these treaties will become international law once they are ratified.

The biodiversity treaty is aimed at preventing the loss of species, habitats, and ecosystems. The treaty requires nations that ratify it to protect endangered species, protect and restore ecosystems, and integrate conservation into economic development. These provisions would allow for using bioresources for tourism, food, and the development of medicines.

One of the most hotly debated issues of this treaty had to do with the ownership and patent rights of technology developed from materials found in "undeveloped" areas, such as tropical rain forests. According to the Citizens Network for Sustainable Development, today the "extractors" of a given country's biological diversity are often white people who work for northern transnational corporations or institutions. Most often the biological materials are extracted from southern rain forests, where indigenous peoples live in their historic homeland. Moreover, says the Citizens Network, for many in the Southern Hemisphere, the "existing economic arrangements regarding biological resources are extremely 'neo-colonial.' "

Language in the treaty referred to the equitable sharing of revenues among source countries and developers of biotechnology, language to which U.S. negotiators objected. In the end, the U.S. became the only major industrialized country to refuse to sign the treaty, making it, according to *The New York Times,* the "focus of international scorn

---

TROPICAL DEFORESTATION

Deforestation has greatly increased in the tropics. Trees are key in recycling water, removing heat-trapping carbon dioxide from the atmosphere and preventing soil erosion. Loss in thousands of hectares (1,000 hectares = about 2,500 acres)

Region	Forest Area 1980	Forest Area 1990
Central America & Mexico	77,000	63,500
Caribbean Subregion	48,800	47,100
Tropical South America	797,100	729,300
South Asia	70,600	66,200
Continental Southeast Asia	83,200	69,700
Insular Southeast Asia	157,000	138,900
West Sahelian Africa	41,900	38,000
East Sahelian Africa	92,300	85,300
West Africa	55,200	43,400
Central Africa	230,100	215,400
Tropical Southern Africa	217,700	206,300
Madagascar	13,200	11,700

Sources: United Nations Food and Agriculture Organization, "Biodiversity" (National Academy Press)

In addition to official negotiations, the summit featured a variety of other festivities. Foremost among these was the 1992 Global Forum, an ecological Woodstock for the estimated 12,000 attendees from nongovernmental organizations, including representatives of human-rights groups and indigenous peoples.

at the Earth Summit." The Clinton administration is now reviewing the treaty with an eye toward signing it, however.

Although the biodiversity treaty was not as strong as it could have been, it at least set up a framework for the international community to work out agreements on tough issues, says Barratt-Brown of the NRDC. "Who owns biotechnology—aboriginals? The country where the genetic material is located? The company that develops it?" she asks. "Preservation is a burning need. There has to be a concerted effort to identify important areas of biological diversity and to help stem the species loss by providing funding early on to preserve these areas."

Clearing the Air

The U.S. did sign the climate-change treaty at the summit, but only after insisting on crucial changes in the wording. "As a result of the United States, it was much weaker than it should have been," says Greenpeace's Curtis.

For instance, the goal of the Climate Change Convention was to stabilize the levels of "greenhouse gases" in the atmosphere; these gases, such as carbon dioxide, trap heat in the atmosphere and may lead to an increase in global temperatures.

Many nations, including the European Community and Japan, wanted to set the year 2000 as the target date for reducing carbon dioxide emissions to 1990 levels. Since the U.S. opposed setting specific targets or timetables, the treaty contains only an ambiguous statement about stabilizing concentrations of greenhouse gases "within a time frame sufficient to allow ecosystems to adapt naturally." However, the 12-member EC voluntarily committed to the goal of limiting carbon dioxide to 1990 levels by the year 2000; in addition, President Clinton has stated he will also commit the U.S. to this goal.

The climate-change treaty requires the ratification of 50 national legislatures in order to take effect. Experts hope that the treaty will be ratified by the end of 1993.

Save the Trees

One final document produced at Rio was the Statement of Forest Principles. Intended to help prevent the destruction of the world's forests, this statement also encountered controversy. Although the Bush administration made an agreement on forest principles its highest priority, poor countries at the summit blocked the agreement because they saw it as an intrusion on their autonomy.

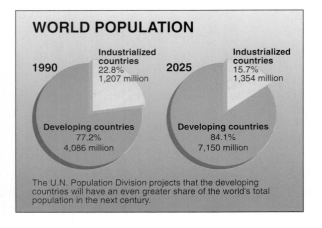

WORLD POPULATION

1990

Industrialized countries
22.8%
1,207 million

Developing countries
77.2%
4,086 million

2025

Industrialized countries
15.7%
1,354 million

Developing countries
84.1%
7,150 million

The U.N. Population Division projects that the developing countries will have an even greater share of the world's total population in the next century.

"Developing countries felt it was a ploy by the U.S. to shift responsibility from fossil fuel and carbon dioxide emissions to forests," says Hajost.

In the end, instead of the convention to produce a treaty that the U.S. had wanted, the Earth Summit produced only a non-binding set of principles that, says Hajost, "some have condemned as worse than the paper it was written on." The principles acknowledge that forests provide habitats for wildlife and help absorb pollution from the atmosphere, at the same time recognizing that poor countries rely on forests for food and fuel. There is no specific call for the creation of a treaty.

The Final Score

In the end—after all the watering down and rewriting to fit the political agenda of each nation—will the work of UNCED have any value? Like almost all who participated, Earth Summit organizers were disappointed but also hopeful: "The current level of commitment is not comparable to the size and gravity of the problems," said U.N. Secretary-General Boutros Boutros-Ghali. However, at the concluding session of the conference, Boutros-Ghali also stated that Agenda 21 will be the "centerpiece of international cooperation for many years to come."

Environmental and citizens' activist groups here in the U.S. and worldwide were encouraged by the interest the conference generated among average citizens. "The fact that so many citizens participated in the Earth Summit was an enormous show of hope," says Barratt-Brown. "We will begin to build a movement that will put pressure on governmental leaders. Plus the fact that you have over 110 heads of state writing speeches on what they thought sustainable development and environmental protection ought to be—[the summit] had benefits that we may not know for a long time."

Rio in Action?

One difficulty with the lofty documents produced in Rio is the actual hands-on work of putting them into practice. For example, every minute, we lose 70 acres (28 hectares) of trees around the world, two-thirds of them in tropical forests. Every year, we lose 45 million acres (18 million hectares) of trees—enough to fill New York State. Since trees produce oxygen, help sustain the ozone layer, provide wildlife habitats, and prevent soil erosion, the destruction of the world's forests poses a serious threat.

To stop this destruction, the documents produced at Rio call for the sustainable development of forests and for the conservation of biological diversity. But, says Richard Donovan, director of the tim-

ber project and Smartwood Certification program for the Rainforest Alliance, "the problem is, how do you implement the darned things?"

For instance, what are the specific steps a forestry operation needs to take in order to conserve biological diversity? Should the operation keep 1,000 acres (400 hectares) in a primitive state for every 10,000 acres (4,000 hectares) they harvest?

Also, what is the mechanism for ensuring that the rights of local, indigenous peoples are acknowledged? The Forest Principles say that operations should understand both the rights of indigenous peoples and the laws of the nation, and should make decisions specific to each country, says Donovan. "But as an example of how that might be done, when someone is doing

At the Indigenous Peoples' Conference on Territory, Environment, and Development, 400 leaders of Brazil's 180 indigenous nations met with 200 representatives from the U.S., Canada, Australia, Latin America, Scandinavia, Asia, and Russia to discuss land rights and self-government issues.

Hajost agrees that citizen action may be the key to future success of the programs set up at Rio. "You can't leave it up to the government; it won't happen. I can say that, having been in the government," he says. "Part of the test of the success of Rio is if groups of all sectors really take this and give it serious attention. It's going to take a substantial effort by nongovernmental organizations all over the world at all levels."

And, notes Hajost, the skeletal framework built at Rio may result in more changes than can be anticipated today, just as the Stockholm conference 20 years ago unexpectedly galvanized the environmental movement: "Stockholm did spur on treaties on the protection of species, the U.N. Environment Programme, ocean dumping," he says. "Let's see if Rio leaves some of the same legacy."

a management plan in a forest region, have they consulted with the local people? If the local people are already using the forest for nuts, resins, and fuel, how are those demands incorporated in the management strategy? Are they even considered?"

Failure to answer these kinds of questions can have catastrophic results, he says. In Somalia, for instance, herders have grazed herds on the same pieces of land for a few months out of the year; they've been doing this every year for thousands of years. Since the herders always came back to the same spot, they were careful to control their impact on the land. Once modern nations came along and stopped them from grazing, however, the herders stopped caring about their impact on the land—after all, they wouldn't be coming back to it.

"It's not just an ethical issue," says Donovan. "It's a resource-management issue."

The Rainforest Alliance is currently working with any group or government interested in determining ways to implement the principles in actual forests. The Alliance's Smartwood Certification program— which "grades" forestry operations according to their impact (or, preferably, lack thereof) on the forest environment—is one method developed to implement the Alliance's principles. Says Donovan: "There have to be real policy changes; there has to be a market which will distinguish between good and bad forestry. There's also a heck of a lot of education that has to be done, because the answers probably lie with the basis we establish now for the next generation."

THE ONSLAUGHT OF ANDREW

by David S. Epstein

Each winter, thousands of Americans leave behind the cold and snowy winters of the North to bask on the sandy beaches of south Florida. Year in and year out, countless other people move to Florida permanently—retirees searching for a balmy climate, engineers and technicians seeking work in the state's high-tech industries, families drawn by the state's relatively low cost of living. To accommodate the ever-increasing numbers of tourists and residents, real-estate developers built houses, apartment buildings, hotels, and condominiums—the closer to the beach, the better.

In the especially desirable Miami-Fort Lauderdale area, the lack of available land led developers to build ever upward. Before long, Florida's Atlantic coast, particularly from Fort Lauderdale southward, was lined with beachfront high rises. First-time visitors to the

Sunshine State often speak of the rush of excitement they feel when those glistening skyscrapers come into sight.

Meteorologists do not—and never did—share in this excitement. For years, they warned of the calamity that would occur should a powerful hurricane set its eye on Miami, Fort Lauderdale, or indeed, anywhere along the state's overdeveloped Atlantic seaboard. The warnings went largely unheeded. The last important hurricane to even threaten southern Florida—David, in 1979—bypassed the Miami area. The last time a major hurricane actually hit the Miami area was way back in 1950. Since then the region's population has grown tenfold. An estimated 90 percent of Floridians had never experienced a minor hurricane, let alone a major one. In August 1992, Hurricane Andrew would change all that with one stunning blow.

Hurricane Andrew may end up being the costliest natural disaster in U.S. history. Fortunately, warnings well in advance of the hurricane's arrival permitted the orderly evacuation of thousands of Floridians (far left). When the storm finally struck south Florida on August 24, 1992, it did so with fury, causing great destruction in several communities, including the town of Homestead (left). Andrew ultimately crossed the Gulf of Mexico and came ashore again in Louisiana (see map below).

THE PATH OF ANDREW

TX
LA
MS
AL
GA
Atlantic Ocean

Baton Rouge
New Orleans

Gulf of Mexico
FL

Storm strikes Florida coast 8/24/92

Andrew reaches Louisiana coast 8/25/92

Miami

Homestead

Worst hit areas

Andrew hits Bahamas 8/23/92

Birth of a Killer

To the untrained eye, the area of cloudiness and thunderstorms over the Caribbean Sea seemed to indicate little more than a typical summertime pattern. But to the experts at the National Hurricane Center (NHC) in Coral Gables, Florida, this apparently minor atmospheric turbulence bore watching. Within a few days, on Monday, August 17, 1992, the disorganized system had developed a characteristic spiral structure and strong enough winds for the Hurricane Center to officially declare it Tropical Storm Andrew, the first of the 1992 hurricane season. Initially, it was a minimal storm forecast to develop slowly. On Wednesday the storm ran into an opposing weather system that sheared apart the top of the storm and caused its center to become disorganized. On Thursday, as Tropical Storm Andrew crossed the northeastern Caribbean, forecasters were calling it a disorganized system that posed little threat to the United States.

But as Andrew continued toward the west, a huge high-pressure area forced the storm to take a more southerly track, one that provided almost ideal conditions for intensification. This high-pressure area to the north remained in place, keeping the storm over warm water and pushing it west toward Florida. By 4:00 A.M. on August 22, as the winds picked up, Tropical Storm Andrew was upgraded to Hurricane Andrew. The storm was now heading toward the Florida coast faster than had been expected. As the National Weather Service tracked the storm's

As part of the post-Andrew relief effort, U.S. Marines set up "tent cities" to shelter people who lost their homes.

position, it was clear that the hurricane did not intend to budge from its westward track —a beeline toward the Miami area.

Complacent No More

As Andrew strengthened out over the sea, south Florida residents kept wary ears open for news. Although some people began storm preparations, many parents and children spent the weekend doing last minute back-to-school shopping in anticipation of the first day of school the following Monday. But by Saturday night, south Florida residents knew that Hurricane Andrew was on its way. The question was: Where would it hit?

Boaters headed for land, filling the south Florida marinas with thousands of boats. People jammed supermarkets to buy bottled water, flashlights, candles, and canned goods. As the storm drew ever closer, hardware stores ran out of plywood, nails, and other items people were using to protect their houses and businesses from the expected onslaught.

By Sunday, Andrew was ranked a category 4 hurricane, and was expected to slam into Florida with sustained winds of 140 miles (225 kilometers) per hour, with higher gusts. Evacuation orders were issued. In an

unprecedented attempt to get people out of the path of a storm, 1 million people were ordered to leave Atlantic coastal areas from the Keys on up to St. Lucie County.

By midafternoon Sunday the storm had grown dangerously strong, with winds clocked at a ferocious 150 miles (240 kilometers) per hour, just 5 miles (8 kilometers) per hour short of a category 5 hurricane, the strongest storm classification. Stores began to sell out of everything. Radio and television stations canceled their regular programming, substituting it with round-the-clock coverage of the impending storm. Governor Lawton Chiles declared a state of emergency, alerted the National Guard, and activated the disaster-response network.

Andrew Hits

In the predawn hours of Monday, August 24, Hurricane Andrew slammed with unprecedented fury into the suburban communities just south of Miami. At the National Hurricane Center in nearby Coral Gables, winds were clocked at 168 miles (270 kilometers) per hour before the anemometer measuring the wind speed was torn off the building. Surveys of the damage after the storm led some scientists to estimate that the gusts of wind had approached 200 miles (320 kilometers) per hour near Homestead, which lay directly in the path of Andrew's 20-mile (32-kilometer)-wide eye.

Indeed, the destruction in Homestead and Florida City (also in the eye's path) was of astronomical proportions. From the air the area looked like the site of a nuclear-bomb detonation. Andrew literally wiped the color green from this characteristically lush area, denuding even those few trees that withstood the hurricane of their leaves and branches. Downed electrical wires, ravaged road signs, and dead or maimed livestock contributed to the postapocalypse look of the area. Fortunately, the massive prestorm evacuation kept the death toll down to 41 people, a comparatively low number considering the magnitude of the storm.

Most of the buildings in Homestead and Florida City consisted of modest homes, trailer parks, camps for migrant agricultural workers, and commercial shopping strip malls. Very few of these buildings survived the storm. The force of Andrew's winds tore the houses apart. Aluminum siding was ripped from the walls, tossed in the air like toys, and dropped to the ground or into any tree strong enough to still be standing.

Homestead Air Force Base, an important part of the local economy, was completely wiped out. Barracks blew apart, and planes were picked up and tossed haphazardly in the wind and the rain. Virtually every structure, including the operations control tower, sustained damage. An Air Force spokesman tersely summed it up: "Homestead Air Force Base no longer exists."

Damage was by no means limited to the communities in the path of Andrew's eye. Fortunately, the brunt of the storm bypassed Miami, doing little damage to the city's skyscrapers. Downtown Miami received only a glancing blow. Just south of the city, however, the glass facades of tall apartment and condominium complexes blew in, exposing the honeycomblike apartments to wind and rain damage. After crossing the Florida peninsula, Andrew reached—in a much weakened form—the bayou country of Louisiana,

Whirlwinds of Destruction

Although the National Hurricane Center (NHC) in Coral Gables, Florida, was unable to measure Hurricane Andrew's winds after its anemometer was torn off in the storm, NHC scientists estimated that there were gusts of up to 200 miles (320 kilometers) per hour. A retired University of Chicago researcher has come up with a theory that says that small whirlwinds within the bigger storm caused much of the extremely severe damage that occurred in sporadic, narrow bands. He theorized that these small whirlwinds spun at about 80 miles (128 kilometers) per hour, but were pushed along at 120 miles (193 kilometers) per hour, the wind speed of the storm, to combine for a total wind strength of 200 miles per hour.

Andrew's winds gusted to an awesome 200 miles per hour. At such velocities, the winds were strong enough to toss boats about as if they were toys.

where it caused heavy rain and flooding. And although damage estimates reached at least $300 million in Louisiana, the storm's legacy belongs to the utterly destroyed communities of south Florida.

Coming Home
For returning Homestead and Florida City residents, the first few days after the storm were almost as terrifying as the storm itself. The homes and businesses lay in ruins. There was no electricity, no water, little food, and no relief workers in sight. Although the National Guard and state and local governments had mobilized for the storm, the lack of communication made it difficult for relief workers to realize the extent of the damage. The Federal Emergency Management Agency (FEMA), which is in charge of coordinating relief efforts, was slow to respond. The people most in need of help were left stranded for about three days, until U.S. Marines were dispatched by the federal government. Then, rapidly, food and water distribution began, and tent cities were erected to shelter those storm victims left homeless.

The economic impact of Hurricane Andrew was enormous. The winds and the 12-foot (3.6-meter) tidal surge that accompanied the storm destroyed many local businesses. The completely leveled Homestead Air Force Base alone represents a loss of $400 million a year to the local economy. The numerous agricultural businesses in the area, world-renowned for growing limes, avocados, tropical fruits, and indoor foliage plants, were virtually annihilated, with plants, greenhouses, and irrigation equipment blown away in the wind. Even the soil was damaged by the salty ocean spray blown inland, making immediate replanting impossible. All told, Andrew left anywhere from 50,000 to 250,000 people homeless, rendered about 86,000 people jobless, and caused over $20 billion in damage in Florida alone.

Environmental Effects
The long-term environmental impact to the area will not be known for years, but the short-term effects could be clearly seen once the hurricane moved through. A huge swath of destruction 20 miles (32 kilometers) wide was cut through the Everglades. Initially, some researchers thought that many of the animals and other wildlife had perished in the storm. Soon, however, the animals emerged from their shelters, and it appears as if most survived quite well.

The effect of Hurricane Andrew on the research being conducted in the Everglades depends on the project. For many botanists, marine scientists, and those in animal husbandry, years of painstaking effort were mostly wiped out during the destructive hours of the storm. Environmental scientists hope that at least one benefit of the hurricane will be to give them an opportunity to

Andrew dealt a heavy blow to many research projects being conducted in the Everglades. On the bright side, heavy rains from the hurricane may have helped flush out much of the Everglades' polluted water.

Next Time Around

As with any natural disaster, Hurricane Andrew provided a wealth of information for the local disaster planners to prepare the emergency-response plans for the next hurricane. They are studying the building codes to determine if the extensive damage to housing could have been prevented, either through stricter enforcement of current codes or by upgrading the codes. Evacuation plans are also being evaluated, to decide if an area that extends farther inland needs to be included in evacuation plans. Experts in disaster research are studying how the densely populated Atlantic coast communities of Miami, Fort Lauderdale, and Palm Beach would have fared if the storm had come ashore 30 miles (48 kilometers) to the north.

Hurricane Andrew was a well-behaved storm in that it stayed on its projected course throughout its life span;

In the wake of the hurricane, Florida's building codes have come under intense scrutiny by state officials and insurance companies.

other storms may not follow this pattern. Disaster planners are concerned that regions are not able to be warned or evacuated safely if a storm changes course quickly. Urban planners and emergency-management teams will be wrestling with these issues for years to come.

better understand how an ecosystem rebuilds after a natural disaster.

Before the storm the Everglades and Florida Bay had become too salty, owing to the diversion of the freshwater supply for human use. In addition the rivers that feed these areas run through both agricultural and industrial regions of the state, acquiring much chemical pollution that finds its way into the Everglades and Florida Bay. Researchers are hoping that the storm was able to flush out at least parts of these areas. Hurricane Andrew moved very quickly across Florida, however, yielding rainfall amounts perhaps not heavy enough to do the job.

At the Metrozoo, just south of Miami, the once-lush landscape was left barren. About 400 animals remain on temporary loan to other zoos while their habitats are rebuilt. A $3 million aviary, the showpiece of the zoo, is in tatters. Aviary keepers dug

about 25 dead birds out of the debris, but 150 others survived, many returning to the aviary on their own. Only five of the zoo's 1,200 mammals lost their lives in Hurricane Andrew—"nothing short of a miracle," according to the assistant curator. "Nature is certainly unbelievable in its instinct to survive," he added.

A Semblance of Normality

South Florida is finally starting to return to some semblance of its former existence. Although some residents have left the area to resettle in other communities, many more are rebuilding. Most of the debris is gone, and some of the trees and plants have started to resprout. School started after a delay. Finally, for many companies and stores, it's business as usual. And a few days before Christmas, four months after Andrew, the Miami Zoo reopened its gates.

The Greening of HOLLYWOOD

by Terry George

By a remarkable coincidence, the 1979 release of The China Syndrome *(left)*, a film about a nuclear meltdown, occurred just 12 days before the accident at Three Mile Island. In 1988, the film Gorillas in the Mist *(below)* focused on endangered species.

Back in 1991 Peter Pan returned, not as the eternal youth in green tights, but as a snarling, mergers-and-acquisitions lawyer and neglectful father, the embodiment of everything that was wrong with the 1980s. At a crucial point in the opening minutes of the $70 million blockbuster *Hook*, Peter (played by Robin Williams) is on his cellular phone, trying to clinch a land deal in a forested area as his noisy kids run by. He screams at them, then turns to the phone. "What do you mean, the Sierra Club report?" he bellows. "I thought there was nothing to report. . . . A what? Cozy blue owl? . . . Well, if they're endangered, maybe there's a reason for it. Since the dawn of time, there have been all sorts of casualties of evolution."

Peter Pan had lost more than his shadow. To show just how much of a rotter he had become, *Hook*'s director, Steven Spielberg, opted for an environmental issue.

The year 1992 marks the greening of Tinseltown. That year, eight major movies with environmental themes crowded the cineplexes or went into production. They stretch from the noble story of Peter Matthiessen's *At Play in the Fields of the Lord* to the satire *The Naked Gun 2½: The Smell of Fear.* In between are David Puttnam's film about martyred Brazilian activist Chico Mendes; *Medicine Man,* a love story about a doctor (Sean Connery) in search of an herbal cure for cancer; *FernGully: The Last Rainforest,* a full-length feature cartoon from Twentieth-Century Fox; *Once Upon a Forest,* another animated feature; *Whalesong,* the story of an injured whale; *Warriors of the Rainbow,* a feature about Greenpeace; and *Amazon,* a Finnish production on the rain forest.

Call it Hollywood Zeitgeist or perverse fate, the question remains: Will this flood of politically correct films produce greenbacks at the box office?

Rain-Forest Morality Play

The most lavish of these projects, *At Play in the Fields of the Lord,* emerged after a 25-year journey through every studio boardroom and a sojourn on nearly every superstar's reading table. Two-time Academy Award winner (*One Flew Over the Cuckoo's Nest* and *Amadeus*) Saul Zaentz's $34 million adaptation of Matthiessen's 1965 novel began its movie life in 1966, when the rights were purchased by MGM for $250,000. The story of its celluloid evolution is at least as captivating as the rain-forest morality tale that made it to the box office. This was, after all, the project that during its quarter-century in the industry limbo called development hell became known as "the most famous unproduced film in Hollywood."

Once MGM had purchased the rights, the project attracted Paul Newman, Jack Nicholson, and Marlon Brando. But by 1970 MGM claimed it was no longer interested in the picture, and offered it to producer Stuart Millar—if he could raise the

Hollywood's environmental awareness reached new heights in 1990 with the blockbuster Dances with Wolves, *a film that combined elements of the traditional Western with eco-sociological themes.*

$500,000 MGM said it had already invested. Millar couldn't raise the money, and MGM kept the rights.

Then John Huston considered doing the movie. Gregory Peck and Kirk Douglas took a look. In 1977 director Bob Rafelson managed to revive MGM's interest. Later that year, he visited the Amazon, and was ready

to start filming, but MGM had reservations about the script. Again production stalled.

Richard Gere was next in the line of superstars involved with the story. In 1983 he tried to bring it to the screen, but his producer, Keith Barish, was a day late in filing an option-renewal check with MGM, and the project was passed to Zaentz, who had been pursuing it for a decade.

In 1986, Zaentz announced that he would soon begin production. But by then his director of choice, Milos Forman, felt he was too old for the rigors of filming in Brazil.

Shooting began in June 1990, with Hector Babenco (*Ironweed*) directing and Tom Berenger playing Lewis Moon, the Cheyenne mercenary who, in a search for self-identity, parachutes into the rain forest to join a Niaruna Indian village. Moon's arrival, along with that of a group of U.S. missionaries who want to convert the Niaruna, ultimately leads to the village's destruction.

At Play opened to mixed reviews. In *Variety*'s roundup of New York and Los Angeles critics' opinions, it scored 12 pro reviews,

An Eco-Movie Time Line

1951 • THE DAY THE EARTH STOOD STILL
Alien freezes the world's power until senseless destruction is stopped.

◄ *Some film historians consider the sci-fi classic* The Day The Earth Stood Still *(left) the first film with a strong environmental message.*

1954 • THEM!
Giant ants are born from atomic fallout—a warning about the side effects of radioactivity.

1959 • ON THE BEACH
The morning after.

1961 • THE DAY THE EARTH CAUGHT FIRE
Two H-bomb explosions tilt the world into global warming.

1972 • FROGS
Pollution breeds killer amphibians.

1979 • THE CHINA SYNDROME
Reporters expose a fault at a nuclear-power plant.

1979 • THE ELECTRIC HORSEMAN
Rodeo-cowboy-turned-activist rescues his horse.

1983 • NEVER CRY WOLF
Scientist sent to condemn wolves in Canada grows to admire them.

1985 • THE EMERALD FOREST
A Brazilian tribe brings up the son of the man who builds a dam in their forest.

1986 • THE MISSION
When Spain and Portugal war over land, natives of the Amazon are the victims.

1986 • STAR TREK IV: THE VOYAGE HOME
Kirk and crew go back in time, beaming up humpback whales to save our future.

10 con, and 13 mixed. Vincent Canby of *The New York Times* described it as a ''rare movie in that it has something on its mind to express,'' while *Rolling Stone* called it ''a plodding washout that trades ideas for cliches.'' The film's distributor, Universal Studios, decided on a phased opening, gradually raising the number of screenings in key movie cities such as Los Angeles, New York, Chicago, and San Francisco, a strategy studios use for movies they perceive as complex films that depend on word of mouth. The plan worked well for Disney's *The Doctor,* a recent film about cancer that built its box office over a period of months and turned a reasonable profit despite its subject matter.

When asked about *At Play*'s apparent timeliness, Zaentz offered the wisdom of a craftsman who has spent more than two decades working on a project: ''You can't think of what the problems of the world are going to be two or three years down the line.''

Catching the Mood

But that's precisely what many producers have done. Despite appearances, the studios analyze political events and social movements closely in hopes of catching the mood of the moviegoing masses.

In previous decades, films with an environmental message generally fell into the science-fiction category. In the heyday of sci-fi movies, during the 1950s, the Earth was often threatened with doom—usually by some sort of atomic catastrophe. In *Them!,* a popular movie of 1954, atomic testing produced giant ants that took refuge in the Los Angeles drainage system. The 1961 British film *The Day the Earth Caught Fire* carried a prophetic warning of global warming after the world had been shifted on its axis by Soviet and U.S. atomic tests.

But in the past 15 years, reality has overtaken sci-fi, and the change was mirrored on the screen. The movie *The China Syndrome*—released in 1979 only 12 days before the nuclear accident at Three Mile Island—marked a turning point. The thriller about a nuclear meltdown was marketed, not as fantasy, but as dramatized reality.

By 1988 the planet seemed on the fast track to self-destruction. Tons of illegally

FernGully: The Last Rainforest *relied on brilliant animation to drive home its environmental themes to younger movie audiences.*

dumped medical waste washed up on beaches along the northeastern seaboard. Out in the Atlantic, seals were being decimated by a mysterious virus. The western United States, plagued by the worst drought in memory, seemed to spontaneously ignite: A third of Yellowstone National Park burned. During the presidential campaign that year, George Bush drowned Michael Dukakis in the polluted waters of Boston Harbor. And two phrases were added to the popular vocabulary: *global warming* and *ozone depletion.* Then, in March 1989, the *Exxon Valdez* spilled 10 million gallons (38 million liters) of crude oil into Alaska's waters.

In 1990, 200 million people worldwide mobilized for Earth Day. That mass demonstration, and the aforementioned environmental catastrophes, provided the impetus for today's flood of green movies. Three Mile Island, Chernobyl, the destruction of the atmosphere, and the Brazilian rain forest—all are real-life disasters that eclipse yester-year's silver-screen melodramas. The end of the world is no longer something that evokes a shiver and a laugh in the darkness.

Struggle and Conflict

Conflict, film-school lecturers will tell you, produces great drama. And it seemed that the Earth was in the greatest of all conflicts —a life-or-death struggle. So Hollywood did what it does best. It passed around the mineral water and took a few meetings. Screenwriter Tom Schulman (Dead Poets Society) met with director John McTiernan (Die Hard, The Hunt for Red October) and laid the groundwork for Medicine Man. The makers of Crocodile Dundee met with a former Disney animator, Bill Kroyer, and plans were made for FernGully: The Last Rainforest.

But while these moviemakers conducted themselves in a businesslike manner, some big names in the industry became embroiled in one of the more reprehensible episodes in Hollywood history.

On the night of December 22, 1988, a Brazilian rubber tapper and activist, Chico Mendes, was gunned down in his backyard by a *pistolero* linked to local ranchers. Mendes, a self-taught environmentalist, had organized the Amazon rubber tappers against the ranchers' slash-and-burn chain-saw gangs. His struggle became the spearhead of the campaign to save the rain forest, and his

An Eco-Movie Time Line

1987 • THE MILAGRO BEANFIELD WAR
Bean farmer versus developer.

1988 • GORILLAS IN THE MIST
Dian Fossey's inner journey to save the endangered mountain gorilla.

1989 • THE BEAR
No dialogue, but lots of water lapping by the star.

1990 • DANCES WITH WOLVES
Bison killing on the American frontier.

1991 • AT PLAY IN THE FIELDS OF THE LORD
Missionaries and Indians in the Amazon.

1991 • THE NAKED GUN 2½: THE SMELL OF FEAR
Even in comedy, there's conscience.

1992 • AMAZON
Doctor recruits daughters to save the Amazon.

1992 • FERNGULLY: THE LAST RAINFOREST
Cartoon pals foil loggers in the rain forest.

◀ *The endangered rain forest—center stage in FernGully (left)—was something of an environmentally correct backdrop in the film Medicine Man.*

1992 • MEDICINE MAN
Love in a deteriorating jungle.

1992 • ONCE UPON A FOREST
Forest fairy tale in full animation.

199? • WARRIORS OF THE RAINBOW
Adventures with Greenpeace.

199? • WHALESONG
A young girl's love for an injured whale.

Endangered humpbacks traveled to the future in the "save the whales" hit Star Trek IV: The Voyage Home.

murder was greeted with shock around the world. In Hollywood the news sparked a race to the Brazilian town of Xapuri by executives vying for the rights to the life of an eco-martyr.

Robert Redford, Ted Turner, David Puttnam, William Shatner, and Warner Brothers producers Peter Guber and Jon Peters were all interested in the story. Mendes's colleagues decided that each of the filmmakers should pitch his deal at a meeting of rubber tappers and indigenous tribes, to be held in the regional capital, Rio Branco. The combatants arrived armed with VCRs and videotapes of their movies. The locals were treated to screenings of *The Mission, Batman, Rain Man,* and *The Color Purple.*

In June of 1989, a Brazilian company, J. N. Films, announced that it had secured the rights from the Chico Mendes Foundation, the group formed to continue Mendes's work. In truth, the J. N. deal had split the foundation right down the middle. Ilzamar Mendes, Chico's widow, and Gilson Pescador, an activist priest, had supported the agreement because they wanted the movie to be made by a Brazilian company. Other members of the CMF claimed that Mendes

had acted unilaterally and that J. N. Films had neither the resources nor the intention to make the movie.

In September 1989, Guber and Peters announced that they had secured the movie rights. By coming up with close to $1 million more for the Brazilian company, the two had quietly walked off with the story.

In the midst of the turmoil, Guber and Peters were secretly negotiating with the Sony Corporation to jump ship from Warner Brothers and take over Columbia Pictures. After their defection, Warner turned over the rights to the British filmmaker David Puttnam.

Films with an Eco-Twist

While some chase the high end of the movie market—namely, a film that reaffirms Hollywood's belief in its own social relevance—others have thrown money into traditional films with an eco-twist.

Disney's contender is a very big-budget romance ($40 million) that plunges Hollywood's most bankable elder statesman, Sean Connery, and Lorraine Bracco (the gangster's wife in *GoodFellas*) into the rain forest. In *Medicine Man,* Connery and

Bracco swing around the jungle canopy in search of an herbal cure for cancer. First they clash. Then they fall in love as, all around them, developers move in to destroy the forest. "It was . . . nice," says Connery, "to take this very serious subject and make a story about two people . . . played with comedy and adventure and still combining the serious theme of the rain forest disappearing at one acre per second."

Nice as it might have been, the shoot was a difficult one, and despite a salary reported at $10 million, Connery got so fed up with the conditions in the Mexican jungle that he left the shoot several days early. "It was getting hotter and hotter," he says. "And if it starts to rain, it can go on five days." Rain-forest blues.

While Connery couldn't get out of the jungle fast enough, the makers of *FernGully: The Last Rainforest* were so determined to re-create every detail of the place that they flew 15 animators and film-crew members to Australia for a month's orientation. "Every flower, every bug, every plant and bird in this film is an actual species," says animator Kroyer. "It was important that our people capture the reality of what is threatened."

FernGully, a children's cartoon, tells how Zak, a teenage member of a logging team, is shrunk to elf size by Crysta, a rain-forest fairy. As Zak sees the destruction unleashed by the loggers, he joins the nymphs in a battle to save the forest.

Kroyer says that this is a consciousness-raising film. Whatever it is, it attracted some of the top talent in Hollywood. The characters' voices are provided by stars such as Christian Slater and Samantha Mathis, Tim Curry and Robin Williams. To put their money where their mouths are, the producers agreed to donate 5 percent of the net profits to rain-forest causes. In return for his very bankable name, Robin Williams, perhaps remembering the old adage "there's no such thing as net profit in Hollywood," negotiated a further 4 percent of the gross for the same charities.

Along with these high-ticket releases are a slew of other projects. Fox, apparently bullish on environmentally correct cartoons, released another animated feature, called *Once Upon a Forest.* Nor has the United States cornered the market. The tiny Finnish film industry blew its biggest budget ever on a rambling, often ludicrous film called *Amazon.* It tells of a widowed Finnish doctor who drags his two daughters off to the depths of the Amazon. Directed by Mika Kaurismaki, it blends the stoic dourness of Scandinavian cinema with the production values of a bad spaghetti Western in a film that may well set the bottom limit in the eco-movie field.

While the Brazilian rain-forest capital of Manaus has taken on the trappings of Hollywood South, a few of Tinseltown's leading figures are looking for plots elsewhere in the environmental movement. Ted Danson *(Three Men and a Baby)* has agreed to star in and executive-produce a $20 million project called *Whalesong*—the story of a young girl's relationship with an injured humpback whale. Miami is the setting.

On the action front, Renny Harlin *(Die Hard 2)* plans a movie about Greenpeace called *Warriors of the Rainbow.* Alas, there is bad news for environmentalists who love cinematic mass destruction. Sly Stallone's contribution to the genre—a toxic-waste sci-fi thriller called *Isobar*—has been put on hold indefinitely.

Sending a Message

Samuel Goldwyn once said, "Messages are for Western Union." But that's a maxim more noted in the breach than in the observance. Hollywood loves to send a message, whether it's about the death of a president or the death of the rain forest. If the cause of the environment becomes the cause of the box office, the industry will campaign to the last seat.

Listen, then, to a quote from *The Naked Gun 2½: The Smell of Fear,* one of the most successful eco-movies yet made. At the end of the film, after Lieutenant Frank Drebin has successfully defeated the triple scourge of the oil, coal, and nuclear lobbies and their respective organizations, SPILL, SMOKE, and KABOOM, he addresses a presidential gathering. "I want a world," he pleads to George and Barbara look-alikes, "where I can eat a sea otter without getting sick."

That's Hollywood.

A WEATHER EYE ON THE WEATHER CHANNEL

by Leonard Ray Teel

The Weather Channel has gained a nation-wide audience for the accuracy of its forecasts, the competence of its meteorologists, and the excellence of its programming.

For nearly 200 years, people have used the term "weather eye" to describe a state of constant alertness to the changes in the weather, and, by extension, to any condition of extreme watchfulness. Today this term finds new applications at one of the television industry's most notable phenomena of the 1980s and 1990s: The Weather Channel (TWC).

Does Anybody Care?
Do television viewers really care about the weather? That's the question that was asked frequently after The Weather Channel's launch in 1982, especially when red ink rained down during its first two years on the air. Now after observing its 11th anniversary on May 2, network executives can answer that question with a resounding "Yes!"

The numbers have quieted the skeptics. Weather Channel executives point proudly to Nielsen ratings that certified that the network has passed the impressive milestone of *50 million* subscribers—the fifth largest weekly audience of *all* cable networks. TWC is being seen in 85 percent of American homes with cable TV, and in 54.7 percent of all households with television.

In 1991 the network was awarded the cable industry's highest honor for its coverage of Hurricane Hugo in September 1989. Tropical-cyclone expert John Hope had tracked the storm from the coast of Africa before Hugo had a name, and meteorologist Dennis Smith met it at Charleston. For this work, The Weather Channel earned the prestigious Golden Ace Award.

By 1992 The Weather Channel had reached the status of an American institution. At the American Meteorological Society convention in Atlanta in January 1992, the network signed an agreement with the Na-

The School Day Forecast (left) helps parents decide how to dress their children. Major storms receive extensive coverage (right), often in the form of special programming.

tional Weather Service designed to improve dissemination of lifesaving weather information, develop new weather services, and create more educational materials like TWC's documentaries.

"This agreement establishes a framework within which a public agency and a private enterprise can work as partners to benefit the general public," said former U.S. Commerce Undersecretary Dr. John A. Knauss, administrator of the National Oceanic and Atmospheric Administration (NOAA) at the time.

The Weather Channel's chief executive officer was all sunshine as he and Knauss signed the agreement at a press conference. "NOAA and The Weather Channel each fill unique and important roles in keeping the public informed about the weather, and we have complemented each other for many years," said Michael J. Eckert, who is also president of TWC's parent company, Landmark Communications' Broadcast and Video Enterprises Division.

National Weather Service head Joe Friday cited The Weather Channel's unique capacity to continuously translate and communicate weather data gathered by the government's vast weather observation, radar, and satellite systems. These government systems are currently undergoing a $3 billion modernization, but the money is earmarked for a new generation of technology, not for weather-information dissemination to the public. Friday likened his agency's role

to that of a "wholesale" distributor, and the role of the media, including The Weather Channel, to that of a "retail" operation.

Another example of The Weather Channel's success in the "retail" weather trade comes from Cox Newspapers, Inc. The Atlanta Constitution and the Atlanta Journal, the chain's flagship dailies, have signed an agreement for The Weather Channel to provide redesigned weather maps and expanded weather coverage. The newspapers have also added a five-day forecast tailored by The Weather Channel for metropolitan Atlanta, a detailed three-day outlook for Georgia, a comprehensive overview of the nation's weather, and a forecast for major overseas destinations served by Atlanta's Hartsfield International Airport.

The Dark Before the Dawn

As The Weather Channel begins its second decade on the air, forecasts indicate fair skies ahead. The dark warnings and dire predictions of the early days have passed. "Some had actually projected the exact date we would close in the summer of 1983," says Eckert. In his office northwest of Atlanta, the 46-year-old chief executive recalls why, like the weather, his network's success proved so hard to predict.

One big problem was the revenue expected from advertising, Eckert recalls. When ABC-TV's "Good Morning America" weatherman John Coleman launched his dream of a weather channel, expectations for

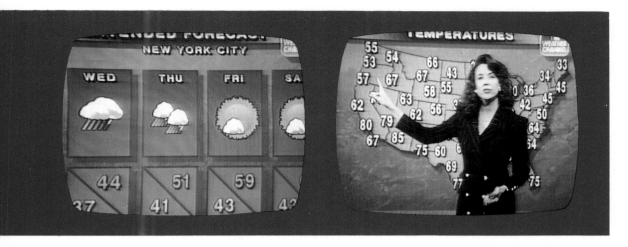

The Weather Channel issues extended forecasts for individual viewing areas (left). The temperatures in selected cities (right) are updated throughout the day.

the whole cable-television industry were generally high.

Eckert, hired by Coleman to sell advertising in the West, already had client contacts and saw big bucks on the horizon. "There was great euphoria," Eckert recalls. "From Michigan Avenue [Chicago] to Madison Avenue [New York City], advertisers said they would support cable networks. They were interested in the new technology, in getting into homes, and in niche opportunities."

Advertisers also believed cable television would give them leverage with the Big Three networks. By threatening to shift advertising dollars to cable, advertisers saw an opportunity to negotiate lower network rates. Amid this euphoria, Eckert was present at the cable-television convention in Las Vegas, Nevada, on May 2, 1982, when the throwing of a ceremonial switch officially launched The Weather Channel.

Advertising sales, however, did not follow as forecast. Part of the reason was the recession of 1981–82, which discouraged investment in the new medium. Then Eckert ran into a stone wall trying to convince clients of the impact of ads on The Weather Channel. "Of the 70 million U.S. television households, we were in 1 to 3 million," he recalls. But he couldn't prove an audience of even 1 million, because the Nielsen firm required a minimum of 13 million homes before it would give a rating. It was a vicious cycle: Eckert had to sell advertising to pay

the bills so the station could survive to get the viewers and the ratings.

"You don't know if anyone is watching?" one cereal company executive asked incredulously. As Eckert recalls, the executive "politely asked me to leave his office."

That was the income problem. On the outgo side, startup costs exceeded expectations. The bottom line was that when Coleman left in 1984, the owner of The Weather Channel, Landmark Communications of Virginia, was reporting losses of $800,000 a month. That was the low point—but there was never a funeral.

Viewer Validation

Eckert became vice president of advertising in 1984, and over the next three years witnessed a dramatic reversal of fortune. Fifty of the largest owners of cable systems, who had been getting TWC free, began paying a fee per subscriber. That fee provided several million dollars to supplement advertising revenues.

This occurred, Eckert explains, because the big cable companies realized that their viewers liked what they'd seen so far. "There was general acceptance," Eckert says, "that The Weather Channel was a viable vehicle."

Next the network invested in viewer research, which showed it should serve different life-styles. So The Weather Channel began tailoring forecasts for housewives, construction workers, truck drivers, business travelers, farmers, private pilots, and sports

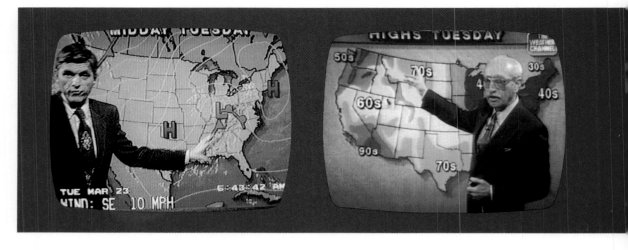

By drawing data from many sources, the meteorologists at The Weather Channel can predict the track of a storm (left) or forecast the nation's temperature profile (right).

fans—including bookmakers, who wanted to know, for example, if it was going to rain at RFK Stadium in Washington, D.C., during the NFL playoff games.

Subscribers were now signing up at the rate of 5 million to 6 million homes per year. That, in turn, boosted advertising income. "We started to get the Nielsen rating, and then we could *prove* we had a large advertising base," Eckert explains.

A Lean, Mean Weather Machine

The two income streams—from advertising and subscriber fees—saved The Weather Channel. By 1987, after five years and several million dollars in losses, the network saw its first profits. By then Eckert had risen to president. "Those who wrote us off," he said in 1987, "never understood our business philosophy. You don't have to generate a lot of revenue to make money in this business. As long as you keep operating expenses relatively low, you can still show a nice profit."

Automation has cut expenses. In the studio, fixed cameras eliminate the need for cameramen. A structured format for standard forecasts does away with writers, directors, and producers. "We're entirely ad-lib in front of the maps and graphics," says Ray Ban, who left the private Accu-Weather forecasting service to join John Coleman's on-camera meteorologists in launching The Weather Channel. Ban's bushy hair of 1982 is now cut short, and he is vice president for

operations, doing his ad-libbing in the halls and in his office, where a monitor is tuned to TWC. Today there are almost 30 Weather Channel weathercasters; about 20 have degrees in meteorology, and the rest have "significant weather experience and can communicate."

Very little has remained the same at The Weather Channel over the past decade. One exception is Coleman's desk, which is now Eckert's. (Coleman took his high-backed chair, Eckert says.) The network has moved to larger headquarters in Atlanta, and has opened new advertising-sales offices in major cities.

Video Actualities

With increasing revenues, network executives decided to improve their weather maps and graphics and expand coverage. Research studies had pointed to the viewer appeal of video "actualities"—footage of some of the more newsworthy weather events from across the country, often relayed from Conus, the Minneapolis-based video network. Through the Conus connection, The Weather Channel also feeds its video reports to TV stations around the country.

Carrying the "actuality" idea a step further, the network created an on-site mobile unit. Dick Byrd, who had four years' experience producing news shows for Ted Turner's CNN network, was hired as executive producer. Byrd now has a pool of six meteorologists on call to travel to the scenes of major

Color radar images (left) allow those in a thunderstorm's path to take the necessary precautions. The overall weather pattern depends upon the position of the jet stream (right).

weather events. Each day, the staff meteorologists alert Byrd to potential trouble spots. "When a major storm occurs and we have to show it to viewers quickly, it gets funny around here. It's our version of election night," says Byrd.

It was Byrd who dispatched staff meteorologist Dennis Smith to meet Hurricane Hugo, work for which TWC won the accolades discussed earlier. On April 15, 1991, Smith and a Weather Channel crew arrived on the scene just minutes after an F5 tornado (winds of 261 to 318 miles—420 to 511 kilometers—per hour) killed 17 people at Andover, Kansas, near Wichita. Their broadcast that Friday evening, while other film crews were still arriving, surprised a local TV newsman, who interviewed them on camera Saturday morning.

"We saw your broadcast last night," said the newsman. "When did you all arrive in town?"

"Friday morning," Smith said.

"Wait a minute. The tornado didn't hit until evening. How did you know?"

As Smith explained it, the decision to go to Kansas was "an extremely good guess." The weather situation indicated the likelihood of a serious tornado outbreak "somewhere in northern Oklahoma or Kansas. The most central location was Wichita. It was a big dartboard, and we got lucky."

Smith was the logical choice for the assignment. He had been the weatherman for Wichita's KARD-TV (now KSNW-TV) before joining The Weather Channel at its startup, so he knew the town. He also grew up in the region, so severe thunderstorms and tornadoes are his specialty. Besides, he's a weather zealot. Back in Atlanta, he's one of the guys with his "nose pressed against the glass" whenever there's violent weather outside. "We want to be there!" he says.

By 4:30 that Friday afternoon, the crew was set to broadcast live from a balcony on the 26th floor of a Wichita hotel. Two hours later, while they were on the air giving a live update, a severe thunderstorm struck. "We actually saw a little 'rope' tornado," recalls Smith. "Sirens behind us were blaring, and we were getting pelted with pea-sized hail. The big killer tornado was on the ground on the backside of the storm, just 9 miles [14.5 kilometers] away. We immediately went out to the scene."

Two people were killed when the tornado overtook their car. Thirteen more died in a trailer park, despite police warnings—which were videotaped. Some of the people on the tape appear to be "nonchalantly going about their business as if nothing was happening—or about to happen," reports Vince Miller, TWC's storm historian.

How did Smith's broadcast differ from the other TV coverage? "We were able," he says, "to explain a little bit more about the damage viewers were seeing and what caused it. Why you should seek the lowest shelter . . . some safety rules. The tornado sucked up everything . . . hymnal books were found 100 miles away; canceled checks were found in southern Iowa. There was tremendous destruction."

Later that night, when Smith entered a restaurant, he was shocked to see a sign that read: *Temporary Morgue.*

The on-site crew took special precautions in approaching Hurricane Hugo. Hugo was a Class 4 hurricane, with winds over 130 miles (210 kilometers) per hour, the worst to hit South Carolina this century. Smith, fellow meteorologist Cindy Preszler (now a weathercaster for a Lexington, Kentucky, television station), and their producers and technical staff were getting continuous advice from John Hope in the Atlanta office. Few knew more about what Hugo could do. Hope had been the National Weather Service's senior hurricane specialist when he joined The Weather Channel in 1982.

"They didn't quite get to downtown Charleston until after Hugo hit," says Hope, who stayed on the air 17 straight hours on September 21, 1989. "We were a little apprehensive. Had that hurricane gone ashore about 20 to 25 miles [32 to 40 kilometers] down the coast from where it did, there probably would have been numerous fatalities in Charleston.

"I can remember how I felt that night. You know, meteorologists sometimes get wild during a major hurricane. It's such an intense thing. There's a lot of adrenaline pumping. I was also a little depressed, though, because I knew what that hurricane was going to do, and I thought a lot of people in that place did not know.

"We followed Hugo from the coast of Africa. It began as a tropical depression off Senegal, southwest of the Cape Verde Islands. We thought it would develop into a hurricane, a very powerful hurricane."

As Hugo neared the U.S. coast, Hope emphasized its dreadful power and the risks. "I knew there was going to be a very high storm surge, 20 feet [6 meters] or so, that would rake the lowlands."

Emergency-management people who had to make evacuation decisions were watching, Hope says. "They got 100 percent evacuation of the barrier islands, which was very difficult to do. I think the fact that we emphasized the very serious nature of the hurricane as it was approaching South Carolina played a role in saving lives. Who can say for sure? A lot of people were watching us. I went into a hotel in Charleston the next summer, and the bellboy told me, 'We're glad to have you in this hotel. You saved a lot of lives in this city.' "

The Weather Channel's role was communication and interpretation. Aside from the on-site team's reports, TWC's forecasters relied on information provided by the National Hurricane Center (NHC) in Miami, Florida. "We just relay their information to the public," Hope explains. "The people at the Hurricane Center will tell you that they are almost totally dependent on the media to get their information out."

Hugo is now featured in Weather Channel educational documentaries. "Force Four" focuses on Hugo and its consequences—loss of life and $10 billion in damage, the most costly hurricane ever to hit the United States. In the 22-minute "Danger's Edge," commentator Hope talks about storm surges and five historic hurricanes—the Galveston hurricane of 1990 that killed more than 6,000 people, the storm in the Florida Keys in 1935 that blew a trainload of World War I veterans into the sea, the infamous 1938 New England hurricane, calamitous Camille in 1969, and Hugo—plus a little bit about Bob at the end.

In 1992 meteorologists from The Weather Channel gave round-the-clock coverage of Hurricane Andrew, which devastated south Florida in August, and the great nor'easter that pounded the Eastern Seaboard in December. When winter arrived, TWC was on the spot to cover the drought-busting storms that battered the West Coast —storms that more often than not subsequently brought snowfall east of the Rocky Mountains. And with another hurricane season just around the corner, The Weather Channel meteorologists have their weather eyes set on the tropics.

Even the most confident meteorologist will tell you that weather is notoriously difficult to predict. But there is one prediction that just about everybody agrees upon: The Weather Channel is here to stay.

THE ECOLOGY OF *Waves*

by Michael Rogers

Clustered together in a bright concrete gallery at the Monterey Bay Aquarium in California, a dozen kids stare at the glass front of a wide tank. The attraction is unclear: The display contains a rocky undersea bluff with droopy sea palms, various algae, a few starfish, anemones, sea urchins, all hanging on, immobile, with water dripping slowly into a small tide pool at the tank's bottom. It seems a drab and static vision to so enthrall the kids, but then suddenly comes the twist they're awaiting. An immense gush of seawater—about 200 gallons (750 liters), nearly a ton in weight—falls from the upper right of the tank and explodes in foam and spray. The kids scream, giggle, fall back. The sea palms and companions stolidly withstand the automated onslaught they endure every 45 seconds, around the clock.

For these tidal denizens, it's just another day's work. And for the children now awaiting another crash of water, it represents, albeit without a beach, a high-tech version of one of our species' oldest forms of relaxation: watching waves.

Mathematics and Language

Waves definitely deserve their quota of contemplation. They represent a continual dance between ocean and air, fueled by sun and wind, shaped by the seafloor, capable of inducing soothing meditation or a cataclysm akin to a small nuclear weapon. Waves can banish a beach overnight, then build a new one the next spring. Waves feed, shelter, and transport myriad life-forms, often nourishing a rockbound filter feeder, then moments

The hypnotic regularity with which ocean swells crash to shore offers only a fleeting glimpse of something that scientists have just now begun to fathom: the immense power hidden within a wave.

later knocking it into the maw of a hungry predator lurking below. Yet in the end, perhaps the most romantic and ineluctable aspect of waves is the perennial fascination they present for humans.

Most have at some point followed John Keats's two-century-old prescription for world-weary eyes: "Feast them upon the wideness of the Sea. . . ." The view is rich and endlessly varied in form, from the wind-blown scatter pattern known technically as "sea," to the immense tsunamis that can wipe entire towns off the Earth. The mathematics of waves are complex, soon plunging into rigorous calculus to describe volumes of water and shapes of swells. The language of waves, however, is simple and often poetic; it seems to mimic the hissing sibilance of the phenomenon itself: surf, set, swell, spume, spray, spill, seiche, spindrift—even the dread tsunami.

Those last, the unquestioned kings of wavedom, are often erroneously dubbed "tidal waves." In fact, they have nothing to do with tides. Tides are massive, large-scale movements generated by the gravitational forces of the Moon and, to a somewhat lesser degree, the Sun. Waves may interact with tides, but are nonetheless almost entirely products of the winds. The only exception: the fierce tsunamis, capable of traveling vast distances at an amazing 500 miles (800 kilometers) per hour, are caused by undersea earthquakes or landslides.

Following the 1964 earthquake in Anchorage, Alaska, a tsunami obliterated the small fishing town of Valdez (which was relocated, later to become the terminus of the Alaska pipeline). Hours later, thousands of miles away, the Alaska-spawned waves were still so strong that they tossed abalone into the streets of Monterey, California, and set that town's bay rocking for days in the standing swell called a seiche.

Keats also praised the "eternal whisperings" of waves. In fact, waves do offer a unique aural mix. The sound is basically a version of what acoustic engineers dub "white noise," of a random mix of frequencies. But unlike the slightly discomfiting hiss of, say, an untuned radio, wave noise has a fairly regular yet slightly random modulation. "It's sufficiently irregular to be fascinating," says Walter Munk, a wave-studies pioneer at Scripps Institute of Oceanography in California. "Not so irregular as to be disturbing, not so regular as to be boring."

Not everyone has a beach at hand, so by now there are at least half a dozen tapes and compact-disk recordings of ocean sounds. One of the first, Nature Recording's five-year-old *The Sea,* sold 44,000 copies last year. You can buy the sound of waves on rocks, waves amid driftwood, waves on sandy beaches, waves punctuated by gulls, bells, or foghorns. There are even competing schools of wave recording: live and unmodified versus a studio-polished mix.

THE ANATOMY OF COASTAL WAVES

When wind blows across the surface of the water, friction causes ripples to form. If wind strength is sufficient and prolonged, the ripples develop into a "sea," wave lengths that last at least 5 seconds. As the forming waves leave the "fetch"—the area of open water over which the wind blows—they become organized into even lines of so-called swells that radiate with increasing speed toward the shore.

A wave crashing over a coral reef increases the oxygen supply to shallow waters. The waves also stir up the food supplies for some coastal creatures and help other animals reproduce by distributing their spores.

Wave Prediction

Curiously, the mathematical understanding of waves came well before scientists really began to study waves in the wild. Victorian scientists were well advanced in the mind-numbing equations of fluid mechanics, but it took another century—and specifically World War II—for wave scientists to get their feet literally wet. In the Pacific, European, and North African theaters, the Allied strategy sometimes depended on amphibious assaults—putting troops ashore with boats or armored landing vehicles. But heavy surf could create death by drowning even before the fighting started, or produce a squad of thoroughly seasick soldiers tumbling into enemy fire.

Researchers like Scripp's Walter Munk were among the first to develop ways to predict waves, based on complex readings of shoreline structure, tides, and weather. Although crude by today's standards, the highly classified wave work was crucial. During the first winter landing in North Africa, scientists managed to pick two relatively calm days on beaches that usually presented

Moving through the ocean, swells cause water particles at the surface to spin in a circular orbit (see circles), helping to maintain wave motion. As the particles go down, their orbit diameter decreases. Heading into shallower areas, the swells slow down and their wave length shortens. The swells also grow steeper in height, since less water is available to fill in the crest and maintain a symmetrical shape. The increasingly shallow bottom prevents the complete rotation of circulating water particles and causes the waves to become unstable. The waves peak and break, crashing near the shore in a foamy surf.

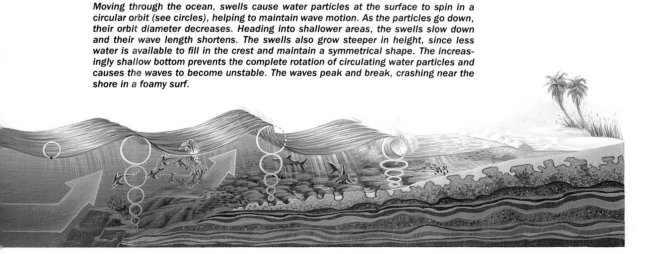

When Surfers Make Waves

For decades the waves that pound the Pacific Coast beaches have drawn surfers like magnets. But in recent times, those waves have carried more than just surfboards back to the beach. In the past three years alone, unusually high bacterial counts and unhealthful pollutant levels in the coastal waters have prompted hundreds of beach closings in California, and thousands more in at least nine other states.

For many ocean lovers, frolicking in such dirty surf is no longer a totally awesome experience. Some have developed health problems, from skin rashes to viral infections. The situation has helped galvanize West Coast surfers, a clan traditionally known more for its free spirits than its social activists. The result: the 15,000-member Surf-

Environmentally active surfers? That unlikely concept has become a reality, especially now that an estimated 60 percent of California's beaches contain potentially health-threatening levels of sewage.

rider Foundation, a nonprofit group based in Southern California that has been kicking up political waves from Orange County all the way to Washington, D.C.

"Those of us who love the ocean have a responsibility to help protect it," says Surfrider executive director Jack Grubb, a

forbidding waves. And for D-Day in Europe, prognosticators warned Dwight Eisenhower that the wave pattern looked rough. But delay appeared even more dangerous, so the invasion went ahead, encountering exactly the tough conditions predicted. By the early 1950s, wave prediction proved a tactical device: Douglas MacArthur's amphibious attack at Inchon during the Korean War surprised an enemy unaware that the tides and waves of that area even permitted such a landing.

To sharpen predictive skills just after World War II, the Navy financed extensive hands-on wave measurement off the Oregon coast by a team of University of California researchers headed by John Isaacs and including Willard Bascom, now a grand old man of oceanography. The team used huge military vehicles called ducks (from their code designation DUKW), built with six

wheels and a propeller, to drive out through the shallows into monstrous breakers to make observations and install measuring devices. The crews would then promptly turn the vehicles around in the pounding water and half-drive, half-surf, always on the verge of swamping, back to the safety of the beach.

"Don't expect us to come rescue you if you get in trouble," the local Coast Guard warned the scientist. "We didn't," Bascom recalls. "We would have drowned long before they could have launched a boat." Indeed, at one nearby lighthouse, winter waves pounded with such force that sometimes the "flung spray" broke windows 135 feet (40 meters) above sea level. "Oregon spray," observes Bascom, "is a thing to be reckoned with." And also something not forgotten: Even nearly 50 years later, Bascom still visits the Oregon coast to videotape long series

43-year-old former graphics publisher who has surfed since he was 11. "By the very nature of the sport, surfers must be environmentalists."

To help discern the severity of the problem, the group has distributed more than 1,000 water-testing kits to volunteers who periodically check near-shore contamination levels at selected sites along the West Coast. The program, headed up by Surfrider environmental director Scott Jenkins, a researcher at Scripps Institute of Oceanography, found that more than 60 percent of the Pacific beaches tested in one study were contaminated by sewage at levels high enough to potentially cause harm to human skin.

The group also has developed a number of materials about coastal ecology and has been taking its message to schoolchildren. "Our goal is to educate people, not make enemies," says Grubb. "We're not a litigious organization, either by design or attitude." Maybe so, but Surfrider has not backed away from confrontation.

Three years ago, after surfers complained that plumes of contaminants were turning waves black in northern Califor-

nia's Humboldt Bay, Surfrider filed suit against the two paper companies responsible for the pollution. The U.S. Environmental Protection Agency (EPA) soon joined in the lawsuit, which was finally settled when the two companies—the Louisiana-Pacific Corporation and the Simpson Paper Company—agreed to pay nearly $6 million in fines and spend some $50 million to reduce toxic discharges.

The settlement, which is one of the largest ever under the federal Clean Water Act, also requires the companies to conduct a series of tests to show that four ocean species—abalone, giant kelp, and certain sea urchins and sand dollars—can all survive in the discharged wastewater. "As far as I know, this is the first time in this country a discharger has been required to reduce toxicity as measured by its effects on organisms," says Grubb.

To help protect its constituents, Surfrider also requested a novel addition to the settlement: The two companies agreed to install solar-heated showers at area beaches so surfers can wash off toxic contamination until the mills clean up their act.

Mark Wexler

of waves. "You find yourself far from bored," he says. "You're caught up in it, always trying to guess what the next one will be like."

Killer Waves

No matter how long you watch, predicting specific waves is maddeningly difficult. Surfers and beachcombers may gain a pretty good sense of what to expect in a given set of waves, but there's rarely an entirely dependable sequence—no magical "seventh wave" or "ninth wave" that's always the biggest. The result of this unruliness can be lethal. Along the rocky coast of northern California, Oregon, and Washington—meccas for devoted wave-watchers—there are invariably three or four deaths a year attributed to "killer" or "sneaker" waves. The killer may appear out of a relatively calm sea, typically on nice days, sweeping the unaware victim out to sea or against the rocks.

Such killers amid apparent calm derive from the fact that few waves are simple. Almost any wave striking the beach is in all likelihood a combination of numerous waves, some far out at sea, their crests and troughs meeting in a process of addition and subtraction. "It may not always be the seventh or ninth wave that's the big one," says Steven Webster, education director of the Monterey Bay Aquarium. "The combinations are so complicated that it may turn out to be the 251st." Which, very rarely, may also turn out to be a killer. Webster's prescription for safety on rough beaches: Sit above the tidal zone and observe the patterns of the waves for 10 minutes or so before proceeding—and never turn your back on the sea.

The biggest sleeper waves of all are found far at sea: full-size ship killers. These leviathan waves, implicated in dozens of mysterious mid-ocean ship disappearances, are again the chance coincidence of various crests and troughs all adding up to one monstrous construct. The towering wall of water can easily sink a ship 500 feet (150 meters) long and weighing many tons. "It's the trough that kills you," says Willard Bascom. "The bow drops away, and then the wave comes straight across and cleans the wheelhouse off the deck." Recall the weight of even the Monterey Aquarium's tiny pet wave; now consider the heft of a wall of water 112 feet (34 meters) high—the tallest wave thus far reliably measured at sea. Experts suspect killer waves that leave no witnesses may be nearly twice that record.

Energy for Ecology

But waves produce far more life than they take. In fact, on shorelines, wave action contributes more energy to the ecosystem than do the direct rays of the Sun (the waves themselves, of course, are indirect results of the Sun). Some of the energy is actually heat —breaking waves do create friction—but because of the volume of water involved, the resulting temperature rise is negligible.

Scientists at the Hydraulics Laboratory in Ottawa, Canada, study waves they generate themselves in an Olympic-sized "wave tank" equipped with fans, paddles, and 1 million gallons of water.

What's not negligible is the sweep and power of the water itself. For denizens of the interface between ocean and shore, waves are an energetic, multifaceted benefactor.

There's meal service: Waves deliver fresh portions of rich ocean broth for the filter-feeding mussels and barnacles. At the same time, a wave may knock a loosened limpet into the center of a hungry sea anemone, or unearth a luckless sand crab for a vigilant shorebird or a surf perch. Waves oxygenate the shallow waters, of course, but they also actually help organize communities. Feather-boa kelp and sea palms, for example, adapted to withstand pounding, grow on the ocean-facing sides of rocks; barnacles and mussels colonize areas of less turbulent backwash. Finally, waves help propagate species—from distributing algae spores to carrying Southern California grunions up a moonlit beach to bury their eggs.

Waves, given their random complexity, are yet only partially understood. Engineers building piers or jetties still run into unpredictable situations: disappearing beaches or sudden new sandbars. Biologists now suspect that migrating whales may ride beneath deepwater swells to speed their passage. The waves off the coast of England have grown taller over the past few decades, for no apparent reason. And the business of wave prediction still has a way to go.

But change is coming. Walter Munk recalls a breakthrough experiment during the 1950s that used ships and land-based stations to track a single wave disturbance for two weeks, from its genesis off the coast of New Zealand to Alaska. "It took forever to do that experiment," he chuckles. Today satellite-based radar can do the same job effortlessly, even through layers of clouds. "In another five years," Munk says, "you'll be able to get a reliable wave forecast for anywhere in the world."

Except, of course, you probably won't want one for that isolated beach where you go to contemplate the universe, soothe some heartache, or simply mull the swells. After all, when the science is over, the final sweet consolation of waves is simple: Each one is slightly different—and there will always be another wave.

JUNGLE POTIONS

by Cathy Sears

A helicopter settles down beside a tiny village in the Amazonian jungle of northeastern Peru. Walter H. Lewis and Memory Elvin-Lewis, plant researchers from Washington University in St. Louis, Missouri, step off it, their backpacks loaded with plant-collecting gear. For years the husband-and-wife team has been studying the plants that the Jivaro Indians, once fierce headhunters, use as medicines. The Jivaro's rain forest is a virtual medicine cabinet, best known as the source of the drug quinine, a malaria remedy the Jivaro shared with Spanish Jesuits in the 1500s, and one that remains widely used today.

The Lewises are part of an international gold rush of scientists who are prospecting the world's jungles and forests for lifesaving

The pace of destruction of the world's rain forests has spurred a race against time to harvest endangered plants that could potentially yield myriad lifesaving drugs. The Amazonian Jivaro Indians are master pharmacists who use a variety of plants for healing purposes.

Tropical forests are being destroyed at the astounding rate of 50 to 100 acres each minute, leading to the extinction of an alarming 17,500 plant and animal species a year. Scientists are concerned that in the time it takes to bring a drug to market—10 to 20 years—a substantial fraction of the world's jungle and forest species will disappear.

plants. The Tropics, which contain two-thirds of the world's estimated 250,000 flowering-plant species and more than 1 million animal and insect species, are a logical place to look, since these species have had to develop complex chemical arsenals to survive against myriad attackers: other insects, fungi, viruses, and bacteria.

Right now the Lewises are on the trail of drugs for hepatitis, cancer, and birth control. A Jivaro treatment derived from tree sap for speeding the healing of wounds is already being tested on animals.

Synthetic Focus

For thousands of years, plants were the primary source of medicine. In the United States, at least one in four prescriptions still comes from plants. But botanical drugs began to fall out of favor in this country after World War II, as American drug companies shifted their focus to synthetics. There was a brief resurgence of interest in the 1960s, with the discovery of the anticancer drugs vincristine and vinblastine in the rosy periwinkle plant, but by the early 1980s, even

the National Cancer Institute (NCI) had virtually scrapped its long-standing program for screening medicinal plants. "During this period, you couldn't give plant extracts away," explains Norman Farnsworth, director of the program for collaborative research in the pharmaceutical sciences at the University of Illinois in Chicago.

Even today the major American drug companies involved in plant and marine research—you can count them on one hand—commit only a tiny fraction of their research-and-development budgets to that area, says Farnsworth. Government funding, while improving, is also tight. One leading researcher aptly refers to current funding levels as "chicken feed."

Part of the blame lies with American medicine's long-held bias against traditional medicine and botanical drugs, which many U.S. doctors view as quaint and/or of marginal effectiveness. Yet, of 121 prescription drugs used widely around the world, "74 percent come from following up folklore claims," according to Farnsworth. "Logic would tell you there are a lot more jackpots out there," he says, adding that the government and drug companies are still reluctant to fund significant research, despite the fact that the U.S. plant-based prescription market was worth $15 billion by 1990. One explanation is that sifting through unidentified plant compounds and identifying the structures of active ingredients can take more time than cooking up already-familiar chemicals, which are also easier to patent, in a laboratory.

Stephen Brewer, manager of bioproducts chemistry for Monsanto, adds that his company's budget for microorganism-based drugs—which are easier to grow in vats in a lab—is five times greater than its budget for collecting and cultivating plants. Still, pharmaceutical researchers admit they can't come up with all the answers in the lab. "Scientists may be able to make any mole-

cule they can imagine on a computer, but Mother Nature—over the course of millions of years—is an infinitely more ingenious and exciting chemist," says organic chemist Monroe Wall of the Research Triangle Institute in Research Triangle Park, North Carolina. Wall pioneered research that led to the development of the anticancer drugs taxol and topotecan, both of which come from the Pacific yew tree, and camptothecin, derived from a subtropical tree from China.

Scientists have only begun to plumb the fast-disappearing biological treasures found in rain forests—an often tedious and time-consuming process. One in four drugs used by doctors today contains ingredients derived from tropical plants. The chart below describes important plant-based prescription drugs currently on the market.

DRUG	USE	DERIVATION
TUBOCURARINE	surgical anesthetic	curare vine (Chondodendron tomentosum), used to make an arrow poison
EPHEDRINE	active ingredient in decongestants	stem of a Chinese shrub (Mahuang), used against colds and asthma attacks
MORPHINE	painkiller	opium poppy (Papaver somniferum), used by the Romans and Greeks as a sedative and sleep aid
CODEINE	analgesic and cough suppressant	opium poppy
PAPAVERINE	muscle relaxant	opium poppy
ASPIRIN	headache remedy and heart-attack medication	willow (Salix alba), used in folk medicine to ease headaches
DIGITALIS	used to treat congestive heart failure	foxglove (Digitalis purpurea), a plant used as a popular folk remedy for stimulating the heart
RESERPINE	tranquilizer and hypertension drug	snakeroot (Rauvolfia serpentina), a plant used in India to calm anxiety
ATROPINE	antispasmodic and pupil dilator	belladonna (Atropa belladonna), used by women in medieval European courts to dilate their eyes
PHYSOSTIGMINE	glaucoma treatment, poison antidote	Calabar bean (Physostigma venenosum), a deadly legume used in ritual trials by ordeal

Due to the lack of money and interest, pharmacognosy programs (the study of biologically active compounds derived from living organisms) have been closing in recent years in many American universities; today there are only half a dozen left. Ethnobotany programs—the study of how cultures use plants—are also facing troubling times. "We have more people studying the surface of Venus than out in the field studying plants," says Paul Alan Cox somewhat incredulously. Cox, a professor of botany at Brigham Young University in Provo, Utah, has been studying American Samoa since 1985.

Making Botanical Drugs a Priority

In sharp contrast, research on plant extracts is booming in such industrialized nations as Germany, Japan, and South Korea. (New Zealand, England, and Sweden also have expanding programs.) A U.S. Department of Agriculture (USDA) report warns that the United States, still the world leader in drug development, is in danger of falling out of the race if these other countries successfully tap forests for new remedies. For example, Japan has made natural-product research a national priority, just as it did with building cars and semiconductors. Today Japan holds fully one-half the world's patents on natural products.

As diseases around the world become resistant to human-made drugs, maintaining a diverse gene pool for new treatments is essential. For example, a synthetic drug called chloroquine replaced quinine when strains of the malaria parasite became quinine-resistant. Now chloroquine is losing its effectiveness. "Once again we've found a cure in nature—*Artemisia annua*, a Chinese plant—that is proving effective," says James Duke, a botanist at the USDA. "But what will happen if we reduce the biodiversity of our planet to the point where we lose the answer to artemisia-resistant malaria?"

Clearly the pace of destruction of the world's jungles and forests has rekindled interest in botanical drugs. In 1986 the NCI reopened its screening program, and soon put an additional focus on compounds that might help in the fight against AIDS. The NCI hired scientists to collect plant samples from jungles and forests in Asia, Africa, and Central and South America, and to speak with healers there about medicinal uses of plants. Researchers are also scouring the

Plant Drugs

Mother Nature is still the most ingenious chemist. The rosy periwinkle plant is the source for anticancer drugs that can't be synthesized in a lab.

Plant-based medicines can be useful in four ways: First, they can provide the actual compound for a drug, such as the anticancer agent vincristine, which cannot be synthesized. Second, they can provide the chemical building blocks that are used to synthesize modified compounds. Oral contraceptives, for example, are derived from steroids found in the wild Mexican yam, dioscorea. Third, plants can provide leads for developing compounds that act in an entirely new way. Snakeroot, for example, originally used as a tranquilizer, was developed into the hypertension drug reserpine. And finally, medicinal plants are also sold directly to consumers in health-food stores in the form of tinctures, pills, tablets, and extracts.

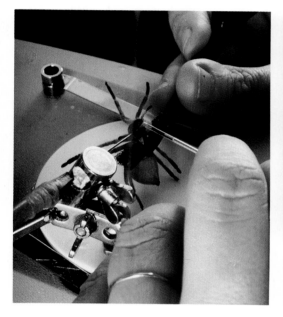

Plants are not the only source of medicinal compounds. Venom extracted from spiders (left) shows promise in blocking stroke-related brain damage. And the toxin from the Central American poison arrow frog (above) may one day be used to increase the rate and force of heart contractions and to treat degenerative muscle diseases.

world's oceans for chemically rich species of algae and sponges.

Advanced technology has made this renaissance possible. Today the NCI uses sophisticated equipment to screen 500 extracts a week from flowers, stems, bark, and roots. The samples are tested against a range of 60 types of human cancer cells and on HIV-infected cells. A decade ago, NCI researchers were simply testing samples on a limited number of mouse tumors.

Even so, the odds of finding a promising compound are long: Of 50,000 extracts recently screened for use against AIDS, so far only three—among them prostratin, found in a Polynesian remedy for the yellow-fever virus—are likely to wind up in clinical trials. Cox discovered prostratin's therapeutic value while documenting the medical practices of an aging Samoan healer. As for cancer, the NCI has screened 33,000 extracts, yielding five compounds for further study. A plant extract discovered at the NCI in the 1960s, and currently exciting great interest, is taxol, touted as potentially the most significant drug of the decade for breast, ovarian, lung, and other cancers. The Food and Drug Administration (FDA) is expected to approve taxol for limited use in the next few months.

Unusual Sources

The NCI is not the only government agency involved in natural medicines. The first congressionally funded natural-product research center is being built at the University of Mississippi in Oxford, where drugs to fight the effects of poison ivy, nausea, and quinine-resistant malaria are being developed. In other areas, Renuk Misra, a natural-product researcher at the National Institute on Aging, is studying plants used as memory boosters in Ayurveda, the traditional system of medicine of India.

Government scientists are also developing drugs from other, unusual tropical sources—among them poisonous frogs. The venom of a tiny Ecuadorian frog contains a newly discovered compound, epibatidine, that affects nerve-cell function. "It's at least 200 times more potent a painkiller than morphine," says organic chemist Tom Spande at the National Institute of Diabetes and Digestive and Kidney Diseases. Another toxin (pumilio) from a Panamanian frog also holds promise for increasing the rate and force of heart contractions and for treating degenerative muscle diseases.

However tentatively, some American drug companies are getting into the act.

Learning from Healers

When ethnobotanist Paul Alan Cox arrived in American Samoa in 1985 to observe Polynesian healers, he thought he would finish his "apprenticeship" quickly. "I knew the language. I thought I'd be back teaching at Brigham Young University within a couple of months."

Seven years later he is still traveling back and forth between Brigham Young and the grass huts of women healers (in Samoan culture, women fill the role of medical healers), taking detailed notes about their ancient medical practices on his laptop computer.

One of Cox's first tutors, 78-year-old Epenesa, still astounds the ethnobotanist. One of her techniques for administering medicine resembles the skin patch, a 20th-century invention. She mixes the plant remedy into coconut oil and slowly massages the mixture into the chest and head of a patient for a slow, sustained release. "Talking to Epenesa is like talking to a Ph.D.," says Cox. "She can identify over 200 medicinal-plant species and their uses. And she understands anatomy."

After Cox spent months observing the aging healer treat patients, Epenesa gave him her prescription for yellow fever. The remedy sat in his fridge along with other plants; he couldn't get a drug company interested in any of them.

In 1986 the National Cancer Institute (NCI) heard about Cox's collection of Samoan plant remedies and asked to screen them. A year later the NCI called with good news and bad news: The yellow-fever remedy was found to be very effective in stopping HIV from invading

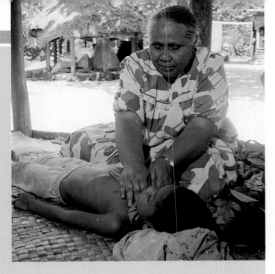

A Samoan medicine healer demonstrates a technique passed down from generation to generation for applying a plant mixture to treat inflammation.

healthy cells, at least in the test tube. But some NCI officials wanted to drop further studies. The remedy, from the tree *Homolanthus nutans,* contained prostratin, which is closely related to some tumor-promoting chemical compounds (phorbol esters). Cox and other NCI scientists convinced the officials to run the compound through cancer screens, which found that prostratin did not promote the development of cancer.

Healers like Epenesa are a dying breed. Many of them have no apprentices—their daughters and granddaughters are more interested in Western culture than in Samoan traditional medicine. With only a handful of healers left on Samoa and the other Polynesian islands, Cox has been racing to transcribe their remedies and medical practices into the Samoan language in the hope of encouraging young islanders to study their own culture.

Merck, the world's largest drug manufacturer, signed a $1 million agreement with INBio (Instituto Nacional de Biodiversidad), a nonprofit group in Costa Rica working to preserve the rain forest. Merck is also providing INBio with lab equipment and is training some of its scientists. In exchange, Merck reserves the right to screen Costa Rica's diverse "library" of plants, insects, and microbes for medicinal use. Another American company, Syntex Pharmaceuticals, has arranged with the Chinese Academy of Sciences to review China's sophisticated pharmacopoeia, which includes thousands of medicinal plants. And researchers at Natural Product Sciences are milking spiders for their venom, which contains compounds that seem to block brain damage associated with stroke. Monsanto and Merck are collaborating with botanical-garden scientists to look for new compounds.

Environmental Politics

But Michael Balick, head of the New York Botanical Garden's Institute of Economic Botany in the Bronx, New York, worries that in the time it takes to bring a drug to market —10 to 20 years—a substantial fraction of the world's jungle and forest species will disappear. In many countries, including India and Sri Lanka, the loss of rain forests is almost complete. Many developing countries are allowing loggers, miners, and ranchers to strip forests while local farmers continue clearing land, too. "What do you say to a farmer who needs to clear-cut the land to plant corn to feed his family?" says Balick, referring to the dilemma.

Balick and others are helping developing countries assess their plants with an eye to harvesting them without depleting the forests. For example, Shaman Pharmaceuticals, a small drug company that gets its leads for new drugs from traditional healers, is committed to buying local plants from tribes in Ecuador for an antiviral drug, rather than synthesizing active ingredients, as most companies do.

Peru, Brazil, Mexico, India, and Rwanda are also mounting modest efforts to record, screen, and sell plant drugs to their citizens, who still rely largely on traditional medicines. A tiny program in Rwanda has been remarkably successful: Together, traditional healers and doctors have produced 15 plant medicines, including an anti-inflammatory ointment, a cough syrup, a disinfectant mouthwash, and a treatment for a parasitic skin disease.

They're also developing a market for tropical-based cosmetics, insecticides, and textile dyes.

Balick and others criticize former President Bush's refusal to sign the biodiversity agreement at the Earth Summit in Brazil in June 1992, which would provide royalties to countries for their indigenous plants.

Poorer countries, says Charles Barber, a research associate at the World Resources Institute, an environmental think tank in

How to Get Involved

Readers interested in learning more about medicinal plants should contact the following groups:

THE RAINFOREST ALLIANCE
Attn: The Periwinkle Project
270 Lafayette St., Suite 512
New York, NY 10012
212-941-1900

THE MISSOURI BOTANICAL GARDEN
P.O. Box 299
St. Louis, MO 63166
314-577-9540

THE NEW YORK BOTANICAL GARDEN
The Institute of Economic Botany
Bronx, NY 10458
212-220-8700

CONSERVATION INTERNATIONAL
1015 18th St. NW, Suite 1000
Washington, DC 20036
202-429-5660

Washington, D.C., "no longer want to be a cut-rate supermarket for raw materials, only to have drugs and other products sold back to them—and at prices they can't afford." Others are not sure that Bush's stance had so great an effect. On its own, for example, Merck has pioneered a royalty agreement in its Costa Rican venture.

Critics and defenders alike agree that the Earth Summit thrust the "green pharmacy" into America's environmental consciousness. And with Albert Gore, now the vice president of the United States, a vocal supporter of environmental causes, biodiversity will undoubtedly come to the forefront.

But, like many natural-product scientists, Walter Lewis believes the issue transcends political ideology, that it's really about survival and quality of life. "If you want better health care, you have to support the preservation and study of nature's pharmacopoeia. Maybe the Jivaro's treatment for speeding the healing of wounds will make a difference," he says optimistically.

Agriculture

CROPS AND LIVESTOCK

Farmers will reduce acreage for corn, rice, and some other crops in 1993 in order to establish some balance after record or near-record harvests in 1992. The combination of high stocks and depressed prices for some crops will lead many growers to take land out of production to qualify for government subsidies. For example, corn yield in 1992 was 10 to 11 bushels per acre (4 to 4.5 bushels per hectare) higher than the previous year, according to the U.S. Department of Agriculture (USDA). Consequently, farmers will need to reduce corn acreage 5 percent in 1993 in order to qualify for benefits. Early estimates suggest that approximately 75 million acres (30 million hectares) of corn would be planted, down about 5 percent from 1992. Rice acreage will also be reduced by 5 percent in 1993.

Conversely, 5 percent more land could be planted with wheat in 1993, and there should be a 2.5 percent jump in cotton acreage. Soybeans, which were harvested in near-record amounts in 1992, are not part of the government acreage-reduction program.

Severe weather during the 1992–93 winter season both helped and hurt the agricultural outlook. In California the mid-March snowpack in the Sierra Nevada Mountains was 150 percent in excess of normal. Since the state's major source of irrigation is the runoff from that snowpack, crop yields should be very good. California produces around 55 percent of the nation's fruits, nuts, and vegetables.

However, cattle prices hit record levels, and hog prices soared during the first quarter of 1993, based in large part on the unusually wet winter in the Midwest. Far greater than normal accumulations of snow turned many feedlots into mud fields, greatly hampering the ability of livestock to feed well enough to gain weight and mature. A severe shortage of fully grown livestock would be reflected in noticeably higher meat prices in supermarkets across the nation.

▶ **The "Storm of the Century."** In March 1993, a powerful storm battered the Southeast with snow, freezing rain, and gale-force winds. Fortunately, the storm came early enough in the season so that farmers in the central and northern parts of Florida could replant some crops. But farther south, in Dade County, one of the largest agricultural counties in the state, 80 to 85 percent of the winter fruit and vegetable market was lost, according to state agencies. The grapefruit and orange crops fared pretty well.

Farther north, the Georgia Department of Agriculture reported that 153 poultry houses were destroyed, 70 were moderately damaged, and 300 sustained minor damage. Over 2 million birds were lost—two-thirds of a typical day's slaughter. The long-term effects on the poultry industry are still not clear. The peach industry in the southern part of the state took heavy losses, perhaps 50 percent or more. About 70 percent of the nurseries reported losses totaling $3 million in structural damage and $5 million in plant losses.

PESTICIDES

In 1993 the U.S. Supreme Court let stand a lower-court ruling disallowing even "negligible" traces of pesticides in the foods Americans eat. The Environmental Protection Agency (EPA) will have to find some way to act on that decision without endangering the yield and diversity of crops.

In a related matter, chemical companies may decide to stop producing pesticides for minor crops—kiwis and artichokes, for example—rather than go through the costly process of reregistering those pesticide products with the EPA. If there is no alternative pesticide available for these smaller crops, or if the crops cannot be produced profitably, then it is likely that both price and availability would be adversely affected.

Neil Springer

Energy

ENERGY POLITICS

In October 1992, Congress passed, and President Bush signed, the National Energy Policy Act. After more than two years of research and debate, this landmark legislation sets a course for the country's energy future for the first time since the oil crises of the 1970s. Before the bill was passed, however, its most controversial aspects—opening the Arctic National Wildlife Refuge for oil drilling, and forcing increased fuel-economy standards for automobiles—were removed. The Energy Policy Act plans to reduce dependence on imported oil by encouraging conservation and increased reliance on and development of domestic energy sources. The U.S. Department of Energy (DOE) estimated that the Act could decrease energy demand by 6 percent, and increase use of alternative fuels by 50 percent, by 2010. Although it proposed tax and fee increases of $5 billion over the next five years (including a $1 billion increase in taxes on ozone-depleting chemicals), the Act also proposed partially offsetting credits of about $4 billion for the same period—like a $2,000 deduction for buying an automobile that runs on alternative fuel, or lower mortgage rates on well-insulated homes.

The bill also encourages increased competition within the electric-utility arena by allowing small energy producers access to transmission lines; streamlines licensing for nuclear-power plants; and creates incentives for increased domestic gas and oil production.

During the 1992 presidential campaign, Bill Clinton and Al Gore stated that they favor conservation over additional energy production and consumption, and made clear their commitment to the environment. Shortly after his inauguration, President Clinton nominated Hazel O'Leary, a Minnesota utility executive, to be his secretary of energy.

President Clinton (left) chats with Hazel O'Leary, a former Minnesota utility executive who now serves the administration as secretary of energy. At right is Richard Riley, the secretary of education.

In February 1993, Clinton proposed a new tax package that included a controversial tax on energy. Opponents of the tax said the president was reneging on his promise not to impose new taxes on the middle class, stating that taxing gasoline and other forms of energy would hit middle- and lower-income persons the hardest because they spend the greatest percentage of their income on energy. The tax could force the average American family to spend more than an additional $100 a year, with gasoline prices increasing 7.5 cents a gallon, and the average monthly home electric bills increasing by $2.25. Some critics said the tax would hurt business and cost jobs. Supporters, however, said the new tax not only would help reduce the budget deficit, but also would encourage greater energy efficiency, reduce pollution, and help make the United States more dependent on its own supplies of energy.

FOSSIL FUELS

Oil. The number of active domestic oil wells continued to decline in 1992, dropping below 700 for the first time in more than 50 years. This compares with 860 in 1991, and 4,500 in 1981. In a disturbing trend, more than 12 percent of the independents, or "wildcatters," have already moved their exploration operations abroad, where costs are far lower. The average U.S. daily demand for oil remained steady in 1992 at 17,006,000 barrels, slightly above 1991's 16,989,000 barrels. The DOE predicted continued growth in demand of 0.4 to 1.6 percent per year, but also projected adequate supplies at least into the next century. Imports supplied more than half of the country's oil at times during the year, but the annual average—46.12 percent—remained in the same range as in the past few years. Oil's contribution to U.S. energy consumption was 40.9 percent, only marginally higher than last year's 40-year low of 40.3 percent. Oil prices remained steady throughout the year, hovering mostly in the $18 to $21 range.

Natural-gas consumption remained steady, supplying 24.3 percent of the nation's energy needs. Prices for producers, which had been in a free-fall for some time, dropped as low as 83 cents per 1,000 cubic feet (28 cubic meters) before steadying and rising to an average of $1.83. Prices to consumers remained nearly constant at $5.90 per 1,000 cubic feet. The low prices for producers continued forcing companies out of business at the rate of one a week, resulting in a decline of 60 percent since 1985. And the wide gap between the price of gas at the wellhead and the price paid by consumers prompted increasing complaints to regulators.

Coal production continued dropping in 1992, to 994.3 million tons, down from last year's 998 million tons. Utilities used coal to produce 56.3 percent of the nation's electricity in 1992, and coal supplied 23 percent of the country's overall energy consumption.

HYDROELECTRIC

Hydroelectric power supplied 8.6 percent of the nation's electricity in 1992, down slightly from 1991's 9.0 percent. After the state of New York canceled its $15 billion contract to purchase electricity from Hydro-Quebec in Canada, the company slowed down work on one phase of the project, moving its completion date back to 2008 from 2000, and its cost up from $40 billion to $54 billion. The project had stirred a storm of criticism from environmentalists and Cree Indians, and remained in some doubt well into 1993.

NUCLEAR POWER

The 109 operating nuclear plants in the United States produced a record 22.1 percent of the nation's electricity in 1992, while the 420 plants operating in 26 countries produced about 16 percent of the world's electricity. Economics became a major issue in 1992, with three U.S. plants permanently shutting down years before their licenses were due to expire, because they had become too expensive to operate. Nuclear power's cost advantage has shrunk in recent years as the industry has been faced with increasing expenditures for plant modifications, an economy mired in recession, and low prices for oil and other alternative sources of electricity. When long-term alternatives become less expensive than existing nuclear plants (or any other kind of plant), it's purely a business decision to retire the more costly generators. Even the proposals of the Energy Act would do little to help existing nuclear plants: As demand growth has remained low throughout the recession, few utilities are planning to order any large generating plants—nuclear or otherwise. With President Clinton's "hands-off" philosophy toward supporting nuclear, this technology faces serious economic challenges in coming years.

In March the Electric Power Research Institute (EPRI) announced it would invest another $12 million to research "cold fusion" because of promising results in an experiment conducted at SRI International in Palo Alto, California. EPRI has already invested about $2 million in this research, and is the only U.S. supporter of such research. The DOE ended its funding in 1989, and the state of Utah pulled out in 1991.

Anthony J. Castagno

Environment

POLITICS AND THE ENVIRONMENT

Politically, 1992 saw two major environmental events: the Earth Summit (see page 58) and the U.S. presidential elections, in which Bill Clinton ultimately defeated the incumbent, George Bush. Clinton's running partner, Senator Albert Gore, is a noted environmental activist.

Upon taking office, Clinton replaced Vice President Quayle's Council on Environmental Quality with the White House Office on Environmental Policy, to coordinate and influence environmental policy at the highest level. He began working with Congress to elevate the Environmental Protection Agency (EPA) to cabinet status, and thereby give it more authority. Several early appointments, including Carol Browner to administrator of the EPA and Bruce Babbit to Secretary of the Interior, were applauded by environmental groups across the nation.

Later on, Clinton took a bold step concerning the management of government land. In a land-use plan to be included in the national budget, the administration proposed tripling the grazing fees on 280 million acres (113 million hectares), and raising the price of timber from national forests. It also proposed royalties on gold, silver, and other minerals mined on federal land.

In April the new administration hosted a Forest Conference to help decide the fate of the Pacific Northwest's last great stands of timber. On behalf of rare fish and wildlife species, court injunctions had limited logging, threatening the jobs of thousands of timber workers. Clinton asked his cabinet to produce, within 60 days, a proposal that would both end the stalemate and ensure the long-term health for the forests and economic stability for timber-dependent communities.

THE OZONE HOLE

The National Aeronautics and Space Administration (NASA) reported that the

Vice President Al Gore's long-held commitment to the environment has been instrumental in making this issue a key focus of the Clinton administration.

ozone hole in the atmosphere above Antarctica in 1992 was the biggest ever—larger in area than the entire continent of North America. For the first time, the edges of the ozone hole crossed the southern tip of South America. In addition, ground and satellite data showed that ozone losses now occur in both hemispheres—not just in the winter, but during spring and summer as well. Amid all this bad news, evidence that ozone loss poses health threats to life on Earth continues to accumulate.

Alarm at worsening ozone depletion quickened chlorofluorocarbon (CFC) phaseout measures. International negotiators met in Copenhagen, Denmark, at the end of November 1992 to redraft the Montreal Protocol, an international agreement that controls the production and consumption of ozone-destroying chemicals. Industrialized nations agreed to completely stop the production of the predominant ozone depleters, like CFCs, by the end of 1995—five years earlier than previously contemplated. The deadline was extended by 10 years for developing nations.

The EPA had already announced guidelines to accelerate the phaseout of

U.S. CFC production and importation in February 1992. But due to the meeting in Denmark, the EPA later proposed harsher measures, a full phaseout of CFC production by 1996, and limited use of methyl bromide and other substances that deplete stratospheric ozone. However, both international and domestic actions allow the continued production of prohibited chemicals for "essential uses," like health and safety, for an indefinite amount of time.

MARINE SANCTUARIES PROGRAM

The year 1992 marked a turning point for the Marine Sanctuaries Program, which was created in 1972 to protect ecologically rich and historically important marine areas.

To begin with, Congress reauthorized the program, designating four new sites for protection, including: Stellwagen Bank Sanctuary in Massachusetts; Hawaiian Humpback Whale Sanctuary in Kahoolawe, Hawaii; Monterey Bay Sanctuary in California; and Flower Garden

Banks in Louisiana and Texas. The total acreage added represents a 50 percent increase. With these new sites, there are now 12 sanctuaries in all. Three others will be considered in 1993, and one in 1994.

In addition, Congress finalized the 1993 Commerce Appropriations Bill, which included a funding increase of $2 million, a 40 percent increase for the program. Administrators say that even this amount falls short of their needs. Still, reauthorization and increased funding will broaden and strengthen the program—a significant victory for the marine environment.

THE CIA SHARES ITS SECRETS

In 1992 the Central Intelligence Agency (CIA) opened wide its guarded doors to a team of about 40 environmental scientists. The idea: use Cold War surveillance satellite data to help fight the world's environmental wars.

In 1990 Al Gore, then a Tennessee senator, began work to declassify 30

**Experiment in the Forest:
New Approaches to Logging**

In traditional forest management, old-growth forest may be clear cut and replaced with tree farms of a single species. A new approach, which the Forestry Service calls "ecosystem management," will test theories of forestry that aim to mimic what happens after natural events. Logging according to historic natural fire patterns would theoretically leave trees behind as a "biological legacy" to foster regeneration. The approach might also leave a network of tree reserves, connected by corridors along river banks, instead of fragmenting the remaining forest.

years' worth of CIA espionage data to qualified researchers. Two years later, President Bush authorized the program. His directive included both the intelligence community and the Pentagon. The hope is that the same satellite images that identify missile silos will also document such environmental data as climate change and desertification.

Already, declassified satellite data has revealed the structure of the world's ocean floors in more detail than ever before seen. Similarly, spy data could be used to produce high-definition maps of Earth's landmasses that would aid ecologists and other researchers. Classified Air Force data on atmospheric composition, and Navy records on the temperature of the world's oceans at various depths, could offer powerful clues as to the condition of this planet.

THE SIERRA CLUB CENTENNIAL

On May 28, 1892, John Muir formed the Sierra Club to "do something for wildness and make the mountains glad." One hundred years later, the mountains have much to be thankful for.

Among other things, Muir's writings galvanized support for the creation of Yosemite, Sequoia, Mount Rainier, Petrified Forest, and Grand Canyon national parks. Today the Sierra Club, based in San Francisco, has 625,000 members and multiple chapters across the country. The club has continued to establish, enlarge, and protect wilderness areas. It has become a potent force in a variety of battles, including the reduction of "greenhouse" gases and the protection of old-growth forests.

RECYCLING BOTTLENECK

Separating clear glass bottles from colored ones and placing newspapers in a bin apart from white paper is fast becoming habit for most Americans. Neighborhood curbside collection programs have skyrocketed from a mere 600 in 1989 to over 4,000 today. Recyclers across the nation are patting themselves on the back and waiting for the positive news about an improved environment.

Sadly, that good news may be delayed. In a turn of events that was probably predicted by economists, but handily buried, so to speak, by environmentalists, it appears that the recycling ritual in many parts of the country has quickly outpaced the capacity to process and manufacture recycled material.

The bottleneck seems to be in two areas—economics and technology. Take newspapers, the most widely recycled material. Facilities to remove ink from newsprint are extraordinarily expensive. Furthermore, the huge glut of recyclable material has depressed an already soft market for used papers.

Another problem is that recycled materials compete with virgin plastics or papers that benefit from government subsidies. Newsprint producers are indirectly subsidized by public-area logging and logging access roads. Depletion allowances for petroleum subsidize makers of oil-based plastics.

Technological innovation is needed to create better processes for removing contaminants. Currently, ubiquitous items—such as clear plastic tape on envelopes, and sticky yellow Post-it notes—tend to gum up recycling systems.

Despite these setbacks, there is money to be made in trash, and ingenious methods are being developed. In Houston, Champion Recycling Corporation, a subsidiary of a leading paper manufacturer, has built an $85 million de-inking plant, and in return, they receive the city's entire collection of old newspapers and magazines. Manufacturers, such as Lever Brothers, have begun to produce super-concentrated versions of their products, packaged in smaller containers that take up less landfill space—a concept that has been shown to attract consumers.

For all its promise, both environmentally and economically, recycling is becoming recognized as only a small solution to the waste-disposal problem. More and more, experts are looking for new ways to reduce waste in the first place and still show a profit.

Beth Livermore

Geology

GIANT CRATER

Geologists discovered what appears to be the remnants of a giant crater, created when a massive meteorite or comet slammed into Earth 65 million years ago, at the end of the Cretaceous period. The impact could explain why the last remaining dinosaurs and many other forms of life went extinct at this time.

The 110-mile (177-kilometer)-wide crater is buried beneath 3,300 feet (1 kilo-

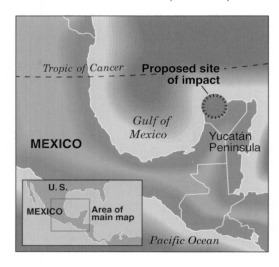

Scientists are close to establishing that a giant crater on the Yucatán peninsula in Mexico (map, left) was created by a mammoth meteorite or comet (above) that slammed into Earth. The impact triggered the mass extinction of dinosaurs and other species 65 million years ago.

meter) of rock at the northern end of Mexico's Yucatán Peninsula. Scientists had identified the crater several years ago, but they were unsure whether it formed at the same time as the Cretaceous extinction. In 1993 researchers from the Institute of Human Origins used radioactive-dating techniques to show that the crater did indeed match the age of the extinctions. Before wrapping up this ancient murder mystery, researchers must drill into the crater to confirm their recent findings. They must also sort out whether factors aside from the Yucatán impact helped trigger the mass extinctions 65 million years ago. Some scientists have suggested that several other meteorites struck the Earth at the same time. Others maintain that more-mundane processes such as climate change played a role in causing the die-offs.

CURRENTS INSIDE THE EARTH

Recent experiments with supercomputers yielded evidence that could help resolve one of the biggest debates in earth sciences. For decades, researchers have argued about the currents of rock flowing slowly through Earth's mantle, the thick region separating the planet's iron core from its outer crust. Many scientists believe that heat inside the planet stirs the entire mantle much like a boiling pot of soup. Yet other investigators have maintained that the mantle is layered into upper and lower parts that do not mix—a situation that would resemble a double boiler. The new computer research suggests that the upper and lower mantle may remain separate most of the time, with occasional leaks developing between the two regions.

Two research groups, led by scientists from the California Institute of Technology (Caltech) and from the University of Hiroshima, simulated the mantle using models run on supercomputers. Judging from these simulations, the scientists think that rock from the upper mantle may sometimes burst through the boundary and flow into the lower mantle. Such currents inside the Earth are important because they ultimately provide the force that shuffles the continents around the planet's surface.

HUGE PREHISTORIC ERUPTION
By matching volcanic deposits on either side of the Atlantic, a team of geologists discovered evidence of what may well have been the largest eruption in the past half billion years of Earth's history. The scientists compared layers of 454 million-year-old volcanic rocks that exist both in Europe and North America. Using rare minerals in the rocks as a "chemical fingerprint," the investigators showed that the deposits on both continents apparently came from the same volcanic blast. The scientists believe a huge eruption spewed dust into the atmosphere, which settled down onto Europe and North America to form the layered volcanic rocks. At the time of the blast, the two continents were closer together than they are today. Such a large eruption would have cooled the climate for several years, the scientists believe.

SEATTLE SEISMICITY
Five teams of geologists uncovered signs indicating that a large earthquake 1,000 years ago ripped through the region that would later become the city of Seattle. The tremor was generated by a newly discovered fault that runs under Seattle's Kingdome stadium. Seismologists think the fault could spawn another earthquake in the future.

The prehistoric quake instantaneously raised land north of the fault by 20 feet (6 meters), triggering a large tsunami wave in Puget Sound that buried marshes along the sound. The shock also set off avalanches in mountains around the city. Prior to this discovery, seismologists had not considered the possibility of a large earthquake occurring directly beneath Seattle. Now they must take that threat into account. At this point, scientists cannot predict when the next quake will hit the Seattle area, though it may not strike for many centuries.

VOLCANO UNDER ANTARCTIC ICE
While flying over Antarctica in an instrument-laden plane, a team of geoscientists discovered an active volcano hidden beneath the thick ice cap that covers 98 percent of the continent. This is the first active volcano found under the ice cap. Led by a geophysicist from the University of Texas, the scientists were studying the sub-ice geology with radar and other instruments, when they detected an isolated mountain under the ice. The investigators believe the mountain is an active volcano because there is a depression in the ice above the peak—an indication that the ice is melting in that spot. Because the volcano is hot enough to melt the ice, it must have erupted within the recent geologic past—sometime during the past few hundred years. The team also discovered hints of other volcanoes in the same part of Antarctica. Because the ice in this region is more than 1 mile (1.6 kilometers) thick, the scientists believe that an eruption of one of these volcanoes would probably not melt clear through the ice cap.

MANTLE ROCKS
A crew of earth scientists achieved the long-sought goal of drilling rocks from the border between Earth's crust and mantle. Since the late 1950s, geoscientists have wanted to drill straight through the 4-mile (6.4-kilometer)-thick ocean crust and into the mantle beneath, but this would take years of expensive work. A team of scientists took a shortcut by drilling directly into the upper mantle in a place where a rip in the ocean crust has exposed this layer. Rocks pulled up by the drill will help scientists study how molten rock from the mantle rises up to form the ocean crust, which covers 60 percent of the planet's surface.

Richard Monastersky

Oceanography

VOLCANO CLUSTER DISCOVERED

Scientists mapping the seafloor 600 miles (965 kilometers) northwest of Easter Island in the South Pacific have found a vast cluster of undersea volcanoes. The finding, announced on February 13, 1993, may include the greatest concentration of active volcanoes on Earth. Peering into the ocean depths with sophisticated sonar scanning devices, oceanographers and geophysicists aboard the research vessel *Melville* were surprised to discover 1,133 volcanic seamounts in a 55,000-square-mile (142,000-square-kilometer) area— about the size of Illinois.

"We thought we would find a few dozen new volcanoes. Instead, we found over 1,000 that had never been mapped before," says Dr. Ken MacDonald, a professor of marine geophysics at the University of California at Santa Barbara, the leading scientist on the project.

The area of intense volcanic activity lies near the East Pacific Rise, a ridge running from north to south where the Pacific and Nazca plates, two of the gigantic plates that comprise the planetary crust, are separating at a rate of 8 inches (20 centimeters) a year—faster than anywhere else on Earth. Some of the volcanoes are almost 7,000 feet (2,135 meters) tall, with their peaks 2,500 to 5,000 feet (760 to 1,520 meters) below the surface. According to MacDonald, "Two or three could be erupting at any given moment."

The finding underscored just how little is known about the ocean depths—the last unexplored frontier on Earth, where less than 5 percent of the sea bottom has been mapped in detail. Although much of the seafloor remains a mystery, scientists have known for some time that there is much more volcanic activity on the ocean floor than on land.

"Ninety percent of all volcanic activity is on the ocean floor," says Haraldur Sigurdsson, Ph.D., professor of oceanography at the University of Rhode Island. Sigurdsson estimates that the quantity of molten rock pushed up into the ocean floor every year is 5 to 10 times greater than that which erupts on land.

The discovery also intensified speculation over whether volcanic activity, pouring huge amounts of heat into the ocean, could change water temperatures enough to affect large-scale weather patterns in the Pacific, such as the El Niño. Concerning such a possibility, many scientists remained skeptical.

"The ocean is so big, and even if you are putting a lot of heat in a few spots, it does not have much effect," says William Chadwick, Ph.D., a volcanologist at the National Oceanic and Atmospheric Administration (NOAA).

Scientists mapping the seafloor northwest of Easter Island discovered a vast cluster of volcanoes covering an area larger than Illinois. At left is a color-enhanced computer scan of three of the volcanic formations.

OCEAN HISS

Crew members in submarines may soon be able to steer a course through a thicket of undersea obstacles by watching an image of nearby surrounding environs generated not by light, but by subtle fluctuations in naturally occurring oceanic sounds. A series of experiments conducted in early 1992 by cooperating scientists at the Scripps Institution of Oceanography in California and Florida Atlantic University at Boca Raton utilized "daylight sound," the acoustic counterpart of natural daylight—to exploit the constant hiss of noise radiating throughout the world's oceans.

The hiss is a consequence of myriad oscillating microscopic bubbles that have been entrained by the breaking waves bursting at the ocean's surface; a similar bursting process causes carbonated soft drinks to fizz. Writing in the March 25, 1992, issue of *Nature*, the scientists reported how the new experimental underwater-detection system reveals oceanic objects—simply by listening to the ambient underwater noise and displaying the objects as a video image.

OCEAN HOT SPOTS

During April 1992, satellite images were able to confirm ground observations of an unusual oceanic hot spot called the Western Pacific warm pool.

"We've been aware of this, but it is the first time vivid images from satellite data agree with observational data gathered at sea level," says Xiao Hai Yan, Ph.D., of the Graduate College of Marine Studies at the University of Delaware. The warm pool appears to be part of a slow, 10-year warming that might be a function of trade winds and global warming, and might also be connected to El Niño (unusual wind and current shifts primarily affecting Pacific waters).

Beginning March 5, 1993, the Central Equatorial Pacific Experiment (CEPEX)—a month-long, comprehensive field investigation of clouds, climate, and the ocean surface—was expected to answer many questions about ocean hot spots in the Pacific Ocean, perhaps shedding scientific light on how the pools are able to maintain themselves and how clouds may help regulate their temperature. CEPEX data were to be collected from a vast oceanic area in the western equatorial Pacific, where an exceptionally large, warm pool of water persists that always exceeds 80° F (26° C), but never goes above 87° F (30° C)—as if by a natural thermostat.

RAPID GLOBAL CLIMATE CHANGE?

Rapid changes in North Atlantic air and sea temperatures—by as much as 2.8° F (5° C) during the past half-century—were confirmed by scientists at the Woods Hole Oceanographic Institution (WHOI) in Massachusetts in May 1992. This finding suggests that greenhouse warming and the melting of snow and ice at the poles may have far more rapid effects on ocean and climate conditions than was previously thought. The temperature changes, caused by sudden shifts in the ocean conveyor-belt circulation system that transports heat from the equator toward the poles—could eventually lead to serious climatic consequences—such as the present climate of Britain and Norway changing suddenly to that of Greenland and northern Canada.

Woods Hole scientists were also able to use miniaturized laboratory modeling of fluid motion and time-lapse photography to gain a better understanding about circulation and temperature patterns in the Arctic Ocean, the least-studied ocean on Earth, and the oxygen-poor Black Sea—without ever having to travel to these locations. After building a Styrofoam ice cap and continental shelves to simulate the topography of the Arctic Ocean, the oceanographers constructed a similar model of the Black Sea with the help of Russian colleagues. By injecting colored dyes into the water and controlling the flow and temperature of water in the test tanks, they were able to track currents, eddies, and other physical features, and photograph the paths of the dyes with still and video cameras over time.

Gode Davis

Seismology

Strong earthquakes struck several parts of the world in 1992, providing a deadly reminder that the Earth's continents are continually shifting their positions in a process known to scientists as plate tectonics.

CALIFORNIA QUAKES

The most powerful shock of the year ripped through the Mojave Desert east of Los Angeles on June 28 near the small town of Landers, California. Measuring 7.6 on the Richter scale, the jolt was the largest to hit California in the past four decades. Despite its strength, the quake caused only one death because it hit a sparsely populated area and sent its most powerful vibrations out into the desert. Seismologists consider the quake to be one of the most important ever, from a scientific perspective, because it occurred in an area with many sensitive instruments that recorded very fine details about the tremor.

The Landers quake concerned scientists because it struck near the San Andreas fault and apparently increased the risk that a major jolt will occur on the infamous fault. The San Andreas runs through highly populated parts of Southern California, so a strong earthquake on this fault will likely cause considerable death and destruction.

Many earth scientists were surprised by the Landers quake because it triggered "swarms" of smaller jolts across much of the western United States, in regions up to 750 miles (1,200 kilometers) away from the epicenter. The swarms occurred in northern California, southern Idaho, southern Nevada, southern Utah, and even in Yellowstone National Park in Wyoming. Scientists are currently trying to understand how a major earthquake in one area can set off small- and medium-sized tremors in very distant regions.

The Landers quake was preceded by several foreshocks, one of which measured 6.5 on the Richter scale. Studies of these preliminary jolts may help scientists recognize when a geologic fault is preparing to unleash a large earthquake. By looking back over records of the pre-Landers tremors, researchers with the United States Geological Survey (USGS) discovered that small earthquakes displayed an unusual type of behavior just before the big shock. Tiny jolts occur often throughout much of California, but the Lander foreshocks were unusual because they were clustered together in a tight space. Seismologists hope that clues such as this might help provide a warning of some large quakes.

Scientists have long held the San Andreas Fault responsible for California's earthquakes. The epicenter of the 1992 Landers quake and those of other recent temblors suggest that a new fault may exist.

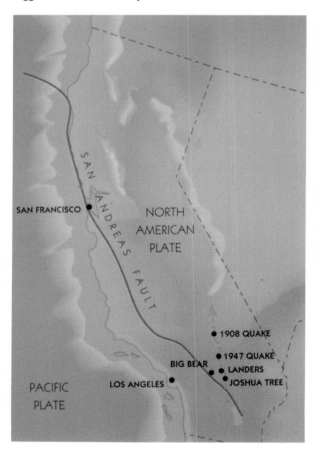

The Landers tremor was caused by stress that accumulates in the Earth's crust as a result of motion between tectonic plates, which are the broken pieces of Earth's outer shell. The planet's surface is divided into a dozen large pieces, or plates, that are continually shifting position. The plates do not move smoothly; instead, their edges stick together until enough stress builds up and the plates jerk forward, creating an earthquake. The Landers shock occurred near the edge of the North American plate where it slides past the plate carrying the Pacific.

Elsewhere in California, several major shocks struck the northern part of the state near the town of Petrolia in April. The main quake measured magnitude 6.9 and was followed by aftershocks of magnitude 6.0 and 6.5.

OTHER NOTABLE QUAKES

The deadliest earthquake in the world last year occurred in the Flores Island region of Indonesia on December 12, 1992, killing at least 2,200 people. The earthquake, measuring 7.5 on the Richter scale, triggered a large tsunami wave up to 80 feet (25 meters) high that devastated coastal areas, causing many of the deaths.

At least 541 persons died as a result of an earthquake near Cairo, Egypt, on October 12, 1992. The earthquake measured 5.9 on the Richter scale, but it caused considerable damage because it occurred near a populated area and leveled many poorly constructed buildings.

A magnitude-7.2 earthquake struck off the Pacific coast of Nicaragua on September 2, triggering a tsunami wave that reached heights of 25 feet (8 meters). The wave killed at least 62 people.

A major earthquake measuring 7.1 on the Richter scale hit northern Japan on January 15, 1993. The quake was centered near the fishing port of Kushiro on the island of Hokkaido and killed two people.

A deadly tremor occurred in Kyrgyzstan in Central Asia on August 19, 1992. The magnitude-7.5 shock caused landslides in the mountainous region, killing at least 42 people.

PREDICTIONS AND THEORY

The end of 1992 marked a major disappointment for seismologists who hope to forecast future earthquakes. The year's end signaled the failure of the first and only official earthquake prediction issued by the USGS. In the mid-1980s, the USGS predicted that a strong earthquake would occur on the San Andreas fault near the small town of Parkfield, California, by the end of 1992; instead, the year passed without the expected jolt. The scientists had based their prediction on the history of Parkfield, with an earthquake having struck the region roughly every 21 years for the past century. The most recent earthquake occurred in 1966. Scientists still believe a Parkfield earthquake will occur soon, but they are less sanguine now about their ability to forecast the potential for strong tremors.

A debate erupted among seismologists last year over one of the central tenets of their science. For several decades, earthquake researchers have subscribed to the seismic-cycle theory, which holds that when a large quake occurs on a fault, the Earth's crust must build up stress over a number of years before that same fault can produce another large quake. In the late 1970s, researchers used the seismic-cycle theory to produce maps of earthquake hazard. According to the theory, areas that had passed many decades since the last big earthquake were judged the most hazardous. They were called seismic gaps. Other areas, which had suffered major quakes in recent decades, were deemed less hazardous.

Two seismologists from the University of California at Los Angeles (UCLA) challenged the idea of seismic gaps, saying it doesn't accurately describe what happens in the real world. They examined a hazard map made in 1979, and found that, contrary to expectations, more large earthquakes occurred in the "less hazardous zones" than in the seismic gaps. The scientific community is now trying to determine whether to amend the seismic-cycle theory.

Richard Monastersky

Volcanology

The Pacific Ring of Fire lived up to its reputation last year as volcanoes erupted on all sides of the Pacific Ocean.

The Mayon Volcano in the Philippines unleashed several blasts in early 1993 that killed at least 72 people and prompted the evacuation of 60,000 others. This 8,077-foot (2,463-meter) mountain is located roughly 200 miles (320 kilometers) southeast of the capital city of Manila. The volcano has erupted a dozen times this century, the last time in September 1984. The dormant volcano reawakened without warning in early February, when it sent a plume of ash 3 miles (4.8 kilometers) into the atmosphere. Volcanologists are predicting that Mayon could produce an even larger eruption soon.

Another Philippine volcano, Mount Pinatubo, killed more than 700 persons when it erupted in 1991 in one of the biggest blasts of the century. Since then the volcanic debris from Pinatubo has remained in the atmosphere as tiny droplets of sulfuric acid that block out sunlight and cool the Earth. The average surface temperature of the planet dropped by roughly 1° F (0.5° C) during 1992, largely as a result of the Pinatubo eruption.

The sulfuric acid droplets from Pinatubo also apparently worsened the Antarctic ozone hole. In October 1992, scientists reported that the ozone concentrations above the South Pole were the lowest ever measured.

A small eruption of the Galeras Volcano in Colombia killed six volcanologists who were studying the crater when the blast occurred on January 14, 1993. Scientists from around the world had traveled to Galeras as part of a workshop on the volcano, one of the most active in historic times. The researchers are particularly concerned about Galeras because the volcano lies only 8 miles (13 kilometers) from the center of the city of Galeras. A large eruption from this mountain could prove disastrous. The scientists were climbing inside and around the Galeras crater when the blast occurred without warning.

In Antarctica a team of researchers attempted to study the inside of an active volcano by lowering a spiderlike robot into the dangerous crater. The project failed, however, when a cable leading to the robot broke. Scientists have tried for many years to explore the crater atop this peak, called Mount Erebus. But previous expeditions inside the crater had proved extremely hazardous because the volcano frequently has small eruptions that send smoldering boulders flying through the air. The researchers had hoped to use the robot, named Dante, to collect gases escaping from a lava lake at the crater bottom. They had also wanted to test the idea of using robots to explore the Moon and Mars.

Richard Monastersky

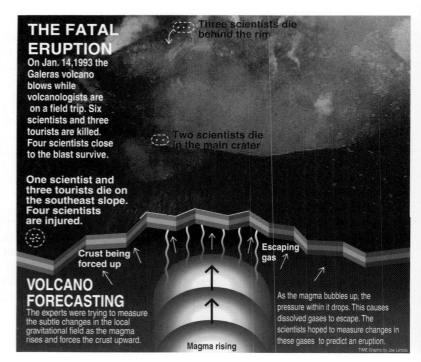

THE FATAL ERUPTION

On Jan. 14,1993 the Galeras volcano blows while volcanologists are on a field trip. Six scientists and three tourists are killed. Four scientists close to the blast survive.

One scientist and three tourists die on the southeast slope. Four scientists are injured.

Three scientists die behind the rim

Two scientists die in the main crater

VOLCANO FORECASTING

The experts were trying to measure the subtle changes in the local gravitational field as the magma rises and forces the crust upward.

Crust being forced up

Escaping gas

As the magma bubbles up, the pressure within it drops. This causes dissolved gases to escape. The scientists hoped to measure changes in these gases to predict an eruption.

Magma rising

TIME Graphic by Joe Lertola

Weather

The downtown area of Montpelier, Vermont, suffered much water damage in 1992 when ice jams caused the Winooski River to overflow its banks.

From the perspective of journalists, 1992 will go down in the record books as the year that Hurricane Andrew devastated Florida, and the year that drought-ending rains arrived in California. Years from now, however, meteorologists may view 1992 as the year that a subtle but significant change in the overall weather patterns brought to an end the years of warmer-than-normal temperatures in much of the country.

Interestingly, the start of 1992 gave no indication that anything approaching a cool-down was in the works. Despite a record-making snowstorm in the Deep South, the entire country had a mild January and February. In fact, the weather through February 1992 was the warmest in the lower 48 states in 98 years of record keeping.

SPRING 1992

Everything changed just as spring should have been getting under way. Cold Arctic air began making frequent excursions southward during March, bringing snow and subfreezing temperatures to northern and eastern areas of the country. Heavy rainfall occurred in California, a harbinger of what would eventually be enough precipitation to end the prolonged drought in the West. In Florida, more-localized weather phenomena made the news. Thunderstorms caused severe damage to the state's citrus industry. A hailstorm in Orlando caused an awesome $60 million worth of damage—the city's worst natural disaster ever. Back in New England, ice jams formed in the Winooski River, causing record flooding in Vermont. In the state capital of Montpelier, the downtown area suffered its worst flooding since the mid-1880s.

In mid-April a blizzard dumped a foot or more of snow from Nebraska across to Iowa and the Dakotas. In May an incredible 60 inches (152 centimeters) of snow fell on Mount Pisgah in North Carolina.

May also witnessed the typical clashes of warm tropical air and cold polar air across the Midwest. As always, these clashes created turbulence. During mid-May alone, 106 reports of severe weather were filed—reports that included 22 tornadoes.

Fortunately, more-tranquil weather reigned in several regions of the country. Miles City, Montana, reached a record high of 72° F (22° C) on May 27. For downright tropical readings, Lake Havasu, Arizona, recorded a sizzling 107° F (42° C) late in the month.

SUMMER 1992

To a meteorologist, 1992's summer weather maps bore an unsettling resemblance to those more typical of late winter. Temperatures in the Mid-Atlantic states dropped to record lows during the third week of June. The cold air in the northern states, and the overriding warm air from the South, caused numerous tornadoes in the Midwest.

By July the jet stream, the river of air high in the atmosphere that steers most of our weather systems, had settled 10 to 15 degrees of latitude farther south than normal for the time of year. Such a jet-stream position left the upper Midwest and the Great Lakes region much colder than normal. In fact, many people drew parallels between the cool summer of 1992 and the summer of 1816, "the year without a summer" in weather lore. These parallels may have had some basis in fact. The cool 1816 summer is thought to have resulted from a distant volcanic eruption that projected much dust in the air. Similarly, many experts blamed dust injected into the atmosphere by the 1991 eruption of Mount Pinatubo in the Philippines for the cool temperatures, a trend reinforced by the reappearance of the El Niño weather pattern in the Pacific.

In December 1992, a powerful storm caused severe beach erosion along the East Coast. Fire Island, New York, was particularly hard hit (above).

Whatever the reasons, many localities registered their coldest July ever. Among the cities to earn this dubious honor were Aberdeen, South Dakota; Albany, New York; Great Falls, Montana; and Sault Ste. Marie, Michigan. At the other end of the thermometer, Daytona Beach, Florida, sweated out a broiling July, with 29 of 31 days above the 90° F (32° C) mark.

As summer waned, the upper peninsula of Michigan saw its earliest foliage-color change in memory. In the Pacific Northwest, dry weather exacerbated for-

est fires. In the Shasta area of northern California, 50,000 acres (20,250 hectares) of forest were destroyed by fire.

FALL 1992

The unprecedented chill of summer paved the way for a colder-than-normal autumn. In the Northeast, snow would fall every month for the rest of 1992 and beyond into April 1993. In New England, some areas saw their earliest snowfalls on record. In the Midwest, cold halted the growing season by the first week of October. A record cold snap swept through the northern tier of states in the second half of October. An early onset of lake-effect snow hinted of a severe winter ahead. Out West, stormy weather in California eased the drought, and some municipalities were able to lift their water restrictions by Christmas.

WINTER 1992–93

For the first time in years, the mountains along both coasts had ideal skiing conditions. Indeed, snow and lots of it fell throughout the country.

A powerful storm moved up the Eastern Seaboard in early December. Beaches from New Jersey northward were buffeted by record tides, high winds, flooding, and severe erosion. In New York City, rapidly rising water levels caused the East River to suddenly inundate the FDR Drive, one of Manhattan's busiest arteries, stranding many motorists in their cars. The high winds caused the city to close down all of its suspension bridges. Farther north, heavy snow fell over much of the New England states, with Worcester, Massachusetts, buried under 3 feet (1 meter) of the white stuff.

Three months later to the day, another monster storm roared north from the Gulf of Mexico. This so-called "Storm of the Century" first spread snow across the Deep South, from Atlanta to Tallahassee, Florida, to Birmingham, Alabama, where 13 inches (33 centimeters) fell—more than in any other entire winter. People from North Carolina to Maine measured snowfall in feet rather than inches,

A blizzard in March 1993 delivered a rare snowfall to Atlanta (above) and other cities in the Deep South.

and gusty winds in excess of hurricane force caused widespread drifting. One indication of the power of the storm was its extremely low barometric pressure: 28.08 inches of mercury—lower than that of Hurricane Hugo when it slammed into South Carolina in 1989.

The severe damage (left) inflicted on Hawaii by Hurricane Iniki in September 1992 was overshadowed by the devastation in south Florida caused by Hurricane Andrew a few weeks earlier.

In California the rain began to wear out its welcome. Floods and landslides hit coastal areas, while record snows and avalanches beset the mountains. Still, by early March, most areas of California had lifted their drought emergencies, as reservoirs found themselves replenished by the winter's almost endless series of storms.

THE 1992 HURRICANE SEASON

With one significant exception, 1992 was a quiet year for the tropics. A tropical depression in late June dumped 24 inches (60 centimeters) of rain on some parts of the Gulf Coast of Florida. In July, precipitation from a Pacific hurricane named Darby caused rare summer rainfall in Southern California and Arizona. The big story of the year—Hurricane Andrew—caused billions of dollars in damage when it hit Florida on August 24 (see the article on page 66). Another Pacific storm—Lester—gave California another extraordinary summer rain in August. In September Hurricane Iniki struck the Hawaiian Islands, causing damage in the $1 billion range. Late in September the remnants of Hurricane Danielle meandered along the mid-Atlantic Coast, bringing heavy rain to the area.

David S. Epstein

HUMAN SCIENCES

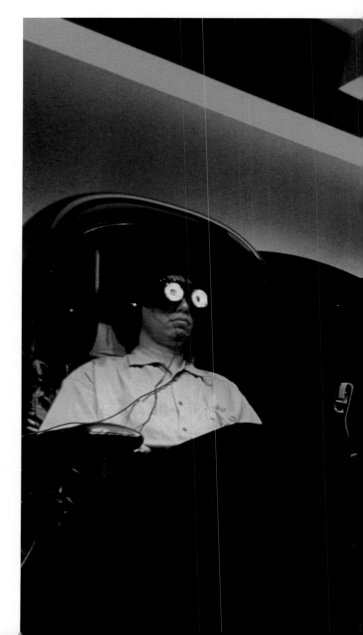

CONTENTS

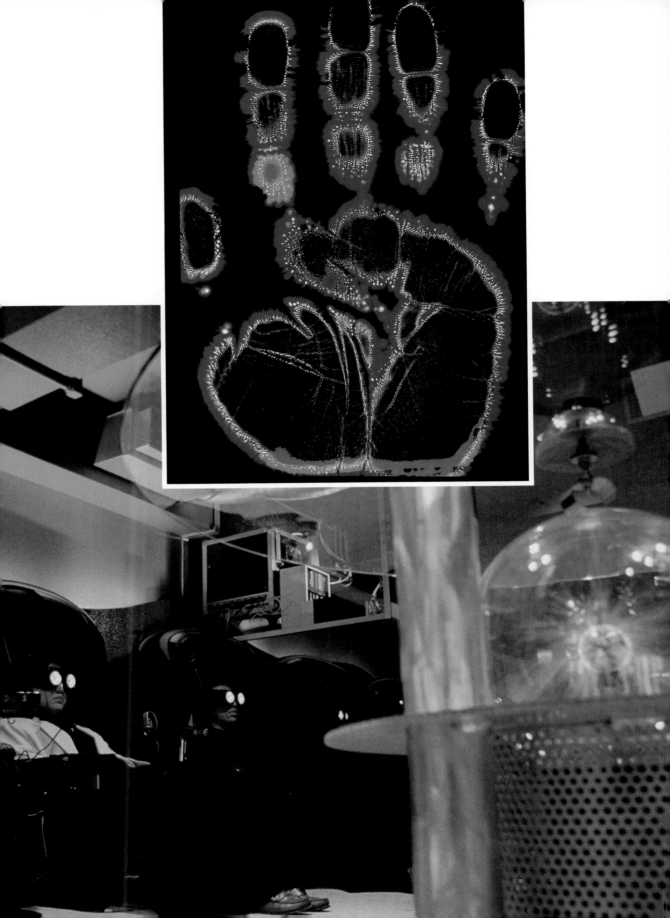

The Power of LIGHT

by Susan Gilbert

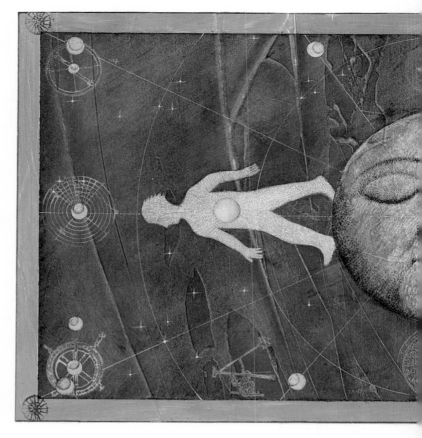

Y ou're flying from New York to Tokyo on business. To ensure that you will arrive rested and alert for a round of meetings, your company has bought a special kind of ticket. This is more than just business class. Your reservation entitles you to sit in an area designed to prevent jet lag. When you board the plane, things seem ordinary enough. You read a little and eat a meal, maybe watch the movie. It's not until well into the flight that something extraordinary occurs. Bright light bathes your face from lamps above your seat. After a few hours, the light goes off, only to come back on several hours later. You don't feel that anything profound is happening, but the light is resetting your internal clock to Tokyo's time zone as rapidly as the plane can fly you there.

Light's Healing Power

For now, this kind of overseas travel exists only in the imagination of people like R. Curtis Graeber, manager of flight-deck research at the Boeing Commercial Airplane Group in Seattle, Washington. He calls it "light-class service," without a trace of humor in his voice. "There are a lot of issues we need to work out," he says—for example, how to design the lighting in such a way that the glare won't be maddening. But one thing is clear: "There is interest on the part of the airline industry in using lights to combat jet lag for flight crews as well as travelers."

Welcome to the dawning of the age of light therapy. This may sound like New Age fluff, but it is serious science, and jet lag is only part of the story. In recent years, while some scientists have warned us to stay out of the Sun to prevent skin cancer and cataracts, other scientists have discovered that bright artificial light free of damaging ultraviolet rays can be harnessed to treat a variety of problems. The healing power of light is a subject of inquiry in clinics and laboratories around the world. In the United States alone, $15.5 million was spent last year by the National Institute of Mental Health (NIMH) on light-therapy experiments. In addition, university researchers and businesses are developing architectural lighting and portable lamps that could one day make light therapy accessible to people in their homes and offices, as well as on airplanes and in hotel rooms.

Science has finally acknowledged something humans have always instinctively known: Light uplifts and energizes. Its magi-

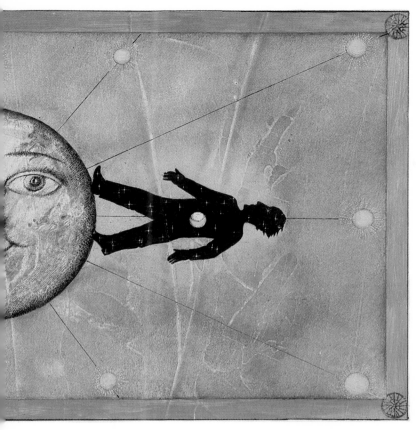

fairly well established. Less familiar is growing evidence that exposure to certain intensities of light at particular times of day and for particular durations can cure some kinds of insomnia, make night workers more productive, and improve the body's immune function. On the frontier of research, there is even speculation that, depending on the timing, light therapy might help treat infertility or act as a natural contraceptive.

The airlines are years away from offering light-class service, but light therapy is beginning to be used outside the laboratories

In ancient times, people worshipped the Sun for its life-giving powers. Today, scientists use light to improve moods, stabilize sleep patterns, and achieve other therapeutic goals.

cal power was recognized by the ancients, who worshiped gods of the Sun, the Moon, and the dawn. Light clarifies what darkness obscures. Light makes us feel good. Given the choice, we'll walk on the sunny side of the street (unless it's a humid day in July). Given the means, we'll head south during the winter. Plants arch toward the Sun, and the human impulse is to follow the light, too.

We now know that the powerful hold that light has on us is less magical than it is biological. It's a matter of hormones coursing through the bloodstream, nerve cells signaling to each other. Although no one knows exactly how, light tinkers with the brain, changing the way it regulates the flow of various chemicals in the body. This bit of knowledge led researchers to discover that properly timed exposure to light affects our mood, our performance, and the hours when we sleep and wake.

The success of light therapy in treating a clinical form of depression known as seasonal affective disorder, or SAD, believed to afflict 6 percent of the people in northern areas of the country during the winter, is

and SAD clinics. Shuttle astronauts now receive light therapy for several days before a launch if it is scheduled at night or if the mission requires nighttime work. According to the National Aeronautics and Space Administration (NASA), the light-therapy regimen works better than sleeping pills in helping astronauts rest during the day and stay alert at night. More down-to-earth applications are on the horizon. This year a venture-capital group formed by Harvard University began a consulting service to advise companies on how to help night-shift workers. Eventually the operation will be expanded to give information over the telephone to individuals interested in practical, scientifically proven advice on which kinds of lights to buy and how to use them.

External Cue
"Light therapy has the potential to really get at the root problem that people have with such things as shift work and sleep disorders," says Charles A. Czeisler, M.D., a pioneering researcher in the field who is director of circadian and sleep-disorders medicine at Brigham and Women's Hospital in Boston, Massachusetts, and an associate professor of medicine at Harvard Medical

Space-shuttle astronauts who participate in a night launching first undergo several days of light therapy. Such a regimen helps the astronauts remain alert during times when they would normally be sleeping.

School in Cambridge, Massachusetts. "Previously, all that we had was palliative, like an aspirin. Light therapy is like an antibiotic."

Other scientists are no less restrained in predicting the impact that light therapy will have on everyday life in the near future. "We're looking at a revolution in the way we will use light," says George Brainard, M.D., a neuroscientist at Thomas Jefferson University in Philadelphia, Pennsylvania, who is investigating the effect of bright light on human performance. "Now we use indoor lights to see things when it's dark, and for aesthetics. But in the future, lighting will be used in architecture and design in a way that we can benefit from it biologically."

"It'll be a technology parallel to indoor heating and air-conditioning," explains Michael Terman, M.D., the director of the winter-depression program at the New York Psychiatric Institute of Columbia-Presbyterian Medical Center.

Light is the single most powerful external cue for the daily fluctuations of the internal chemistry and temperature of virtually all living beings. After entering the retina, light passes right to a part of the brain known as the biological clock, or suprachiasmatic nucleus, and makes sure it adheres to a daily schedule. It is no coincidence that we typically wake up shortly after dawn and wind down after dusk, and that, like the Earth revolving around the Sun, our internal rhythms follow a 24-hour cycle. Sunrise and sunset prompt the clock to orchestrate the undulations in body temperature, blood pressure, hormones, and nervous-system chemicals that determine how alert or tired we feel at different times of day. Although there are "day" people and "night" people, the rise and fall of the Sun governs everybody's schedule to a great degree.

Seasonal variations in sunlight can also have significant effects. Studies show that sexual activity and conception vary from season to season, with peaks occurring in the spring. One theory is that the lengthening and brightening of the days after the spring equinox may increase levels of reproductive hormones.

Roughly a decade ago, psychiatrists identified SAD as a legitimate clinical condi-

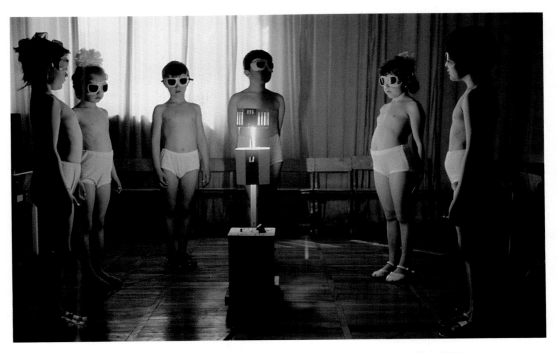

Children suffering from sunlight deficiency sometimes benefit from exposure to intense artificial light.

tion rather than a manifestation of hypochondria. Sufferers tend to begin feeling depressed in the fall, and, as winter draws closer, they become progressively more lethargic. Researchers don't know why people become depressed during the low-light months, but there is evidence that the relatively short days and dimness of the sunshine affect the body's production of or ability to use serotonin, a brain chemical that helps induce feelings of calmness and happiness.

Most SAD symptoms can be eliminated with two hours a day of light therapy from fall until spring. This therapy consists of exposure to 10,000 lux of white fluorescent lights, usually contained in a box. It's not necessary to look directly at the lights— which are as bright as sunlight at sunrise— to benefit, nor is it recommended, since that can irritate the eyes. People can even sleep through the therapy, because light can reach the retina through closed eyelids. "The best studies show 75 to 80 percent clinical remission of winter depression," says Terman. "We're talking about days before you see benefits, rather than weeks or even months with proven medication for depression."

Exploring Therapeutic Options

His kind of success has prompted researchers to explore other therapeutic applications. Dr. Czeisler's team at Harvard was the first to show that artificial light can reset a person's biological clock. In the course of exposing volunteers to bright light at different times of the day and night, he discovered something akin to an international date line on the biological clock. The time is between 4:00 A.M. and 5:00 A.M.—the exact moment varies with the individual, depending on when body temperature is generally lowest. Bright light prior to this time turns the clock back. Exposure after this time pushes it ahead. The closer to this sensitive point the light is turned on, the more pronounced the effect.

"Properly timed exposure to light can shift us to the equivalent of any time zone or reset the biological clock by up to 12 hours in just two to three days," says Dr. Czeisler. "With just a single light exposure of three to four hours around breakfast time, we can get you halfway to Europe." For someone flying west, to an earlier time zone, a light-therapy session the evening before a flight would help prevent jet lag. A dose of bright light in

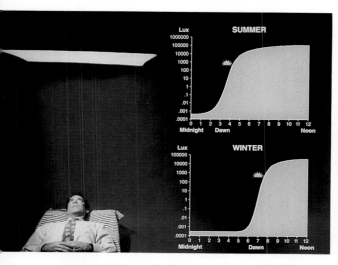

Dr. Michael Terman basks in the simulated dawn light of his Naturalistic Bedroom Illuminator. The panels at right compare the dawn profiles at a latitude of 45 degrees (as in Minneapolis, Minnesota). The Sun icon indicates the moment of sunrise, when the solar disk has risen halfway over the horizon.

the morning would be appropriate for someone who needs to be on a later schedule, such as a person who wants to fall asleep by midnight, but routinely thrashes in bed until 2:00 A.M. Several hours of brighter-than-normal lighting in the workplace would enable employees on the graveyard shift to stay alert all night long and sleep during the day.

For many people, light therapy on the job is already a reality. Light Sciences, a new company formed by investors associated with Harvard, is selling light fixtures to be used in conjunction with software that controls when light levels go from normal to bright. The timing depends on dozens of variables, including the length of a company's shifts, how often the shifts rotate, and the age of the company's employees. The proprietary software, based on calculations by Dr. Czeisler and Richard E. Kronauer, a Harvard mathematician, is designed to incorporate these variables and come up with one lighting schedule to accommodate everybody on a shift.

Light Sciences says that its clients include a chemical company and a public utility, but will reveal little else. "This is a very competitive business," says Leah R. McIntosh, vice president of consumer products and applications. She adds that the company is developing software and lights for use against jet lag.

Too Time-consuming?

Some scientists believe that using light therapy for anything other than treating SAD is premature. "I don't think we have a reliable strategy to guarantee a jet lag-free trip yet," says Margaret Moline, M.D., from the department of psychiatry at New York Hospital-Cornell Medical Center in White Plains. "More studies need to be done to see if the treatment is equally effective for all travelers." She points out, for example, that no studies on light therapy and jet lag have included women.

Some of the latest investigations of light therapy do involve women. In 1991, Hugh McGrath, Jr., M.D., a rheumatologist at Louisiana State University, got encouraging results by using artificial light to relieve the symptoms of women with lupus, an incurable disease in which the immune system assaults the body's tissues. Lupus patients have traditionally been told to stay out of the Sun because it can induce an autoimmune response. But in earlier animal experiments, Dr. McGrath found that this response was induced by only two components of sunlight —the ultraviolet light known as UVB and UVA-2, the primary causes of sunburn and skin cancer. To his surprise, a third component of sunlight, UVA-1, mitigated the immune system's attack on the body.

Ten women with lupus were exposed to a combination of UVA-1 and ordinary fluorescent light for about 10 minutes a day, five days a week, for three weeks. Nine out of 10 patients had less joint pain, fatigue, and other symptoms.

In one of the odder experiments with light, Daniel F. Kripke, M.D., a psychiatrist at the University of California, San Diego, found that exposure to a 100-watt incandescent light could normalize the menstrual cycles of women whose cycles were too long, spanning as many as 54 days. The light, about a tenth the intensity required for treating SAD or shifting the biological clock, was left burning all night by the women's bedsides for five days in mid-cycle. Dr. Kripke does not know why the light worked, although he thinks it somehow affected the levels of various hormones involved in the menstrual cycle.

Photo above by Eve Vagg, graphic design by Rachel Yarmolinsky, N.Y. State Psychiatric Institute at Columbia-Presbyterian Medical Center. Apparatus courtesy of Medic-Light, Inc., Lake Hopatcong, N.J., (201) 663-1214.

"Some women with long menstrual cycles have trouble conceiving," he says. "The findings are very preliminary, but it might be possible to treat infertility this way. Also, different timing and intensity of light might inhibit ovulation and be used as a natural contraceptive."

Even if clinical trials show that light therapy is effective for a variety of difficulties, there is an obstacle to its widespread use. "The biggest problem is compliance," says Scott S. Campbell, a psychologist at New York Hospital-Cornell Medical Center. It is hard to imagine light boxes the size of window fans becoming part of the furniture. And how many people would spend two or more hours each day in the presence of bright lights?

Some people find it easier to wear their light-therapy equipment. Dr. George Brainard (below) invented a form of headgear in which bulbs hidden beneath a visor shine light onto the wearer.

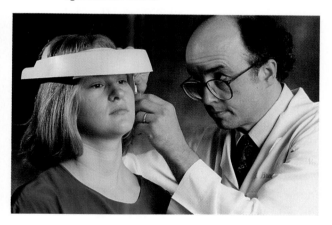

The answer is not encouraging. After 10 days of light therapy for insomnia under Campbell's supervision last year, 12 elderly people slept better. "But they weren't happy enough with the results to continue with the therapy," he says. "They had to sit in front of the light boxes for two hours at night, from 8:00 P.M. to 10:00 P.M. The bottom line is, it took too much time."

Some researchers believe that the solution is to make the technology less intrusive. George Brainard of Thomas Jefferson University helped invent a kind of cap with bulbs beneath the visor. Although he doubts that anyone would want to wear it in public, he envisions the cap being used at home and in the privacy of hotel rooms. "People can wear the hat while getting the kids ready for school," he says.

The Dawn Simulator

Perhaps the most appealing invention is a dawn simulator, designed to be placed next to a bed. Devised by New York Psychiatric Institute's Michael Terman, it looks like an ordinary lamp, but is controlled by a microprocessor that can be programmed to reproduce the intensity of a sunrise in any location on Earth at any time of year. If you live in New York and suffer from SAD, you can program the lamp to simulate dawn on a beach in Fiji in July. In mimicking dawn, the simulator shines for about two hours before you want to wake up. No further exposure is required. "The therapeutic response for people with SAD is the same as with regular bright-light therapy," Terman says.

The dawn simulator can also help people who have trouble falling asleep at night and waking up in the morning. "For someone who isn't waking up until noon, we'll say, 'Set the lights for 11:30 A.M., grab your bagel, and stay under the lights,' " says Terman. "This advances the internal clock, so it becomes easier for that person to go to sleep a little earlier and wake up a little earlier the next morning. Within a week or two, the sleep schedule can be normalized."

The results are promising. But even if simulating sunrise on a beach in Fiji is the next best thing to being there, it's not good enough. Terman concedes that his product needs some fine-tuning before people will embrace it.

A prototype of the dawn simulator recently failed an informal marketing test. An elderly woman who had been awakening in the middle of the night used the device in the evenings. She had one in her apartment in Manhattan and another in her country house. Even though the lights helped her sleep better, she stopped using them because they were too much of a bother.

The woman was Terman's mother.

THE HUMAN GENOME PROJECT

by Thomas H. Maugh II

Within one month in 1993, researchers announced the discovery of five important genes. Two of the genes were major jewels: the defective genes that cause Huntington's disease and amyotrophic lateral sclerosis (ALS), better known as Lou Gehrig's disease. Damage to the third of these genes, called neurofibromatosis 2, plays a key role in the development of brain tumors. The last two were genes that play a role in the development of high blood pressure. These discoveries have great significance. Within two months of the identification of the ALS gene, physicians had begun clinical trials of the first treatments for the disorder. Identification of the other four genes is also likely to provide the key to new treatments that can delay the onset of Huntington's, prevent cancer, and hold down blood pressure.

Five genes, all with the potential to ease human suffering. But humans are prey to more than 3,000 distinct genetic disorders, and human cells contain more than 100,000 genes. Imagine what physicians could do if they knew the identity of each of those genes, its function, and whether it was defective in each individual. That is the goal of the largest biology-research project ever undertaken: the Human Genome Project, a $3 billion, 15-year project designed not only to identify each of those 100,000 genes, but also to obtain the precise identities and order (sequence) of the 3 billion individual chemicals that compose the human genome. The objective, according to Nobel laureate James D. Watson, the molecular biologist who headed the project in its first two years, "is, to say the least, heroic. It's to find out what being human is." Adds another Nobel laureate, Walter Gilbert of Harvard University: That objective is the "ultimate answer to the commandment 'Know thyself.' "

Now in its third year, the project is—to the surprise of many observers—proceeding on pace. Workers have made the first "maps" of the genome and are slowly refining them.

Others are building the tools, both mechanical and biological, that will be used during the final 5 to 10 years of the project, when each of the 3 billion individual chemicals that compose the genome will be identified. Nonetheless, the project has suffered controversy: Most researchers have criticized the government for trying to patent large segments of genes that have been isolated during the initial stages of the project, while the government has been critical of the industrial links of many federally funded researchers. Biologists not affiliated with the mammoth project have argued that it is draining funds away from more-fundamental research. And finally, critics fear that the vast new knowledge obtained through the project will be misapplied. Insurance companies, for example, might charge higher rates to individuals who are genetically fragile, perhaps even denying them coverage altogether. Both scientists and administrators are now going through the difficult process of working out all these potential problems.

Genetic Blueprint

The human genome is a genetic blueprint, "the complete set of instructions for making a human being," according to biochemist Robert Sinsheimer of the University of California at Santa Barbara. That information is hidden away in the nucleus of each of the body's 3 trillion cells (except red blood cells, which have no nucleus), in coiled strands of deoxyribonucleic acid, better known as DNA. DNA, whose structure was deciphered by James Watson and Francis Crick a mere 40 years ago, is organized into small units called genes, each of which serves as the blueprint for a specific protein that is crucial to the structure or functioning of the cell. Genes are bundled into chromosomes. The estimated 100,000 human genes are contained in 23 pairs of chromosomes.

DNA is composed of four distinct chemicals called bases. These bases are joined

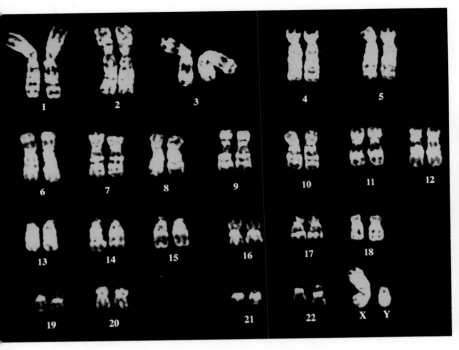
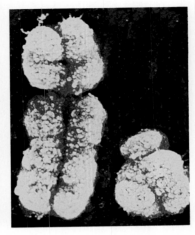

Of the 46 human chromosomes, the Y chromosome (right lower corner above and far right) confers maleness. The development of more sophisticated maps of the Y chromosome is paving the way to understanding sex-linked illnesses.

together like pearls on a string to form genes and chromosomes in the same fashion that letters of the alphabet combine to form words, sentences, and books. If the DNA in a single cell could be uncoiled and stretched out in a straight line, it would be about 8 feet (2.4 meters) long. But if it were enlarged so that each base was the size of a letter in this article, the DNA would form a huge sentence 4,700 miles (7,562 kilometers) long, enough to fill a book with 1 million pages, or to stretch from Los Angeles, California, to Jacksonville, Florida, and back again. So far, researchers have determined the structure of about 3,800 genes.

Researchers have two main goals in the Human Genome Project: to create a "physical map" of the genome, and to create a sequence. A physical map would be a collection of perhaps 100,000 DNA fragments that would span the entire genome and whose positions on the chromosome are known. A sequence would give the order of each of the 3 billion bases. Mapping might be compared to the process of locating every city and town in the United States, while sequencing would be the equivalent of pre-paring a detailed street map of the entire country.

Some critics, such as geneticist Francisco Ayala of the University of California at Irvine, argue that mapping may be sufficient for locating all human genes, and that only the genes themselves—which account for less than 10 percent of the genome—should be sequenced. Sequencing the other 90 percent, he says, "is not likely to provide meaningful insights." But Nobel laureate Paul Berg, a molecular biologist at Stanford University in California, disagrees: "I call it arrogance to assume we already know what is junk and what is real. My premise is . . . that information which now appears to be undecipherable will, in fact, have substantial meaning in terms of understanding how the human genome works." This "nongenetic DNA" may also yield information about the evolution of the human genome.

Advances in Mapping and Sequencing

Researchers reported three major advancements toward the development of a physical map in October 1992. Molecular geneticist Helen Donis-Keller of the Washington University School of Medicine in St. Louis reported the development of the first high-resolution map of the genome, a significant step, but still far from the map researchers eventually hope to prepare. Compiled from data submitted by researchers around the world, the map contains information that identifies 1,416 DNA fragments spanning more than 90 percent of the genome. For this new map, markers on most chromosomes are spaced about 5 million bases apart, a distance researchers eventually hope to narrow to between 100,000 and 500,000 bases. The utility of such fragments and markers lies in the fact that family studies can usually place the gene that causes a genetic disease on the DNA fragment between two markers, greatly narrowing the search for the cause of the disorder. The closer the markers, the easier the task of finding a particular gene.

At the same time, French and American researchers reported the construction of more sophisticated maps of chromosome 21, which carries the gene for Down syndrome and perhaps Alzheimer's disease, and the Y chromosome, the chromosome that confers maleness on an infant and that carries genes for a variety of sex-linked illnesses. These were good starting points because they are the smallest of the 23 chromosomes. These projects involved more than the simple accumulation of data. In both cases the researchers broke the chromosome down into a large number of overlapping pieces of DNA and inserted each fragment into the genome of yeast, so that the fragments could be both easily manipulated and produced in large quantities. These assemblages are called yeast artificial chromosomes (YACs).

In the case of the Y chromosome, the researchers produced 196 overlapping DNA fragments that covered the entire chromosome; for chromosome 21 the number was 810. The researchers also identified the order in which each fragment occurs in its chromosome, as well as unique markers at the end of each fragment. When researchers begin the actual process of sequencing the human genome, they will work with the DNA in these YACs. As data is assembled for each fragment, the information can then be confidently combined to yield the sequence for the entire chromosome.

Researchers at other institutions are preparing similar physical maps of the other 21 chromosomes; their goal is to have them all finished by 1995. "That goal now appears realistic," says molecular biologist David C. Page of the Massachusetts Institute of Technology (MIT) in Cambridge, who led the team that mapped the Y chromosome. "I'm not sure it would have a year ago." The mapping of the two chromosomes in itself has not led to the discovery of any genes linked to inherited diseases—yet. But it has produced some new insights. Page and his colleagues found, for example, that there are

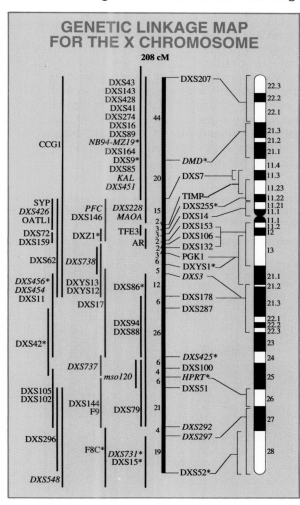

GENETIC LINKAGE MAP FOR THE X CHROMOSOME

ADVANCES IN HUMAN GENE MAPPING

The table at right describes the state of progress of human genome mapping as of July 15, 1992. Since the same time in 1991, 214 new genes and 806 new polymorphic markers had been added. Polymorphic markers are used for linkage studies and mapping of disease genes. Sequence tagged sites (STSs) are useful for physical mapping. The number of polymorphic STS markers, helpful in genetic mapping, have almost doubled in the last year.

Chromosome	1	2	3	4	5	6
Genes						
Mapped	247	138	85	99	92	131
Disease-related	55	24	24	25	23	27
Markers						
Total	466	266	1351	495	491	396
Polymorphic	197	121	442	179	242	185
HET >0.7	24	14	14	19	28	17
STS markers						
Total	63	66	93	81	51	41
Polymorphic	44	36	45	39	46	26
Number of sequenced loci	144	92	52	148	53	87
Number of sequences	734	515	263	342	211	750
Kilobases of sequence	760	867	204	379	302	778

far more genes on the Y chromosome than anyone had expected, and determining their functions may lead to new understandings about gender.

Picking up the Pace

Sequencing the genome is a more formidable function. Five years ago, it was not unusual for a researcher to spend his whole career sequencing one gene, which could be anywhere from 1,000 to 100,000 bases long, at a cost that was usually around $3 to $5 per base. At that rate, 400,000 human-years would be required to sequence just the 10 percent of the genome that contains genes. The cost would be prohibitive. But researchers in Japan and the United States have been developing automated machinery that can perform the task much faster and at lower cost. Chemist Richard A. Mathies and molecular biologist Alexander N. Glazer of the University of California at Berkeley, for example, reported in July 1992 that they had developed a technique capable of sequencing 10,000 bases per hour, and they predicted that refinements could raise the rate to 100,000 bases per hour.

The backbreaking work of sequencing the genome is most likely to be conducted either in large government centers or in an industrial setting. Already a number of companies have been established, both to market the new technologies for sequencing genes and to carry out the sequencing itself. Other start-up companies are avidly recruiting the principal genome researchers in hopes of producing drugs and other therapies based on the gene discoveries. In addition to the estimated $170 million that the federal government was expected to pour into the genome project in 1993, private companies were expected to spend an impressive $85 million.

The year 1993 marks the fortieth anniversary of the decipherment of the double-stranded helical structure of DNA by James D. Watson (left) and Francis Crick.

7	8	9	10	11	12	13	14	15	16	17	18	19	20	21	22	X	Y
132	64	77	79	154	136	31	66	65	83	135	28	132	45	39	71	225	18
25	22	26	15	46	23	13	19	17	17	26	9	24	14	8	17	111	1
608	308	242	263	1059	218	200	133	173	516	932	79	364	143	322	360	1382	406
226	178	105	116	320	82	87	63	76	163	268	51	97	66	110	140	290	32
12	12	34	11	23	7	11	11	8	19	23	14	11	21	19	6	21	2
34	34	65	29	67	27	28	20	33	59	269	25	42	42	62	16	136	180
23	26	57	21	48	19	25	13	29	37	46	19	31	38	44	11	52	0
83	41	50	43	84	93	12	49	31	53	98	17	86	24	23	41	72	10
697	170	219	262	432	360	114	881	109	151	288	90	412	129	141	196	435	17
598	229	277	303	815	648	113	686	183	429	860	137	487	175	95	402	553	24

The large industrial contribution has led to charges of conflict of interest based on fears that researchers would be using government funds to discover genetic information that they would turn over to private companies. In fact, Watson resigned as head of the genome project in 1992 at least in part because of his holdings in such companies. Researchers are taking great pains to ensure that such conflicts do not occur, including measures such as publishing results quickly and not providing insider access to their data. And many researchers believe that there is a real need for such corporate involvement. "On the one hand, you've got universities, research institutes, and government laboratories, where all the work is going on," says venture capitalist Kevin Kinsella of Avalon Medical Partners. "On the other hand, you've got pharmaceutical companies, which make all the drugs, but don't know anything about genes. We see the role of a [genome-related] company as a bridge between them."

Ethical Questions

Perhaps the most ticklish aspect of the Human Genome Project involves the ethical questions that it raises. The prospect of having a complete library of the genome raises in some minds the specter of eugenic improvements—the manipulation of the genome to alter the characteristics of humans.

The vast majority of molecular biologists agree, however, that the human genome must not be tampered with, even if it could be done safely. Moreover, the chance that it could be done safely, says biologist John Maddox, editor of the British journal *Nature,* "is so small that this fear is . . . more distant than it may appear."

Perhaps more immediately threatening is the possibility that information gleaned from the genome project will be used to assess each individual's risk of contracting not only the common genetic diseases, which are typically caused by defects in a single gene, but also diseases caused by a more complex interaction of genes, such as diabetes, heart disease, and cancer. If these propensities can be identified, critics charge, the individuals might have to pay much more for health or life insurance or might be denied it altogether. They might also be prevented from working in industries in which the environment could be foreseen to interact adversely with their genetic propensities. Advocates, however, note that such actions already occur on a more limited scale, and they argue that safeguards can be enacted to prevent the most severe inequities. These questions have yet to be resolved, but they are being vigorously addressed: At the instigation of Watson, the project is setting aside 3 percent of each year's budget for exploration of such ethical considerations.

SIAMESE TWINS

by James A. Blackman, M.D.

Twins who are joined together physically are said to be "conjoined." The term "Siamese twins" arose from Chang and Eng (above), a famous pair of conjoined twins born in Siam (now Thailand).

Have you ever had a friend or family member with whom you so enjoyed spending time and sharing experiences that people said the two of you were inseparable?

But what if you were truly inseparable, not by choice, but by a fluke of nature, joined physically rather than intellectually or emotionally? This describes the rare phenomenon of conjoined twins—or Siamese twins, as they are commonly called. Siamese twins frequently share essential body organs but typically have separate and individual personalities. With present surgical methods, early separation seems advisable for the sake of optimal development of each twin. However, separation may not be feasible without sacrificing the life of one of the twins. In such cases, parents and their doctors must make difficult decisions regarding the best course of action. Although some Siamese twins have lived for many decades and have led productive lives, most have died in early childhood.

Causes and Frequency

Siamese twins are always of the same sex and have identical chromosome patterns. In essence, they are regular identical twins except for the fact that parts of their bodies are physically joined. In the womb, they begin as a single fertilized egg until, sometime between the 13th and 15th day after fertilization, the inner cell mass splits into equal, but not fully separated, halves. The twins may share a simple bridge of skin, or major organs such as the heart, liver, or brain.

Siamese twins occur once in every 50,000 to 80,000 births. For unknown reasons, they are much more common among black infants. Over half of Siamese twins, usually boys, die before birth. Scientists describe the various types of Siamese twins with the Greek word *pagus*, which means "fixed," used as a suffix, with the prefix for the connecting point of the body. For example, connection at the skull is called *cranio*pagus; chest, *thoraco*pagus; abdomen, *omphalo*pagus; pelvis, *ischio*pagus; and sacrum, *pyo*pagus. Thoracopagus twins are most common (40 percent), craniopagus the least common (2 percent). Twins joined at the chest and abdomen often share hearts, livers, and possibly intestinal tracts. Brain tissue may be connected or totally separate in craniopagus twins. The extent to which vital organs are shared determines whether surgical separation is possible without the death of one or both twins.

History of Siamese Twins

The earliest record of Siamese twins (though they were not called that then) concerns the Biddenden maids, Mary and Eliza Chulkhurst. Joined at the hips and shoulders, they were born to wealthy English parents in A.D. 1100. After 34 years, Mary suddenly became ill and died. It was proposed that Eliza should be separated from her sister's corpse by a surgical operation. She refused, uttering the words: "As we came together, we will also go together." Legend holds that she died six hours later.

In their will the Chulkhurst twins directed that the rent from their land provide for an annual distribution of food to the poor. To this day, on every Easter morning in the small Kentish village of Biddenden, tea, cheese, and loaves of bread are given to widows and the elderly.

The common name for conjoined twins arose from the most famous of all Siamese twins, Chang and Eng. They were born in Siam (now called Thailand) in 1811, united by a band of tissue at the lower chest and abdomen. In infancy the connecting band was very short, forcing them to lie facing one another. Later the connecting band stretched so that they could stand side by side. Their parents sold them to an American ship captain, who brought them to the United States in 1829.

Chang and Eng Bunker, as they chose to be called, earned a living for a time working in P. T. Barnum's exhibitions around the world. Eventually they became wealthy landowners in North Carolina. At age 42, they married twin daughters of a clergyman and raised large families. Chang became the father of 10 children, Eng of 11. At first the Bunker twins lived together with their wives, Sally and Adelaide, until the wives began to quarrel. Chang and Eng were forced to maintain separate houses some distance apart for the women. The men arranged to spend three days in the home of one, and the next three days in the home of the other.

At age 63, Chang became ill with pneumonia and died, possibly of a blood clot in the brain. Always the stronger and healthier of the two, Eng died two hours later. An autopsy performed in Philadelphia showed that the only internal connection was a thin band of liver tissue containing a few small blood vessels. They apparently had no major blood connections.

The head-to-head connection of the McCarthur twins (right) made surgical separation impossible. The twins were set to graduate from nursing school when they died suddenly at age 43.

Surgical Separation

The history of surgical separation of conjoined twins dates from the 15th century. Two sisters born in Germany in 1495, joined at their heads, lived for 10 years before one died. After the one twin's death, the two were separated. Unfortunately, the other one died shortly thereafter. The first successful separation, with both twins surviving, was recorded in 1689. The twins were joined at the abdomen, and the division was accomplished by tying off the connecting band of tissue between the children with a constricting string.

Because of greatly improved surgical techniques over the past decade, Mark Hoyle, M.D., of the University of Texas Health Science Center at Dallas, believes that, even in difficult cases of conjoined twinning, separation should be considered, with rare exceptions.

The separation of twins joined at the head presents special problems. It took 22 hours to successfully separate a pair of twins, Benjamin and Patrick Binder, who were joined at the back of the head. Though they had separate brains, the twins shared a large area of the skull and the major vein that drains blood from the brain.

Yvette and Yvonne McCarthur, whose connection at the top of their heads was thought inoperable, followed the performing tradition of Siamese twins by singing gospel tunes around the country during the 1970s. Their mother forbade surgery to separate them because of the risk that one would die. (They had separate personalities, but a common circulatory system.) At age 38, they enrolled at Compton College in Los Angeles, California to study nursing. Arrangements were made for round tables and two armless chairs to be moved into classrooms where the twins would sit, replacing the conventional desks. Unfortunately, just as they were about to graduate, they died unexpectedly of natural causes at age 43.

Psychological Development

David Smith, Ed.D., professor of education at the University of South Carolina in Columbia, says that even though Siamese twins share the same genetic makeup and nearly identical environments, they have distinctly separate and individual personalities. Apparently, duplicate minds can view the same landscape, hear the same symphony, or feel the same wind, and yet interpret these experiences differently.

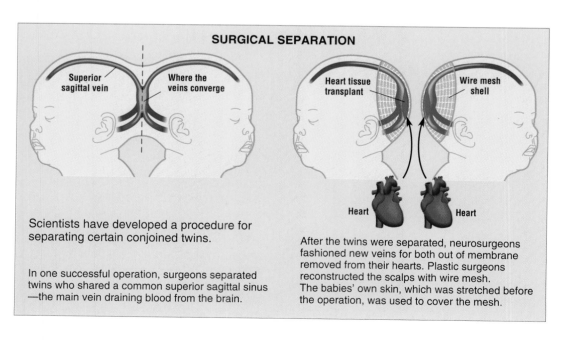

SURGICAL SEPARATION

Superior sagittal vein

Where the veins converge

Heart tissue transplant

Wire mesh shell

Heart   Heart

Scientists have developed a procedure for separating certain conjoined twins.

In one successful operation, surgeons separated twins who shared a common superior sagittal sinus —the main vein draining blood from the brain.

After the twins were separated, neurosurgeons fashioned new veins for both out of membrane removed from their hearts. Plastic surgeons reconstructed the scalps with wire mesh. The babies' own skin, which was stretched before the operation, was used to cover the mesh.

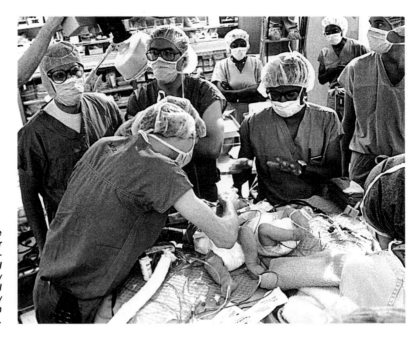

Over half of all Siamese twins die before birth. For those that survive, the prospect for successful surgical separation depends largely on where the twins' physical connection occurs. Surgery is usually performed when the twins are still very young.

Even when reared in the most understanding and tolerant environment, it is difficult for unseparated Siamese twins to live anything approaching a normal existence. Most authorities agree that early experiences influence the evolution of personality and self-concept.

Conjoined twins have few opportunities to master the developmental stages of becoming independent persons. Unseparated, they may find outright rejection by significant adults, have difficulty establishing healthy relationships, and find no role in society other than that of curiosities. Parents must be counseled about these issues when making the decision about separation.

Scientists gained new insights into the plight of Siamese twins through a battery of psychological tests given to a pair of 12-year-old males joined at the abdomen. These youngsters, of average intelligence, were very much aware of their abnormal appearance and of the revulsion it evoked in others. The testing revealed that unconsciously they thought of themselves as freaks or monsters. Their mood was one of despair. They longed to be freed from the imprisonment of their joined bodies.

These tests supported the prevailing belief that, from a psychological viewpoint at least, separation should be accomplished during the first years of life—before an awareness of the unusual condition occurs, and before the development of personality is established.

Ethical and Moral Issues

Parents and physicians face difficult ethical and moral dilemmas when it comes to making decisions about separating Siamese twins. In many cases, Siamese twins share vital organs, such as the heart; only the twin who receives the heart can live. Catholic, Protestant, and Jewish theologians have debated whether it is permissible to sacrifice one infant so that the other can live. Legal and ethical scholars tend to support reasonable medical attempts to separate Siamese twins with conjoined hearts. Notwithstanding all the discussion among lawyers, religious experts, and judges, the greatest burden of decision-making falls on parents, who never find it easy favoring one child over another. Such a difficult decision is illustrated by a recent case from Ireland reported on European and American television.

The Boy of Bengal

Perhaps the most unusual type of conjoined twins are dicephalus dipus (one body, two heads) and ''parasitic'' twins, in which one partially formed body depends completely on the other for life-giving sustenance, including food and oxygen.

One famous case of dicephalus dipus was the two-headed boy of Bengal. He was born in May 1783 in Mundul Gaut, a village in Bengal, India, to a poor farming family. Immediately after the birth, the midwife—terrified by his strange appearance—tried to destroy the twin-headed baby by throwing him into a fire. He was saved from the flames, but an eye and an ear on the upper head were badly burned. The parents soon recovered from their initial shock enough to see the possibility of earning money by exhibiting the child. They left their village and went to Calcutta to ensure a large audience.

The boy immediately attracted the attention of curious crowds in Calcutta. He gained such celebrity, in fact, that his parents had to cover him with sheets between shows to protect him from overexposure. His fame soon spread throughout India. By the time the two-headed boy was four years old, he enjoyed generally good health except for emaciation. One day his mother left him to fetch water. When she returned, her son was dead from the bite of a cobra. Today his two-headed skull is on display at the Royal College of Surgeons' Hunterian Museum in London.

In the Ireland case, the parents of Siamese twin daughters joined at the abdomen struggled with a London surgeon's advice to separate them. Though the odds for survival of both were thought to be excellent, there were no guarantees that either or both would live. Finally, the parents decided it would be best physically and psychologically for the twins to be separated. One twin died shortly after the operation, but an autopsy revealed that she had a very weak heart and would have died anyway, killing her sister as well. The parents, for all their grief, felt they had made a good—and the right—decision.

In an American case, a pair of Siamese twins, Ruthie and Verena, were rushed to Denver for intensive care soon after their birth. They shared a heart and parts of the intestines. The doctors said that they would probably live only a couple of months. Their parents could not bear the thought of surgical separation. According to their mother, "If Ruthie and Verena survive another few years, and it does become possible to separate them, it won't be up to my husband and me. Ruthie and Verena would have to say, 'We understand that medical technology would allow it, and we think that we want to be separate.' " Of course, the twins were too young to make such decisions. Ruthie and Verena died at age 7 of heart and respiratory problems.

The decision about Siamese twins named Emily and Francesca was somewhat easier. Joined at the front, they had separate hearts, but shared a massive liver. While separation promised to be a grueling, 10-hour procedure, the chance of survival for both was good. It took many months before both twins were home, but each twin survived, and their parents were pleased with their decision to proceed with separation in the first few days of life.

An entirely different approach was taken in the case of the Mueller twins in Illinois. Charges of attempted murder and neglect were brought against their parents and doctor for withholding food and medical treatment. The twins were joined at the waist and shared the lower digestive tract and a common leg. They were not expected to live for more than a few days. Although the prosecutor decided not to pursue the case, the decision whether to treat the twins was taken away from their parents. Eventually the twins were returned to their parents and were surgically separated. One of the pair ultimately died from a serious heart ailment.

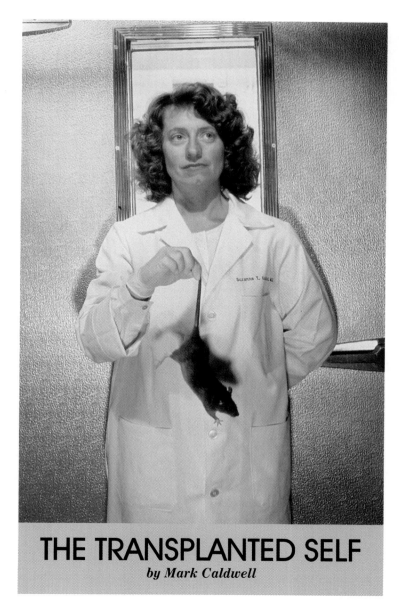

THE TRANSPLANTED SELF
by Mark Caldwell

For 10 years, Suzanne Ildstad trained to be a pediatric transplant surgeon, and she doesn't underestimate the seductions of her calling. "I like using my hands, and I like taking care of patients," she says, "and it's nice to see immediate results. Parts of surgery are really exciting." For example, she says, you always know right away when you've got a successful kidney transplant. "We always hook the vessels up first to restore the blood supply," Ildstad says. "Then if the kidney is healthy, you know immediately because it makes urine and squirts it out. That's one of

the greatest feelings I know as a surgeon: the urine popping out as soon as the blood starts profusing through." Not a knockout image, perhaps. But it is a sign of new life for a patient who 50 years ago would have been given up for dead.

Yet despite the adrenaline charge and the rewards of transplant surgery, Ildstad has

Surgeon Suzanne Ildstad has left the operating room for the laboratory to hunt for ways to improve human organ transplants. The mice in her lab, which have been transplanted with rat cells, may hold the secret.

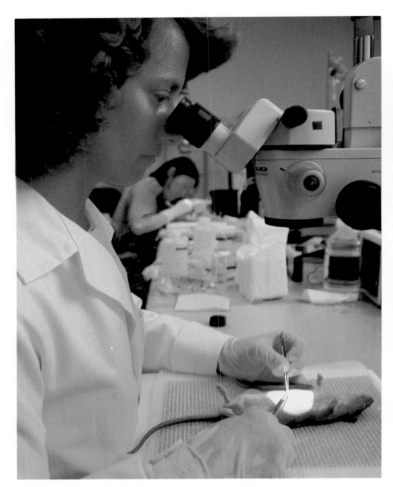

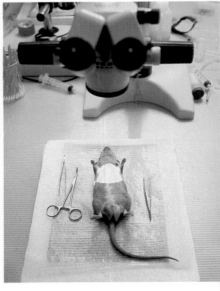

There is a critical shortage of organs for human transplants. The skin transplant experiments in rats (above and left) may one day give way to successful human transplants using organs from pigs or sheep.

largely forsworn them. In June 1991, she decided to take a detour, a temporary turn away from her clinical practice at the University of Pittsburgh School of Medicine and into the research lab. The move, she says, came because miracles in the operating room have their limits. "By the 1980s we'd developed operations to treat almost any surgical problem," Ildstad remarks. "Yet the patients who undergo transplants often suffer potentially devastating complications."

Overcoming Rejection

Kidneys, for example, are the most compliantly transplantable of major organs, yet one in every five kidney transplants from cadavers—the most common source—fails within a year. The kidney is rejected by the transplant recipient's immune system, a vast, intricate array of cells, antibodies, and proteins designed to seek out and destroy anything in the body that's foreign. Usually the foreign interloper is a harmful virus or bacterium; in a transplant, unfortunately, it's a donated vital organ. And this happens even though surgeons carefully select the organs, using only those whose cells barely provoke the recipient's immune system in preoperative tests. The search for such an organ can be long and desperate. There aren't a lot available. Some people undergo kidney dialysis for years. Others die waiting.

Drugs have been developed to help combat rejection, but they do so by undermining the immune system, exposing patients to the very infections and malignancies that the system is designed to attack. And these immunosuppressive drugs don't work forever: the four out of every five transplanted cadaver kidneys that aren't immediately rejected typically succumb within 10 years.

That, Ildstad argues, is simply not good enough. "Our goal is to have everything be perfect: permanent graft acceptance without drugs, no complications, no transplant rejection. The reason I went into research is that I think our next major advances—the breakthroughs that will let us solve the clinical limitations of transplant surgery—won't come from the operating room but from the lab, from research at the cellular and molecular level."

Making Hybrids

So while Ildstad still spends some time on call as a surgeon, these days she's more often found amid a large population of experimental rodents at the university's labyrinthine animal-research facility. When you visit her mice, snuffling about contentedly in their clear plastic cages, they don't look like pioneers on the transplant-surgery frontier. They appear to be utterly normal specimens: stubby, black-furred, companionably beady-eyed. Inside, though, they're made of different stuff. They're part rat.

These animals have been reoutfitted as hybrids, harboring a mixed collection of rat and mouse immune-system cells. On their backs and sides, some of these mice sport thumbtack-size, healthy patches of skin transplanted from the same rats whose immune cells they carry. "I've got pictures of some that've been in place for up to 180 days," Ildstad reports. That's nearly forever for a mouse, which has a life span of at most a year and a half. "And they look just beautiful," she adds. Skin grafts, Ildstad explains, particularly interest transplant researchers because transplanted skin provokes an extremely strong immune response. It's always been used as a standard for graft tolerance.

Ildstad and her colleagues have given these creatures a new biological identity, reshuffling the body's sense of self to enable it to accept a new organ as its own, yet retain a vigilant defense force against invading diseases.

They may also have created—if this success can be repeated in humans—a new future for organ transplants, without exasperating, often futile hunts for donors with tissue types similar enough to the recipient's. This is a future in which children won't die of liver or kidney disease while waiting for a suitable organ, a future in which organs can move freely between people and even between species without being rejected.

It is also a future ripe with the promise of radical new treatments for old diseases. Skin isn't the only body tissue that Ildstad and her co-workers have transplanted. "We've done thyroids, parathyroids, adrenal medullas, cortices; we did some animals where we put multiple endocrine tissues in," she notes with some pride. Recently her lab successfully transplanted clusters of insulin-making cells into diabetic mice.

Learning Immunity

Ildstad completed her medical training at the Mayo Medical School in Rochester, Minnesota, in 1978, and went on to Massachusetts General Hospital for a residency in general surgery (where she was the seventh woman to complete the program).

In 1982 she accepted a research fellowship at the National Institutes of Health (NIH) in Bethesda, Maryland, where she joined David Sachs, a transplant immunologist. The two began collaborating on ways—without resorting to immunosuppressive drugs—to subdue, or even eliminate, the immune system's inveterate hostility to grafts. Their search began, as does the immune system itself, with bones.

Biologists had known for some time that all immune cells originate in bone marrow. The cells that do the immune system's scut work—engulfing and devouring—travel directly to various body tissues, where they mature and proliferate. But in humans and mice (and many other animals as well), the immature cells that will eventually control this activity—deciding *when* to engulf and devour—first migrate into the thymus gland, which lies under the breastbone. The thymus is a finishing school of sorts: there the cells learn what's self and what isn't, and they mature and begin circulating through the bloodstream.

The most important of these cells, for Ildstad's work, are the T cells. From the moment a surgeon implants a donated organ and it makes a first, tentative acquaintance

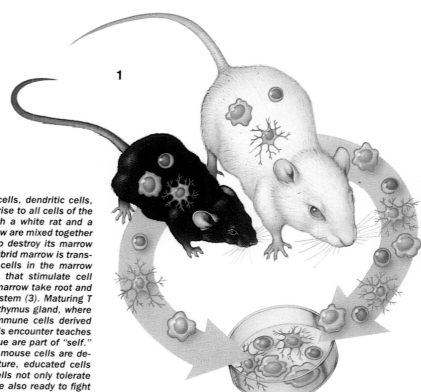

TRANSPLANTED SELF

Bone marrow containing mature T cells, dendritic cells, as well as immature cells that give rise to all cells of the immune system, is taken from both a white rat and a black mouse. The two types of marrow are mixed together (1). The mouse is then irradiated to destroy its marrow and immune system cells (2). The hybrid marrow is transplanted into the mouse. Dendritic cells in the marrow secrete proteins, called cytokines, that stimulate cell growth and may also help the new marrow take root and begin to generate a new immune system (3). Maturing T cells travel from the marrow to the thymus gland, where they encounter other burgeoning immune cells derived from both the rat and the mouse. This encounter teaches T cells that both rat and mouse tissue are part of "self." Any T cells that attack either rat or mouse cells are destroyed in the thymus (4). The mature, educated cells disperse through the mouse. The cells not only tolerate skin grafts from the rat, but they are also ready to fight against any other foreign invaders, such as bacteria (5).

with your arteries and veins, your T cells bolt into a frenzy of activity, drafting other kinds of immune cells into action and launching an ingenious, coordinated, and all-too-often victorious campaign to exterminate the intruder. To complicate matters further, any of the organ donor's T cells that are left lurking in the transplanted organ will spring into action against the recipient—a response called graft-versus-host disease.

More than three decades earlier, Nobel prizewinner Peter Medawar and other researchers had transplanted bone-marrow cells from one animal into a newborn animal of the same species, and shown that the recipient would afterward tend to accept any additional tissue transplanted from the donor. To Sachs and Ildstad, that seemed a promising lead. But there was a big snag with these bone-marrow transplants: They didn't work in adult animals, which rejected the grafts. There appears to be a privileged time before an immune system matures when it can be retrained to accept foreign-marrow grafts. In people the window of opportunity is the first 16 weeks of gestation—not a very useful opening.

Other researchers tried to work around this obstacle by completely obliterating the recipient's immune system with toxic chemicals or radiation. Their idea was to let the donor marrow then provide the recipient with a whole new immune system—the donor's—that wouldn't attack the graft. And indeed, the experimental animals tolerated grafts well. But they developed another, more serious problem. Their immune systems became incompetent—vulnerable to all sorts of infections and malignancies.

The problem, the researchers would eventually learn, was a matter of improper protocol. Like any large bureaucracy, the immune system has procedures that must be followed, and these experiments seriously breached them. T cells can't recognize invaders on their own. An invader must first be grabbed by another type of immune cell called a macrophage, and then hustled into the presence of a T cell. The T cell will react to the invader only if it recognizes that the macrophage making the introduction is an ally from the same body. However, in the experimental animals, that recognition wasn't forthcoming.

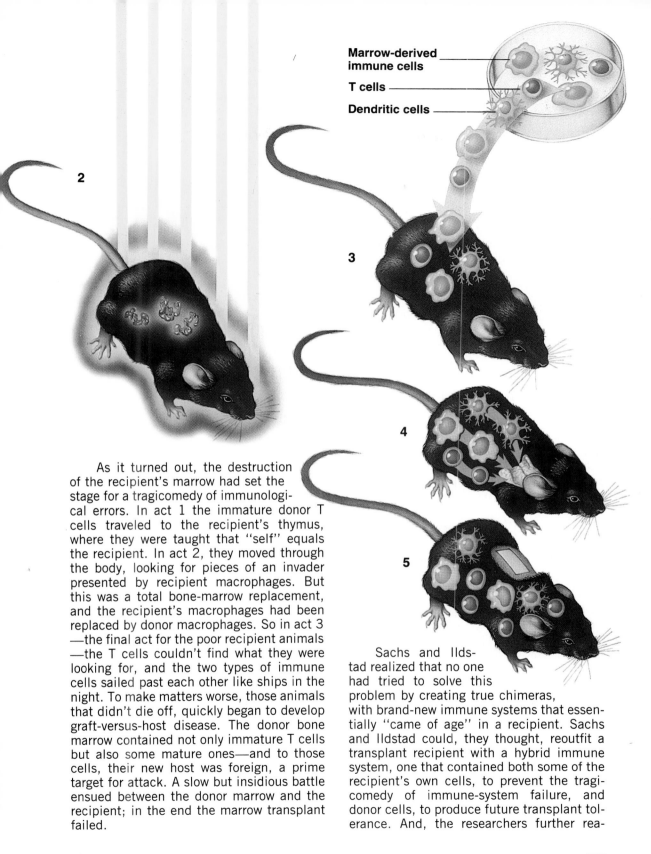

Marrow-derived immune cells

T cells

Dendritic cells

2

3

4

5

As it turned out, the destruction of the recipient's marrow had set the stage for a tragicomedy of immunological errors. In act 1 the immature donor T cells traveled to the recipient's thymus, where they were taught that "self" equals the recipient. In act 2, they moved through the body, looking for pieces of an invader presented by recipient macrophages. But this was a total bone-marrow replacement, and the recipient's macrophages had been replaced by donor macrophages. So in act 3 —the final act for the poor recipient animals —the T cells couldn't find what they were looking for, and the two types of immune cells sailed past each other like ships in the night. To make matters worse, those animals that didn't die off, quickly began to develop graft-versus-host disease. The donor bone marrow contained not only immature T cells but also some mature ones—and to those cells, their new host was foreign, a prime target for attack. A slow but insidious battle ensued between the donor marrow and the recipient; in the end the marrow transplant failed.

Sachs and Ildstad realized that no one had tried to solve this problem by creating true chimeras, with brand-new immune systems that essentially "came of age" in a recipient. Sachs and Ildstad could, they thought, reoutfit a transplant recipient with a hybrid immune system, one that contained both some of the recipient's own cells, to prevent the tragicomedy of immune-system failure, and donor cells, to produce future transplant tolerance. And, the researchers further reasoned, if they removed mature T cells from the donor marrow, they could keep the graft from attacking the host.

Mixing Marrow

The duo used mice as recipients, and either rats or other mice as donors. First they removed some marrow cells from a mouse transplant recipient. Then they did the same to the donor animal. Afterward they filtered mature T cells out of both types of marrow, leaving only the immature cells to be educated in the recipient's thymus. Next they mixed the two types of marrow together. Then they irradiated the recipient mouse to destroy its native immune system, making it a blank slate, immunologically speaking. Finally they transplanted the hybrid bone marrow back into the irradiated mouse.

What Sachs and Ildstad hoped would happen was that as various kinds of immune-system cells from both animals matured, they would travel to the thymus along with the maturing T cells. In the thymus the T cells would bump into mouse-derived cells and learn that those cells are part of self. They would also bump into donor-derived cells, which would teach them a similar lesson. Any T cell that didn't learn both lessons —in other words, any T cell that attacked any of the other cells—would get destroyed in the thymus; it would flunk out of immune school. Any T cell that made it through, however, would recognize both donor- and recipient-derived immune-system cells— such as the all-important macrophages— as allies, and would combine with them for an assault on anything else traveling around the body, which would appear foreign.

And that's pretty much what happened, at least with Sachs and Ildstad's mouse-mouse crossbreeds. Some of them accepted later tissue transplants from their bone-marrow donors, while vigorously rejecting grafts from third parties. There was also no sign of a graft-versus-host reaction. "It was a breakthrough," Ildstad says.

Yet the rat-mouse transplants didn't work. For some reason the hybrid marrow grafts never fully "took" in the mice, and, as a result, the mice rarely reconstituted immune systems containing rat-derived cells.

"The greater the degree of genetic difference between two animals," Ildstad says, "the harder it is to get engraftment. My major goal at the NIH was a permanent, stable graft between species, and I never achieved it."

For Ildstad, this was a major disappointment. "There's a critical shortage of organs. I had a lot of patients who died while they were on the waiting list for a liver transplant. And with children, in addition to the shortage of donors, there's a size limitation that makes it difficult to find livers and hearts." But if a cross-species transplant were possible, she knew, it would make available a whole new source of lifesaving donors—pigs and sheep, for example. With other animals to choose from, the chances of finding an organ that fits increase tremendously. However, the poor results Ildstad achieved made these chances look like a very long shot.

Curiously, though, despite the apparent failure of the marrow graft, some rat-skin transplants did stay on a recipient mouse for brief periods before being rejected. Yet Sachs and Ildstad couldn't figure out why. To do so, they had to track any rat-derived immune cells that might be roaming around inside the mouse, making up some kind of chimerical immune system allowing that animal to accept such skin transplants, if only for a while. To find these cells, they were using monoclonal antibodies, which are like molecular dog tags that clip onto and identify a particular class of cells; Ildstad would take antibodies that attach to rat-derived immune cells, and stain them with a fluorescent dye. If any rat cells were surviving in the mouse, she would see them light up. But few of the cells turned up in these tests, and thus the researchers could not tell if the moderately successful skin transplants were the result of a moderate case of chimerism. "We couldn't track the cells, so we just stopped," says Ildstad.

In 1985 she returned to her surgery residency at Mass General. A year later she moved on to Children's Hospital Medical Center in Cincinnati, where she established her specialty as a pediatric surgeon. Then, in 1988, when Ildstad came to the University of Pittsburgh, she saw a chance to pick up where she had left off.

Serendipity at Work

Ildstad decided to begin her transplant research again, using essentially the same procedure she and Sachs had developed, with rats as the donors and mice as the recipients. But this time, as she was setting up her new lab, she experienced a stroke of luck that has provided momentum for her research ever since. "It was," she recalls, "total serendipity. Because the lab was still unpacking its scientific luggage and settling into quarters, we didn't have an agent to deplete T cells from the rat marrow. So we decided we'd transplant it anyway, just to see what would happen."

In theory, not much should have: it remained the conventional view, remember, that you had to remove all the mature T cells from donated bone marrow, or else they'd attack the recipient, and the marrow graft would fail. But Ildstad was flabbergasted to find her T cell-saturated rat-marrow cells growing like crabgrass in her chimerical mice —much better than they'd ever done when the T cells were removed. "We found unbelievable levels of detectable rat cells," she recalls.

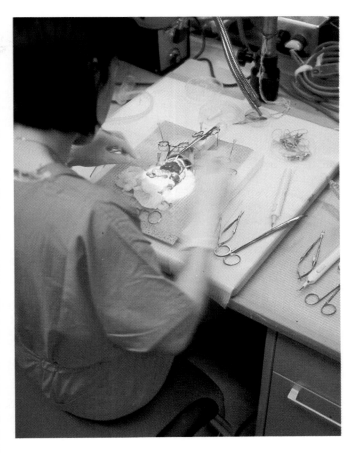

Still more exciting was that marrow transplants also improved the later acceptance of rat tissue. Skin grafts took root with extraordinary alacrity in three-month-old mice, and they lasted until the rodents died a natural death at the ripe old mouse age of a year and a half. "We use tail skin because it's very easy to see," Ildstad explains. "A healthy tail-skin graft is nice and smooth and soft, doesn't have any scabbed-over areas, and doesn't have any petechiae, which are like bruises. If there is rejection, you start seeing areas of redness. Then, if it's acute rejection, over a few days it becomes scabbed over, dried up, and crusted."

Ildstad put three grafts on each animal —one from the donor rat, another from a genetically different rat, and a third from a different mouse. The donor rat's skin survived, while skin taken from third-party rats and even from other mice came off immediately. The mice also showed no sign of immunoincompetence, and they valiantly fought off infections.

What exotic ingredient in this transplanted T-cell-rich rat bone marrow could suddenly render a mouse so receptive to the donor rat's tissue? Presumably it couldn't be a T cell; there was no reason to think a cell designed to react against foreign tissue would suddenly work to ensure a warm welcome for it. Yet whatever it was, this cell, chemically speaking, had to look very much like a T cell; otherwise it wouldn't have been eliminated by treatments designed to recognize and remove T cells. Ildstad suspected the existence of a cell that somehow both helps the rat's bone marrow adapt to the mouse and makes the mouse tolerant of rat tissue as well.

It took nearly two years to confirm those suspicions, but in December 1991, Sherry Wren, a postdoctoral student working in Ildstad's lab, finally put a face on this mystery cell: it was a dendritic cell. These cells are distributed throughout the body and ingest invaders that float by. Dendritic cells originate in the bone marrow, like the T cell, and share many of the T cell's chemical markers. What's more important is that dendritic cells release proteins called cytokines, which, among other things, stimulate colonies of cells to grow. Ildstad isn't sure exactly what happens, but she thinks the cytokines floating around the immediate area of the graft stimulate the marrow cells to take root.

Flushed with this success, Ildstad began to turn to different animals. In addition to mice transformed partway into rats, she reversed the process and gave rats part of the immune systems of mice. She's also used hamsters as donors, grafting part of the hamster immune system into mice.

Ultimately, Ildstad believes, the dendritic cell could become a powerful and invaluable aid in transplants. Indeed, this spring, doctors at the M. D. Anderson Cancer Center in Houston, Texas, are going to use Ildstad's mix-and-match technique to transplant bone marrow into leukemia patients. But as Ildstad points out, there is a wide range of transplants that could be helped by such a procedure. After her stunning success with skin, Ildstad began to explore other types of transplants with the same technique, moving various rat glands into her mice without rejection.

Focus on the Pancreas

Last year she turned her attention to the pancreas, where damage leads to juvenile-onset diabetes. "Nearly 1.5 million people in the United States have diabetes. Half of them go blind, lose their limbs, or suffer kidney failure and need a transplant," Ildstad says. All this devastation is the result of the failure of a small number of pancreatic cells, called islet cells, which produce the hormone insulin to regulate sugar levels in the body. "If we could find a benign way to modify the immune system of people prone to diabetes, we might be able to induce tolerance to either pancreas- or islet-cell grafts. Insulin-producing islets represent less than 2 percent of the pancreatic tissue anyway, so that's all you'd need to cure diabetes and prevent further complications."

At least in Ildstad's mice, that's already a reality. "Our rat islets have survived in mice for 10 months now," she says. And they don't merely survive, they work, controlling sugar levels even in mice with previously induced diabetes. "We've given glucose-tolerance tests to these animals; within 15 minutes, their blood sugar peaks; and within 30 minutes, they're back down to normal."

The human applications of these experiments to diabetes or any other disease are still years away. Research needs to be conducted on larger, more human-size animals, with organs that might prove practical in an interspecies transplant, but with immune systems that might behave very differently from those of rats and mice. And ethical issues need to be settled, such as the moral implications of sacrificing animals for transplants. Ildstad points out that no one is considering primates for the purpose; the likely candidates are pigs or sheep, already used for human consumption. Beyond that, a few ethicists are troubled by the prospect of people infiltrated by inhuman cells and solid organs; they wonder if that could constitute a subtle metaphysical threat to our humanity.

Though Ildstad acknowledges the importance of these issues, the metaphysics involved in being two things at once doesn't trouble her to excess. After all, she's something of a chimera herself, part surgeon and part immunologist. And that, she thinks, is the reason for her success. "People question whether surgeons can do research or whether it's an oxymoron," she says. "But I really think we get a unique perspective through interacting with our patients and seeing the suffering that people sometimes go through. That in turn brings new ideas into the lab with the hope that they can be applied clinically. There are some things we can't fix in surgery. But I think we're close to making a major contribution through research."

Directing Your Dreams

by Rosalind Cartwright, Ph.D.,
and Lynne Lamberg

Although dreams often puzzle us, it's worth the effort to capture and decipher them, for they show us what we otherwise may not see. They help us uncover truths about ourselves that our waking minds may know yet deny, or that, awake, we may not be able to articulate clearly. They do so especially in times of intense emotional upheaval, as when going through a divorce, facing major surgery, or when confronting the fear that some perceived flaw might keep one from getting married.

Few of us go through life without encountering such crises. Indeed, times of crisis highlight the important functions that dreams serve in our lives. Events we perceive as both positive and negative, beginnings and endings, pluses and minuses, all place heavy demands on the dream system. When we gain or lose a job, a mate, a home, or when we undergo any major change in our lives, our internal picture of who we are and our sense of security are called into question.

At such times, our dreams go into high gear. In our dreams, we search through our life story to find memories that can help us cope. We sleep more lightly and awaken more often. Dreams are more apt to stick with us when we're troubled than when we're happy.

The Chinese symbol for crisis includes the characters for both danger and opportunity. The danger we face during a crisis is from the potential shattering of the program on which we run—the present self. The opportunity is in expanding that picture, reshaping how we see ourselves, constructing a new, better-functioning persona.

Dreams during crises show how that equation is working out. Is the danger overwhelming us, or are we dealing with the opportunity to assume new roles? Are these roles positive or negative? "Are you a good witch or a bad witch?" the Munchkins asked Dorothy, who, when threatened with the loss of her beloved Toto, dreamed she was blown out of Kansas and into the land of Oz.

From Bad Dreams to Good

Our dreams serve, Cincinnati psychiatrist and dream researcher Milton Kramer, M.D., suggests, as an emotional thermostat. A bad dream, like an elevated temperature, is a symptom that something is wrong. It is a distress signal, a message from our sleeping mind to our waking mind that is risky to ignore. Kramer's studies show that the mind-set we have during our dreams affects our attitudes and behavior the following day. After a bad dream, we awaken more discouraged than we were at bedtime. After a good one, we tend to feel more optimistic.

Dreams also offer a shortcut to understanding and overcoming the emotional stumbling blocks of people in crisis. The times that try our souls are precisely those when we most need to shift our dreams into action, to turn from simply recognizing the danger to our present self to accepting the opportunity to invent or devise a change, and then to rehearse and work on it.

The crisis-dreaming method—which addresses malfunctioning dreams directly to try to change those that reveal a continuing, underlying, poor identity pattern—offers an opportunity to direct our dreams toward more-positive results. Once you discover the trend of what is happening from dream to dream, you can identify more easily those that are unproductive, and you can start to work on dream repair that very night.

Useful in many situations, the crisis-dreaming method aims to change negative, hurtful dreams to positive, healing ones. It enables people to stop bad dreams while they are in progress, and to rewrite their scripts. In this way, we can redirect dreaming to perform its proper function: to update our inner narrative, our sense of identity—first by recognizing those aspects of our present crisis that are negative and demoralizing, and next by finding images of strength already filed in our memory banks. We can then activate these images to change our waking attitudes.

The crisis-dreaming method has value in good times as well as bad: It shows how we can use dreams not only to understand ourselves better, but also to foster desired changes in our waking lives. A good dream

system enables us to reorganize our sense of ourselves internally, to make the necessary transformations in point of view when circumstances change, to create a new, self-respecting version of who we are.

The first step is to pay more attention to your dreams. You may be skeptical. The conventional wisdom is that most of us recover from a crisis by changing our waking lives, not our dreams. Can it work the other way around? Many of those with whom I have worked tell me that dreams provided them with both the insight on how the present connects to the past, and the impetus to change the program of the self in order to create a better fit with their present lives. They have found that changing the endings of their dreams can be a giant step toward these goals. This concept shocks traditional psychoanalysts. It challenges the basic idea of the nature of the relationship between the waking mind and dreams.

Understanding Dream Dimensions

Key questions to ask yourself about any dream you recall are: "Why this dream?" and "Why dream this dream now?" For that, you need to look for the underlying themes that bind together your dreams and waking life, the emotional issues that prompt your sleeping self to declare: "This is what I am feeling. This is what day-to-day events remind me of. This is what needs more attention."

We build our dream stories to express these underlying themes using various dream dimensions—distinctions that we make to define and categorize our experiences. Dimensions, which reflect opposing states or qualities, constitute our own unique and habitual way of organizing the world we live in. We start in infancy to make big evaluative discriminations: This feels good; that feels bad. This is warm; that is cold. In dreams the specific images, along with their opposites, show how we see people and events and express our innermost feelings about them.

The idea of our using a system of opposites in our dreams is one adapted from the work of the noted anthropologist Claude Lévi-Strauss, who used this approach to analyze the characteristic ways of thinking and

mythmaking in tribal cultures. He suggested that the mind works on problems by dividing key issues into pairs, and then by juggling these elements until they fit the needs of the person telling the story.

Studies of waking memory show that we mentally file new information with bits of similar information as well as with their opposites; that is, we sort a new fact, not only by what it is, but also by what it is not. Thus, it's no surprise that we dream about winning the lottery when we're worried about not being able to pay the bills, or about striking out at bat at age 10 just after receiving a big promotion. The following list comprises some of the most common dimensions we see in dreams in our lab:

- Safety versus danger
- Helplessness versus competence
- Pride versus shame
- Activity versus passivity
- Independence versus dependence
- Trust versus mistrust
- Defiance versus compliance

While each dream in a series may have several dimensions, the same few dimen-

sions are usually expressed repeatedly in a single night. Some parts of a dream may reflect the positive pole, and others the negative side. A dream expressing feelings of danger, for example, may precede or follow one expressing feelings of safety. Homing in on any one concept that a dream presents, and asking, "What is its opposite?" may open our eyes to important issues.

Rewriting Dream Scripts: The RISC Method

If bad dream scripts make you awaken discouraged and downhearted, rewriting the scripts to improve the endings should lead to better moods. Dream therapy has just four steps, illustrated by the sequence of drawings in this article:

- **R**ecognize when you are having a bad dream, the kind that leaves you feeling helpless, guilty, or upset the next morning. You need to become aware while you are dreaming that the dream is not going well.
- **I**dentify what it is about the dream that makes you feel bad. Locate the dimensions that portray you in a negative light—as, for example, weak rather than strong, inept

Helping Children to Learn from Their Dreams

Common problems we all face in growing up—learning to protect ourselves from physical injury, coping with hostility, learning to live with our own upsetting feelings—fuels the creation of our early dream images. We cannot shield children from accidents, injuries, or pain, but we can be alert to the dreams these events leave behind, the stories that clearly need to have better endings.

We can encourage children to share their dreams, especially their nightmares. We can familiarize ourselves with how they handle the problems that we expect to show up in the dreams. If they are falling, we can help them learn to fly like an eagle; if monsters chase them, we can teach ways to tame or even befriend the savage beasts.

Some children relate their dreams better by drawing them. In her book *Nightmare Help,* Anne Wiseman suggests using drawings to help your child master the conflicts the dream expresses. Once the monster is down on paper, you can encourage the child to draw a cage around it, for example. Children need to know they don't have to face their troubles—even in their dreams—all alone.

Wiseman reports that her own son had bad dreams about their house burning down. Her husband suggested to the boy that next time he had one of these dreams, he put the fire out. One morning the boy came to breakfast full of excitement. He did it. He put the fire out! When his parents asked him how, he said proudly, "I peed on it." They may have traded one nighttime problem for another!

rather than capable, or out-of-control rather than in control.

• **S**top any bad dream. You do not have to let it continue. You are in charge. Most people are surprised to find that telling themselves to recognize when a bad dream is in progress is often all it takes to empower them to stop such dreams.

• **C**hange negative dream dimensions into their opposite, positive sides. At first, you may need to wake up and devise a new conclusion before returning to sleep. With practice, you will be able to instruct yourself to change the action while remaining asleep.

The first letter of each step forms the acronym "RISC" to help you remember that the idea is to "risk" stepping in to change the endings of your dreams and to work toward a more positive self-image.

Altering the outcome of a dream is a tall order, but our studies show it's an achievable goal. There's an active give-and-take between the conscious and the sleeping mind. Even if you don't change a particular dream while asleep, your waking exploration of the depressive elements of your dreams, and your awareness of what you can and should change, may have a payoff. People who devise several possible solutions to familiar dream dilemmas report that they often manage to incorporate some of these new waking attitudes into their dreams.

The definition of a "better" ending sometimes proves surprising. One meek young man often had a nightmare in which a bus ran him down. When he changed this dream, he gave himself a machine gun so that he could attack the bus and shoot its driver. He felt much better afterward and never had the dream again. Why not? In his dream, he could retaliate aggressively against the bullies who had picked on him when he was a youngster, and destroy them. His dream success rebuilt some pride he sorely needed.

Such success reverberates with waking life. Becoming more active in dreams helps people to become more positive about the future. A successful night of dreaming produces immediate benefits for mood in the morning. Stopping a bad dream and changing it lifts the spirits. People gain a sense of empowerment from knowing they are not at the mercy of their bad dreams. Then, as they begin to change the image of a rejected, helpless self to one that is more in control, waking behavior begins to improve. They start to try out the new roles, the underdeveloped, better aspects of themselves, that they first practice in dreams.

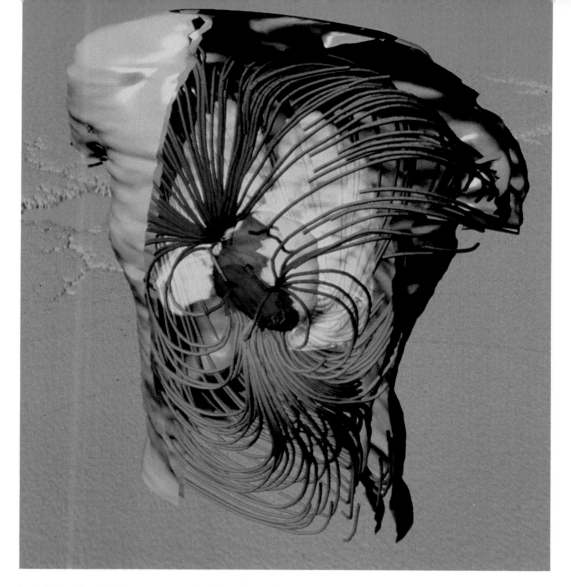

THE BODY ELECTRIC   *by Carl Zimmer*

Without electricity, we would perish. We could learn to do without the flow of electrons that powers VCRs and food processors, but the currents inside our bodies are vital. The brain needs electricity to issue its commands from neuron to neuron. When these signals reach a muscle, they set up a wave of electrical excitation in the fibers, which in turn triggers the chemical reactions that make the fibers contract or relax. The most important muscle is the heart; it shudders under a wave of electricity about once each second.

Inferring Heart Function
The heart's electric field radiates out into the chest cavity, sending clues about the heart's function toward the skin. Cardiologists can get a peek at the heart by taping electrodes to a person's torso; each electrode produces a familiar squiggle on an electrocardiogram (EKG) that shows how the voltage changes at

A computer-generated model (above) of the electric pulses that make the heart beat may one day be used as a noninvasive method to gain insight into the heart's proper function and health.

that single point on the body. Cardiologists spend years learning to infer heart function from these signals, and to recognize in EKG readings the telltale signs of dangerous heart conditions.

But looking at an EKG is like watching a hurricane from a porthole. The 6 to 12 electrodes on a patient's torso and limbs provide only limited information about the heart. So cardiologists also use isotope imaging, echocardiograms, and other methods to gather data. In rare cases, surgeons even have to open the chest to find the problem.

Chris Johnson wants to figure out how to see a heart without cracking open a rib cage. This University of Utah computer scientist has been working to visualize the electric field created by a heartbeat. Ultimately Johnson hopes that he will help create an entirely new diagnostic tool—to let cardiologists, as it were, watch the hurricane from space.

Johnson has been developing his methods for making these calculations since 1985. Last year, when he was ready to use some real heart measurements, he wanted to get them from a heart with clear electrical abnormalities. A condition known as Wolff-Parkinson-White syndrome was a perfect candidate. This birth defect gives its victims an extra patch of cardiac cells that sometimes sets off the contraction of the ventricles—the main pumping chambers—too quickly, throwing off the heart's rhythm. To measure the electrical activity of the heart, surgeons would put a nylon sock covered with electrodes over the heart's ventricles, and measure the voltage on its surface.

Johnson got hold of a set of these measurements and put them into his computer. All he needed was a body to carry the heart, because the shape of the chest influences the way the heart's electric field reaches the skin.

"One of the professors' sons happened to be around one day, looking hungry," Johnson says, "so we offered him a free lunch if he would lie in this magnetic resonance imaging [MRI] chamber for several hours." Radiologists use MRI to measure the magnetic fields of different kinds of tissue and convert those measurements into images of

A HUMAN BLUEPRINT

To create a computer-generated model of the body's electric field, computer scientist Chris Johnson begins with a magnetic resonance image (MRI) of a slice of the torso. By hand, he marks the boundaries between different parts of the body on a computer screen — computers are not yet capable of making such distinctions, although Johnson is working to automate this process. Above, red indicates the heart and yellow the lungs. The two sets of gray balls mark the inner and outer borders of the muscle, and the brown balls mark the skin.

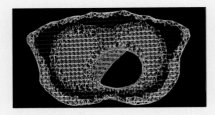

Each ball represents a boundary point Johnson uses to create a mesh. The mesh above shows body fat in yellow, muscle in red, and the chest cavity in blue; the lungs are purple. The mesh produces many small pyramid-shaped building blocks that are used to calculate the heart's electric field. After the computer connects the points in one layer, it joins them to the neighboring layers. As Johnson adds more MRI torso slices to the picture, the organs slowly begin to take shape. Below, the lung is formed in yellow. Eventually Johnson constructs an entire torso, using 130,000 building blocks.

HAVE A HEART

Once the torso is constructed, heart analysis can begin. These four images show how the electric field on the surface of the heart changes during a single beat. The bar at far right shows the color scale indicating maximum to minimum voltage, and the box in the lower left corner of each image shows the electrocardiogram reading at that moment from an electrode placed on the chest directly over the heart. Most cardiac cells at rest maintain a strong negative voltage. When stimulated by electrical impulses, the voltage of the cardiac cells becomes more positive (1). The change creates a small electric current that excites the adjacent cells, creating a wave about one-twentieth of an inch wide that travels over the heart at about three feet per second (2). For a cell sitting in front of the wave, the current creates a positive voltage that gets stronger until the wave breaks over it; then the voltage switches over to negative at the peak of the heartbeat (3). At the end of the beat (4), the heart resumes its neutral to negative charge. Electric waves begin inside the heart, so the centers of positive voltage seen here are where they occur when about to break out onto the surface.

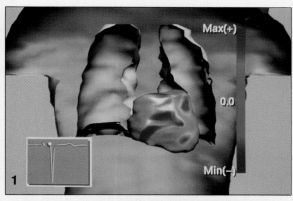

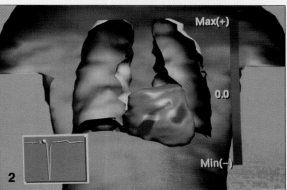

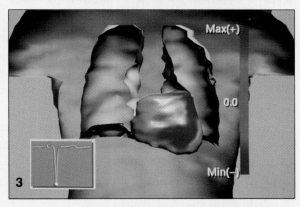

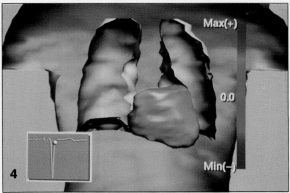

slices of the body. Johnson made 200 scans of the man's torso and turned them into a three-dimensional image in his computer. As the pictures at right show, he was able to combine the heart and the torso to create remarkable views of the body.

A Sock for the Heart

Johnson will be measuring more people's heartbeats in the near future; a Utah colleague is now developing a nylon sock that will fit over the entire heart. Armed with the electrical information from the sock and his computer torso, he'll be able to watch as the heart's activity shows up as voltage changes on the skin. As he generates more pictures of body surfaces, he will learn how to reverse mathematically the entire process—to infer the electric field on the heart from the way it appears on the skin, eliminating the need for the sock. With this information in hand, a cardiologist will be able, not only to diagnose Wolff-Parkinson-White syndrome, but to locate the errant cells as well. But it's a tricky problem. "The body acts as a kind of filter, smoothing things out," says Johnson. When there are two centers of positive voltage on the surface of the heart, for example, the skin has only one.

Johnson foresees a time when patients with heart trouble could put on an electrode jacket to measure the voltages on their skin. "The ideal thing would be to measure the voltage on their body," he says, "zip them through a magnetic resonance imager, and within the afternoon calculate the voltage on the heart itself, pop it up on the computer, and say, 'Oh well, here's the abnormality. We can change this with drug therapy or use a catheter.'"

For now, though, Johnson is trying to perfect the computer program, incorporating complicated features such as the large arteries around the heart. And since the size and shape of a patient affect the electric currents, he's learning how to stretch his standard model to fit the infinite variety of human geometry. "In four or five years, we're going to use a real patient, localize the abnormality, tell the surgeons where it is, and they'll go and find it," says Johnson. "And then they'll tell us how well we did."

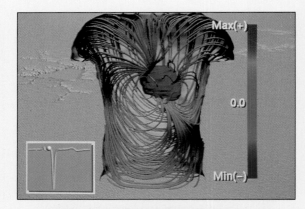
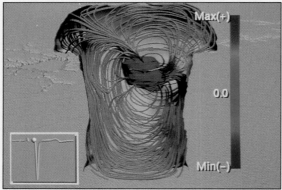
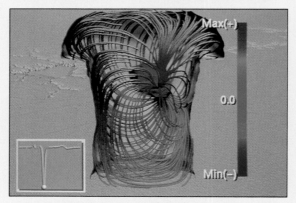
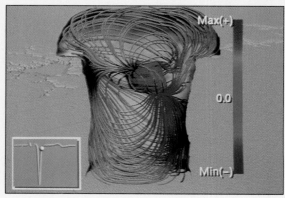

CURRENT CONDITIONS

The next step is to calculate the electric field inside the body at each stage of the heartbeat (opposite page). The loops show the flow of electric current over the course of a single heartbeat. (Chris Johnson creates a total of 500 images per heartbeat; these images show the interior fields that correspond to the surface changes shown in the previous photographs.) Says Johnson, "No one knows what normal looks like. Once we map the normal current pathways we can start looking for distinctions between normal and abnormal." A potential use for this information would be to improve the defibrillating paddles used on cardiac patients. "All of the studies on these paddles have been done on pigs," explains Johnson. "No one has been able to show with any certainty just how big the paddles should be or exactly where they should be placed." Computing a precise map of internal electric fields during a normal heartbeat could put researchers closer to developing a better defibrillator.

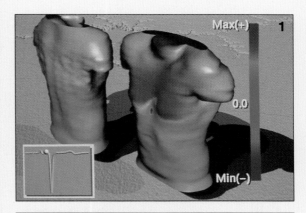

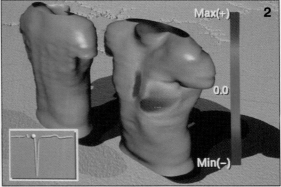

MORE THAN SKIN DEEP

Once the electric field is simulated inside the body, Johnson can use the computer to see what the field looks like from the outside (right). At the beginning of the heartbeat, an area of positive voltage (in red) appears where the wave is about to appear (1). It gets pushed out of the way by a burst of negative voltage shown in blue (2). By the peak of the heartbeat (3), the negative voltage has spread over the chest and, by the conclusion of a single beat, the positive charge has reappeared on the chest (4).

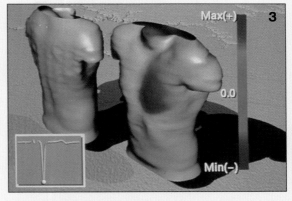

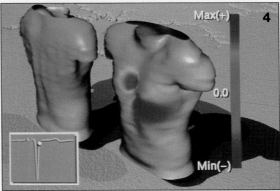

LSD

Makes a Comeback

by Jim Urban

Three decades after LSD defined the flower-child generation, the mind-altering hallucinogenic drug has seized a new set of users in college dorms and suburban high schools.

That's right. Acid. The drug that causes some people to hear colors, see sounds, and jump out of windows thinking they can fly is making a return trip with a generation that never heard of LSD maven Timothy Leary. "Some people will try it and have bad trips and not try it again, but most people like it," says John, 18, a University of Pittsburgh freshman who wanted his last name withheld. He saw LSD at his suburban Philadelphia high school last year, and even more when he got to college.

"It's become more acceptable, not as much as pot, but it is heading in that direction," John says.

Most Drug Use Down

National studies show alcohol, marijuana, and cocaine use by young adults has decreased, while LSD use has increased. The most startling figure pertains to male high school students in predominantly white suburban neighborhoods, where hallucinogen use in 1991 was 18.9 percent, or about one of every five, according to the Parents' Resource Institute for Drug Education, or PRIDE.

"The word is out that there are advantages to taking LSD that are not present with cocaine," says Henry Abraham, M.D., direc-

Use of the hallucinogenic drug LSD has risen at an alarming rate among young Americans. At all-night dances called "raves" (above), some party-goers allegedly "drop acid" to create a more memorable experience.

tor of psychiatric research at St. Elizabeth's Hospital in Boston, who has studied disastrous LSD cases since 1971.

"It's not addictive, and it's cheap," Abraham says. "These kids think, 'Hey, it's not cocaine, you can really party with this drug.' And you don't have to go into hock or become a hooker to feed a habit."

A 1991 survey of 15,000 high school seniors found 8.8 percent had experimented with LSD, up from 7.2 in 1986. Cocaine usage fell from 16.9 percent in 1986 to 7.8 percent in 1991, below that of LSD.

The same study, conducted jointly by the University of Michigan and the National Institute for Drug Abuse (NIDA), found 5.1

percent of college students had used LSD in 1990. That compared with 3.4 percent in 1988. The survey had a margin of error of plus or minus 1 percentage point.

The PRIDE study of 26,000 high school students, released in 1992, showed drops in marijuana and cocaine use, but a 20 percent jump in LSD use since 1989.

Twisted Perceptions

Today's doses, known as hits, are about half as powerful as those in the '60s, researchers say. But many users develop a tolerance and soon increase the number of hits to attain the LSD high that some say blows away cocaine and marijuana.

"It just twists every perception," says Alice Holopirek, a former user who now counsels young users in Larned, Kansas.

"You see sounds, you see music, you hear colors," Holopirek says. "It's a complete distortion of the way we receive things in our senses. Some people told me they would see things that weren't there, but I never had that. I would tend to see things there that were very distorted—like a wall melting. Or I'd look across the street and see waves in the asphalt."

That's what makes the drug dangerous, Holopirek says. She knows of a user who hallucinated having bugs under his skin, and tried to cut them out with a knife.

There are also side effects like flashbacks and panic attacks. And some users suffer from prolonged psychoses in which they lose touch with reality for two to three days and sometimes weeks, says Abraham, the Boston researcher.

"Use of LSD is really like playing Russian roulette with chemicals," Abraham says. "You can spin the chamber, and maybe five times you get away with it. Maybe the sixth time you blow your brains out. Instead of bullets, they are using drugs."

What is LSD?

LSD, or lysergic acid diethylamide, is derived from a fungus that grows on rye and other grains. It is easily produced in clandestine labs, mostly in California, and commonly disseminated in the form of drug-permeated blotter paper decorated with cartoon characters.

"There is a degree of expertise needed for manufacturing, but you don't have to be a scientist," says Ken Jones, who heads a team of narcotics investigators for the Postal Inspection Service in Pittsburgh.

The distribution network is much like those for other drugs. Street dealers buy in bulk from mid-level suppliers, who buy in even greater bulk from the manufacturers. Missing are the drug cartels or Mob families that dominate cocaine distribution.

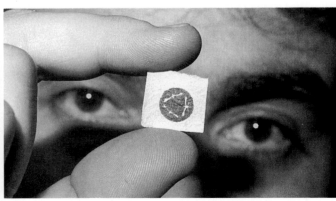

One $5 dose, or "hit," of LSD (above) usually provides a high that lasts about six hours. Most LSD is illicitly manufactured in California and distributed on blotter paper permeated with the drug. The paper is frequently decorated with cartoon characters (left).

During the 1960s, the LSD experience represented a virtual rite of passage for advocates of the "anti-Establishment" movement. In recent years, '60s fashions have seen something of a revival (above), a wave of nostalgia that may have played a role in the renewed interest in LSD and other hallucinogenic drugs.

LSD is an entrepreneurial drug with big money at stake, Jones says. Users place a hit on their tongue or chew on the paper to get a high that lasts about six hours. The experience is called "tripping."

During the '60s, tripping was practically a rite of passage for many who smoked marijuana, tuned in Jimi Hendrix and the Doors, protested the Vietnam War, and became sexually liberated. By tripping, they showed disdain for the Establishment.

"A lot of people in the '60s who were at all into the drug scene dropped acid sooner or later. There is no question about that," remembers Joe White, an associate professor of history at the University of Pittsburgh who spent 1962 through 1967 at a counterculture hotbed, the University of California at Berkeley.

Today experts say a lack of spending money, the passing of time, and marketing of the '60s counterculture have combined to fuel LSD's resurgence.

LSD hits run about $3 to $5 in most areas of the country. The price is higher in some places, but it's still more cost-effective than the other "party drugs" such as marijuana and cocaine. A gram of cocaine will run about $100. Marijuana costs about $65 for a quarter-ounce.

"When you compare one hit of LSD to a gram of coke, LSD gives you a much longer high," says Jones, the postal inspector. "You don't have the intense high for a short time and then a falloff. You get a lot more bang for your buck with LSD."

Another reason for increased LSD usage is the focus on cocaine and its derivative, crack, in the drug war. LSD has been out of the public eye for years, and so have the horror stories about bad trips. But young kids have been bombarded with commercials that assail cocaine and crack use.

"It's been long enough now that LSD doesn't have the same reputation among young people as it did in the '60s or '70s, when young people were reading articles about people flying out of windows while tripping on LSD," says Jonathan Caulkins, a Carnegie-Mellon University professor who researches drug trafficking.

Nostalgia for the '60s

LSD use is also part of the '60s nostalgia being revived today. Tie-dyed clothing and psychedelic colors and patterns reemerged

While maintaining their focus on crack and cocaine, police and federal agents are starting to investigate LSD sales and are buoyed by tough federal sentencing guidelines. If acid and the blotter paper it is on weighs more than 10 grams, about the weight of 10 paper clips, a dealer faces a minimum of 10 years in prison.

LSD's popularity seems greatest in sleepy suburban communities that are a far cry from hip San Francisco or Los Angeles.

Last year, investigators broke up an LSD ring in two Medford, New Jersey, high schools. Prosecutors obtained a guilty plea from a Lafayette, Indiana, teenager who spiked his teacher's soda with acid. And a 21-year-old suburban Pittsburgh man, convicted in December of drug trafficking, had received 20,000 hits of LSD each week from California, postal inspectors said.

Some researchers remain unconvinced that LSD use is a major problem today. Saul Shiffman, a professor of psychology at Pittsburgh, says agencies should remain focused on efforts to contain drinking and marijuana smoking among young people.

"You have to pick your battles," Shiffman says.

The one thing missing from this generation of LSD users is a charismatic spiritual leader—a '90s version of Leary, the former Harvard professor who promoted LSD in the '60s by telling followers to, "Tune in, turn on, drop out."

Leary still stands by the drug, saying bad trips gave it a bad rap.

He defends most of what he has done. He needed to take drugs, he says, to learn about the mysteries of the mind that could be gained no other way.

But Abraham, the Boston researcher, says LSD users who say the drug hasn't harmed their minds are in for a surprise.

"It's not that they are lying. They just haven't looked closely enough for a side effect," Abraham says.

several years ago. Bell-bottoms were the rage in this year's spring fashion shows. The Doors, the innovative '60s rock band with outrageous front man and drug addict Jim Morrison, was the subject of a 1991 Oliver Stone film.

"The rise in LSD and other hallucinogen use is part of the '60s psychedelia culture being heavily marketed once again to young people," says Marsha Keith Schuchard, M.D., a PRIDE researcher.

Unlike marijuana and cocaine, LSD is hard to spot because students can carry hits around in textbooks or send it to friends inside greeting cards.

Drug-sniffing dogs have a hard time detecting it, so suppliers traffic it to street dealers via the U.S. Mail.

"If you're in California and are considering sending 5,000 hits of LSD to Pittsburgh, you're spending $1 postage, and within two or three days, it is hand-delivered," says Postal Inspector Andy Weber.

AMERICAN SIGN LANGUAGE

by Richard Wolkomir

In a darkened laboratory at the Salk Institute in San Diego, a deaf woman is signing. Tiny lights attached to her sleeves and fingers trace the motions of her hands, while two special video cameras whir.

Computers will process her hands' videotaped arabesques and pirouettes into mathematically precise three-dimensional images. Neurologists and linguists will study these stunning patterns for insight into how the human brain produces language.

Only in the past 20 years have linguists realized that signed languages are unique—a speech of the hand. They offer a new way to probe how the brain generates and understands language, and they throw new light on an old scientific controversy: whether language, complete with grammar, is innate in our species, or whether it is a learned behavior. The current interest in sign language has roots in the pioneering work of one renegade teacher at Gallaudet University in Washington, D.C., the world's only liberal arts university for deaf people.

When Bill Stokoe went to Gallaudet to teach English, the school enrolled him in a course in signing. But Stokoe noticed something odd: among themselves, students signed differently from his classroom teacher.

"Hand Talk" a Genuine Language

Stokoe had been taught a sort of gestural code, each movement of the hands representing a word in English. At the time, American Sign Language (ASL) was thought to be no more than a form of pidgin English. But Stokoe believed the "hand talk" his students used looked richer. He wondered: Might deaf people actually have a genuine language? And could that language be unlike any other on Earth? It was 1955, when even deaf people dismissed their signing as "slang." Stokoe's idea was academic heresy.

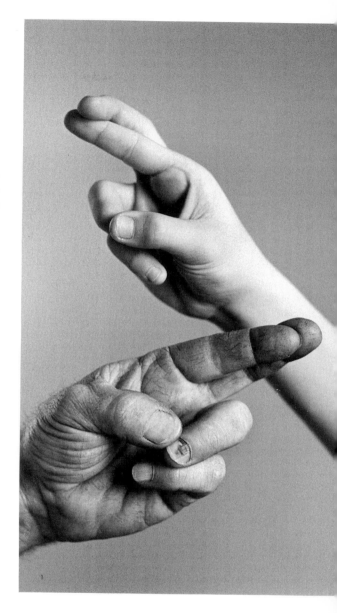

It is 37 years later. Stokoe—now devoting his time to writing and editing books and journals and to producing video materials on ASL and the deaf culture—is having lunch at a cafe near the Gallaudet campus and explaining how he started a revolution. For decades, educators fought his idea that signed languages are natural languages like English, French, and Japanese. They assumed language must be based on speech, the modulation of sound. But sign language is

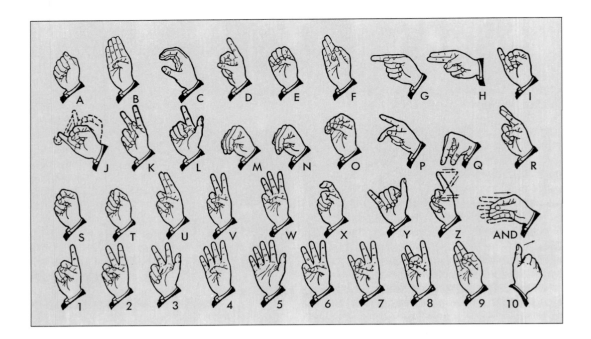

based on the movement of hands, the modulation of space. "What I said," Stokoe explains, "is that language is not mouth stuff —it's brain stuff."

It has been a long road from the mouth to the brain. Linguists have had to redefine language. Deaf people's self-esteem has been at stake, and so has the ticklish issue of their education.

"My own contribution was to turn around the thinking of academics," says Stokoe. "When I came to Gallaudet, the teachers were trained with two books, and the jokers who wrote them gave only a paragraph to sign language, calling it a vague system of gestures that looked like the ideas they were supposed to represent."

Deaf education in the 1950s irked him. "I didn't like to see how the hearing teachers treated their deaf pupils—their expectations were low," he says. "I was amazed at how many of my students were brilliant." Meanwhile, he was reading the work of anthropological linguists like George Trager and Henry Lee Smith, Jr. "They said you couldn't study language without studying the culture, and when I had been at Gallaudet a short time, I realized that deaf people had a culture of their own."

When Stokoe analyzed his students' signing, he found it was like spoken languages, which combine bits of sound—each meaningless by itself—into meaningful words. Signers, following similar rules, combine individually meaningless hand and body movements into words. They choose from a palette of hand shapes, such as a fist or a pointing index finger. They also choose where to make a sign; for example, on the face or on the chest. They choose how to

The "Total Communication" policy at Western Pennsylvania School for the Deaf strives to make children proficient in both speech and sign language.

orient the hand and arm. And each sign has a movement—it might begin at the cheek and finish at the chin. A shaped hand executing a particular motion creates a word. A common underlying structure of both spoken and signed language is thus at the level of the smallest units used to form words.

Stokoe explained his findings on the structure of ASL in a book published in 1960. "The faculty then had a special meeting, and I got up and said my piece," he says. "Nobody threw eggs or old vegetables, but I was bombarded by hostility." Later the university's president told Stokoe his research was "causing too much trouble" because his insistence that ASL was indeed a *language* threatened the English-based system for teaching the deaf. But Stokoe persisted. Five years later he came out with the first dictionary of American Sign Language based on linguistic principles. And he's been slowly winning converts ever since.

The History of Sign

Just as no one can pinpoint the origins of spoken language in prehistory, the roots of sign language remain hidden from view. What linguists do know is that sign languages have sprung up independently in many different places. Signing probably began with simple gestures, but then evolved into a true language with structured grammar. "In every place we've ever found deaf people, there's sign," says anthropological linguist Bob Johnson. But it's not the same language. "I went to a Mayan village where, out of 400 people, 13 were deaf, and they had their own Mayan Sign—I'd guess it's been maintained for thousands of years." Today at least 50 native sign languages are "spoken" worldwide, all mutually incomprehensible.

Not until the 1700s, in France, did people who could hear pay serious attention to deaf people and their language. Religion had something to do with it. "They believed that without speech, you couldn't go to Heaven," says Johnson.

For the Abbé de l'Epée, a French priest born into a wealthy family in 1712, the issue was his own soul: he feared he would lose it unless he overcame the stigma of his privi-

leged youth by devoting himself to the poor. In his history of the deaf, *When The Mind Hears,* Northeastern University psychologist Harlan Lane notes that, in his 50s, de l'Epée met two deaf girls on one of his forays into the Paris slums and decided to dedicate himself to their education.

The priest's problem was abstraction: he could show the girls a piece of bread and the printed French word for "bread." But how could he show them "God" or "goodness"? He decided to learn their sign language as a teaching medium. However, he attempted to impose French grammar onto the signs.

"Methodical signing," as de l'Epée called his invention, was an ugly hybrid. But

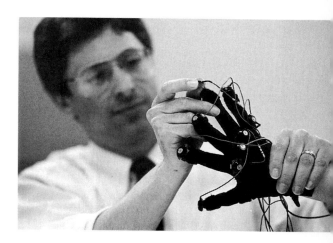

In order to study the nuances of sign language, researchers fit the signer with a fiber-optic glove that records the hand movements on computer.

he did teach his pupils to read French, opening the door to education, and today he is a hero to deaf people. As his pupils and disciples proliferated, satellite schools sprouted throughout Europe. De l'Epée died happily destitute in 1789, surrounded by his students in his Paris school, which became the National Institution for Deaf-Mutes under the new republic.

Other teachers kept de l'Epée's school alive. And one graduate, Laurent Clerc, brought the French method of teaching in sign to the United States. It was the early

1800s; in Hartford, Connecticut, the Reverend Thomas Hopkins Gallaudet was watching children at play. He noticed that one girl, Alice Cogswell, did not join in. She was deaf. Her father, a surgeon, persuaded Gallaudet to find a European teacher and create the first permanent school for the deaf in the United States. Gallaudet then traveled to England, where the "oral" method was supreme, the idea being to teach deaf children to speak. The method was almost cruel, since children born deaf—they heard no voices, including their own—could have no concept of speech. It rarely worked. Besides, the teachers said their method was "secret." And so Gallaudet visited the Institution for Deaf-Mutes in Paris and persuaded Laurent Clerc to come home with him.

During their 52-day voyage across the Atlantic, Gallaudet helped Clerc improve his English, and Clerc taught him French Sign Language. On April 15, 1817, in Hartford, they established a school that became the American School for the Deaf. Teaching in French Sign Language and a version of de

l'Epée's methodical sign, Clerc trained many students who became teachers, who further helped spread the language.

Clerc's French Sign was to mingle with various "home" signs that had sprung up in other places. On Martha's Vineyard, Massachusetts, for example, a large portion of the population was genetically deaf, and virtually all the islanders used an indigenous sign language, the hearing switching back and forth between speech and sign with bilingual ease. Eventually pure French Sign would blend with such local argots and evolve into today's American Sign Language.

Signers vs. Oralists

After Clerc died in 1869, much of the work done since the time of de l'Epée to teach the deaf in their own language crumbled under the weight of Victorian intolerance. Anti-Signers argued that ASL let the deaf "talk" only to the deaf; they must learn to speak and to lip-read. Pro-Signers pointed out that through sign the deaf learned to read and write English. The pros also noted that lip-reading is a skill that few master. (Studies estimate that 93 percent of deaf school-children who were either born deaf or lost their hearing in early childhood can lip-read only one in 10 everyday sentences in English.) And pro-signers argue that the time required to teach a deaf child to mimic speech should be spent on real education.

"Oralists" like Horace Mann lobbied to stop schools from teaching in ASL, then the method of instruction in all schools for the deaf. None was more fervent than Alexander Graham Bell, inventor of the telephone and husband of a woman who denied her own deafness. The president of the National Association of the Deaf called Bell the "most to be feared enemy of the American deaf." In 1880, at an international meeting of educators of the deaf in Milan, where deaf teachers were absent, the use of sign language in schools was proscribed.

After that, as deaf people see it, came the Dark Ages. Retired Gallaudet sociolinguist Barbara Kannapell, who is co-founder of Deafpride, a Washington, D.C., advocacy group, is the deaf daughter of deaf parents from Kentucky. Starting at age 4, she at-

At a pep rally prior to a Gallaudet University sporting event, cheerleaders generate audience participation by signing their cheers to the all-deaf student body.

The football huddle originated at Gallaudet in the 1890s so that players could hide their signs from the other team. Players know that when they feel the vibrations of a drumbeat (right), it's time to start the next play.

tended an "oral" school, where signing was outlawed. "Whenever the teacher turned her back to work on the blackboard, we'd sign," signs Kannapell. "If the teacher caught us, she'd use a ruler on our hands."

Kannapell has tried to see oralism from the viewpoint of hearing parents of deaf children. "They'll do anything to make their child like themselves," she signs. "But, from a deaf adult's perspective, I want them to learn sign, to communicate with their child."

In the 1970s a new federal law mandated "mainstreaming." "Parents could thus keep children home instead of sending them off to special boarding schools, but many public schools didn't know what to do with deaf kids," signs Kannapell. "Many of these children think they're the only deaf kids in the world."

Gallaudet's admissions director, James Tucker, an exuberant 32-year-old, is a product of the 1970s mainstreaming. "I'd sit in the back, doing work the teacher gave me

and minding my own business," he signs. "Did I like it? Hell no! I was lonely—for years I thought I was an introvert." Deaf children have a right to learn ASL and to live in an ASL-speaking community, he asserts. "We learn sign for obvious reasons—our eyes aren't broken," he signs. Tucker adds: "Deaf culture is a group of people sharing similar values, outlook, and frustrations, and the main thing, of course, is sharing the same language."

Today most teachers of deaf pupils are "hearies" who speak as they sign. "Simultaneous Communication," as it is called, is really signed English, and not ASL. "It looks grotesque to the eye," signs Tucker, adding that it makes signs too "marked," a linguistic term meaning equally stressed. Hand movements can be exaggerated or poorly executed. As Tucker puts it: "We have zealous educators trying to impose weird hand shapes." Moreover, since the languages have entirely different sentence structures, the effect can be bewildering. It's like having Japanese spoken to English-speaking students with an interpreter shouting occasional English words at them.

ASL in Infancy
New scientific findings support the efforts of linguists such as Bob Johnson, who are calling for an education system for deaf students based on ASL, starting in infancy. Research by Helen Neville at the Salk Institute shows that children must learn a language—any language—during their first five years or so, before the brain's neural connections are locked in place, or risk permanent linguistic impairment. "What suffers is the ability to learn grammar," she says. As children mature, their brain organization becomes increasingly rigid. By puberty, it is largely complete. This spells trouble because most deaf youngsters learn language late; their parents are hearing and do not know ASL, and the children have little or no contact with deaf people when young.

Bob Johnson notes that more than 90 percent of all deaf children have hearing parents. Unlike deaf children of deaf parents, who get ASL instruction early, they learn a language late and lag educationally. "The average deaf 12th grader reads at the 4th-grade level," says Johnson. He believes deaf children should start learning ASL in the crib, with schools teaching in ASL. English, he argues, should be a second language, for reading and writing: "All evidence says they'll learn English better." It's been an up-

hill battle. Of the several hundred school programs for the deaf in this country, only six are moving toward ASL-based instruction. And the vast majority of deaf students are still in mainstream schools where there are few teachers who are fluent in ASL.

Meanwhile, researchers are finding that ASL is a living language, still evolving. Sociolinguist James Woodward from Memphis, Tennessee, who has a black belt in karate, had planned to study Chinese dialects, but switched to sign when he came to Gallaudet in 1969. "I spent every night for two years at the Rathskeller, a student hangout, learning by observing," he says. "I began to see great variation in the way people signed."

Woodward later concentrated on regional, social, and ethnic dialects of ASL. Visiting deaf homes and social clubs in the South, he found that Southerners, particularly Southern blacks, use older forms of ASL signs than Northerners do.

Over time, signs tend to change. For instance, "home" originally was the sign for "eat" (touching the mouth) combined with the sign for "sleep" (the palm pillowing the cheek). Now it has evolved into two taps on the cheek. Also, signs formerly made at the center of the face have migrated toward its perimeter. One reason is that it is easier to see both signs and changes in facial expressions in this way, since deaf people focus on a signer's face—which provides crucial linguistic information—taking in the hands with peripheral vision.

Signers use certain facial expressions as grammatical markers. These linguistic expressions range from pursed lips to the expression that results from enunciating the sound "th." Linguist Scott Liddell at Gallaudet has noted that certain hand movements translate as "Bill drove to John's." If the signer tilts his head forward and raises his eyebrows while signing, he makes the sentence a question: "Did Bill drive to John's?" If he also makes the "th" expression as he signs, he modifies the verb with an adverb: "Did Bill drive to John's inattentively?"

Sociolinguists have investigated why this unique language was for so long virtually a secret. Partly, Woodward thinks, it was because deaf people wanted it that way. He says that when deaf people sign to the hearing, they switch to English-like signing. "It allows hearing people to be identified as outsiders, and to be treated carefully before allowing any interaction that could have a negative effect on the deaf community," he says. By keeping ASL to themselves, deaf people—whom Woodward regards as an ethnic group—maintain "social identity and group solidarity."

A Key Language Ingredient: Grammar

The "secret" nature of ASL is changing rapidly as it is being examined under the scientific microscope. At the Salk Institute, a futuristic complex of concrete labs poised on a San Diego cliff above the Pacific, pioneer ASL investigator Ursula Bellugi directs the Laboratory for Cognitive Neuroscience, where researchers use ASL to probe the brain's capacity for language. It was here that Bellugi and associates found that ASL has a grammar to regulate its flow. For example, a signer might make the sign for "Joe" at an arbitrary spot in space. Now that spot stands for "Joe." By pointing to it, the signer creates the pronoun "he" or "him," meaning "Joe." A sign moving toward the spot means something done *to* "him." A sign moving away from the spot means an action *by* Joe.

In the 1970s Bellugi's team concentrated on several key questions that have been of central concern ever since MIT professor Noam Chomsky's groundbreaking work of the 1950s. Is language capability innate, as Chomsky and his followers believe? Or is it acquired from our environment? The question gets to the basics of humanity, since our language capacity is part of our unique endowment as a species. And language lets us accumulate lore and

A deaf aerobics teacher can effectively lead a class by signing.

pass it on to succeeding generations. Bellugi's team reasoned that if ASL is a true language, unconnected to speech, then our penchant for language must be built in at birth, whether we express it with our tongue or hands. As Bellugi puts it: "I had to keep asking myself, 'What does it mean to be a language?'"

A key issue was "iconicity." Linguistics has long held that one of the properties of all natural languages is that their words are arbitrary. In English, to illustrate, there is no relation between the sound of the word "cat" and a cat itself, and onomatopoeic words like "slurp" are few and far between. Similarly, if ASL follows the same principles, its words should not be pictures or mime. But ASL does have many words with transparent meanings. In ASL, "tree" is an arm upright from the elbow, representing a trunk, with the fingers spread to show the crown. In Danish Sign the signer's two hands outline a tree in the air. Sign languages are rife with pantomimes. But Bellugi wondered: Do deaf people perceive such signs as iconic?

One day a deaf mother visited the lab with her deaf daughter, not yet 2. At that age, hearing children fumble pronouns, which is why parents say, "Mommy is getting Tammy juice." The deaf child, equally confused by pronouns, signed "you" when she meant "I." The mother corrected the child by turning the girl's hand so that she pointed at herself. Nothing could be clearer. Yet the child continued to point to her mother when she meant "I."

Poetry in Sign

Bellugi's work revealed that deaf toddlers have no trouble pointing. But pointing in ASL is linguistic, not gestural. Deaf toddlers in the "don't-understand-pronouns" stage do not see a pointing finger. They see a confusing, abstract word. ASL's roots may be mimetic, but—embedded in the flow of language—the signs lose their iconicity.

By the 1980s most linguists had accepted sign languages as natural languages on an equal footing with English, Italian, Hindi, and others of the world. Signed languages like ASL were as powerful, subtle, and intricately structured as spoken ones.

The parallels become especially striking in wordplay and poetry. Signers creatively combine hand shapes and movements to create puns and other humorous alterations of words. A typical pun in sign goes like this: a fist near the forehead and a flip of the index finger upward means that one understands. But if the little finger is flipped, it's a joke meaning one understands a little. Clayton Valli at Gallaudet has made an extensive study of poetry in ASL. He finds that maintenance or repetition of hand shape provides rhyming, while meter occurs in the timing and type of movement. Research with the American Theater of the Deaf reveals a variety of individual techniques and styles. Some performers create designs in space with a freer movement of the arms than in ordinary signing. With others, rhythm and tempo are more important than spatial considerations. Hands may be alternated so that there is a balance and symmetry in the structure. Or signs may be made to flow into one another, creating a lyricism in the passage. The possibilities for this new art form in sign seem bounded only by the imagination within the community itself.

The special nature of sign language provides unprecedented opportunities to observe how the brain is organized to generate and understand language. Spoken languages are produced by largely unobservable movements of the vocal apparatus, and are received through the brain's auditory system. Signed languages, by contrast, are delivered through highly visible movements of the arms, hands, and face, and are received through the brain's visual system. Engagement of these different brain systems in language use makes it possible to test different ideas about the biological basis of language.

The prevailing view of neurologists is that the brain's left hemisphere is the seat of language, while the right controls our perception of visual space. But since signed languages are expressed spatially, it was unclear where they might be centered.

To find out, Bellugi and her colleagues studied lifelong deaf signers who had suffered brain damage as adults. When the damage had occurred in their left hemisphere, the signers could shrug, point, shake their head, and make other gestures, but they lost the ability to sign. As happens with hearing people who suffer left-hemisphere damage, some of them lost words, while others lost the ability to organize grammatical sentences, depending on precisely where the damage had occurred.

Conversely, signers with right-hemisphere damage signed as well as ever, but spatial arrangements confused them. One of Bellugi's right-hemisphere subjects could no longer perceive things to her left. Asked to describe a room, she reported all the furnishings as being on the right, leaving the room's left side a void. Yet she signed perfectly, including signs formed on the left side. She had lost her sense of *topographic* space, a right-hemisphere function, but retained control of *linguistic* space, centered in the left hemisphere. These findings support the conclusion that language—visual or spoken—is controlled by the left hemisphere.

One of the Salk group's current efforts is to see if learning language in a particular modality changes the brain's ability to perform other kinds of tasks. Researchers showed children a moving light tracing a pattern in space, and then asked them to draw what they saw. "Deaf kids were way ahead of hearing kids," says Bellugi. Other tests, she adds, back up the finding that learning sign language improves the mind's ability to grasp patterns in space.

Thinking and Dreaming in Signs

Salk linguist Karen Emmorey says the lab also has found that deaf people are better at generating and manipulating mental images. "We found a striking difference in ability to generate mental images and to tell if one object is the same as another but rotated in

American Sign Language (ASL) allows families in which all members are deaf (left) to communicate with as much vigor as any hearing family. Deaf children taught completely in ASL (below) face a much brighter educational future than have generations of deaf kids before them.

space, or is a mirror image of the first," she says, noting that signers seem to be better at discriminating between faces, too. "The question is, does the language you know affect your other cognitive abilities?"

Freda Norman, formerly an actress with the National Theater of the Deaf and now a Salk research associate, puts it like this: "English is very linear, but ASL lets you see everything at the same time."

"The deaf *think* in signs," says Bellugi. "They *dream* in signs. And little children sign to themselves."

At McGill University in Montreal, psychologist Laura Ann Petitto recently found that deaf babies of deaf parents babble in sign. Hearing infants create nonsense sounds like "babababa," first attempts at language. So do deaf babies, but with their hands. Petitto watched deaf infants moving their hands and fingers in systematic ways that hearing children not exposed to sign never do. The movements, she says, were their way of exploring the linguistic units that will be the building blocks of language —their language.

Deaf children today face a brighter future than the generation of deaf children before them. Instruction in ASL, particularly in residential schools, should accelerate. New technologies, such as the TDD (Telecommunications Device for the Deaf) for communicating over telephones, relay services and

video programs for language instruction, and the recent Americans with Disabilities Act all point the way to a more supportive environment. Deaf people are moving into professional jobs, such as law and accounting, and more recently into computer-related work. But it is not surprising that outside of their work, they prefer one another's company. Life can be especially rewarding for those within the ASL community. Here they form their own literary clubs, bowling leagues, and gourmet groups.

As the Salk laboratory's Freda Norman signs: "I love to read books, but ASL is my first language." She adds, smiling: "Sometimes I forget that the hearing are different."

Behavioral Sciences

INFANT ARITHMETIC

A psychologist reported that by about five months of age, babies can already add and subtract very small numbers of items. The human brain may contain an innate mathematical ability that operates in the absence of explicit counting and helps infants to categorize things in the world, asserted Karen Wynn of the University of Arizona in Tucson.

The finding extends previous indications that infants realize when a small number of drumbeats matches an equal number of objects shown on a slide, and also notice changes in the number of a set of items.

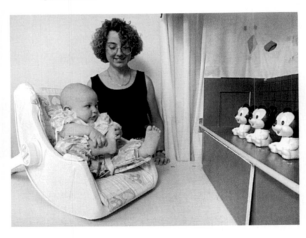

Humans may have an innate mathematical ability from infancy that enables them to use simple reasoning to categorize objects.

Wynn's experiments relied on the well-established tendency of babies to look longer at new or unexpected objects than at familiar objects.

One group of five-month-olds watched an adult place a rubber doll on a table, cover it with a screen, and then place a second doll behind the screen. A second group saw two dolls placed on a table and then hidden by a screen, followed by removal of one doll. The screen was then lifted to show either the correct number of dolls or an incorrect number corresponding to "$1 + 1 = 1$" or "$2 - 1 = 2$."

Babies looked much longer at the incorrect number of dolls, signifying that they expected to see a different number. On another test employing rubber dolls, infants looked longer at the toys when the trial corresponded to "$1 + 1 = 3$" rather than "$1 + 1 = 2$."

Infants undoubtedly cannot deal with amounts much bigger than those in the doll experiments, Wynn said. But they still make use of a simple form of mathematical reasoning, she contended.

TEENAGE SUICIDE ATTEMPTS

A survey of high school students in South Carolina concludes that suicide-prevention efforts may need to concentrate not only on depressed teenagers, but on highly aggressive and alcohol-abusing adolescents as well.

Carol Z. Garrison, an epidemiologist at the University of South Carolina in Columbia, directed the analysis of questionnaires completed by 3,764 youngsters in grades 9 through 12. The students attended public schools throughout South Carolina.

Three-quarters of this group reported no thoughts of or attempts at suicide. About 1 in 10 students cited serious thoughts of suicide, more than 6 percent acknowledged having made a specific plan to kill themselves, and 7.5 percent—nearly 1 in 12 students—reported having made a suicide attempt.

Teens who engaged in the most aggressive behavior, such as getting in fights and carrying weapons, stood the greatest chance of thinking about, planning for, or attempting suicide. Those students who smoked cigarettes or used alcohol or illicit drugs displayed a weaker, but statistically significant, association with suicide thoughts and attempts.

Highly aggressive teenagers may more often act on suicidal thoughts and plans when depressed, frustrated, or scared, Garrison's team suggests. However, the number of severely depressed students in the sample was not established.

At this point, the researchers say, the findings justify an expansion of suicide-prevention efforts from a focus solely on depressed teens to including highly aggressive and alcohol-abusing youngsters.

DEPRESSION RATES RISE

The number of people suffering from severe depression has increased from one generation to the next since 1915, according to researchers who organized the first international study of how depression rates have changed over time.

The study brought together information on approximately 43,000 people surveyed in cities in Canada, France, Germany, Italy, Lebanon, New Zealand, Puerto Rico, Taiwan, and the United States.

In each country the frequency of depression almost always increased in succeedingly younger generations, peaking among those born after 1955. One exception occurred in Los Angeles, where older and younger Hispanic individuals suffered about the same rate of depression.

Large peaks and valleys in depression appeared among some residents of Beirut, Lebanon. The 1950s and 1970s witnessed surges in depression among young adults in their 20s who had to contend with long periods of warfare and social chaos. Fewer cases of depression among young adults occurred during the relative calm of the 1960s.

CIGARETTES AND MEMORY

A recent controversial study suggests that cigarette smoking may interfere in some ways with memory and judgment, perhaps because of the still poorly understood effects of nicotine on the brain. Smoking seems most likely to lower the amount of relevant information a person can recall when confronted with an especially complicated task, such as navigating a car on a wet road after getting a flat

Studies suggest that cigarette smoking may affect a driver's judgment during such complicated tasks as emergency car maneuvers.

tire, said psychologist George J. Spilich of Washington College in Chestertown, Maryland.

Spilich studied regular smokers who had either just smoked a cigarette or who had abstained for three hours, as well as nonsmokers.

All three groups did equally well on basic tests of perception, such as picking out the letter X from an array of other letters. But both groups of smokers, and particularly those who had just smoked, performed more poorly on a test of reading comprehension and had more rear-end collisions while operating a computerized driving simulator.

Some researchers contest these findings, citing much evidence for modest boosts in attention and memory among smokers performing relatively simple tasks, such as remembering briefly studied words. Nicotine may spark these improvements by increasing the flow of a critical chemical messenger in the brain. Critics also charge that a general tendency to take risks in all situations, rather than nicotine effects, makes smokers more accident-prone while driving.

Others theorize that nicotine, like other addictive drugs, enhances mental functioning at low doses and on basic tasks, but has negative effects at high doses and on challenging endeavors.

Bruce Bower

Food and Population

The world supply of cereal grains, the basic staple of most diets, is estimated by the United Nations (U.N.) Food and Agriculture Organization (FAO) to have increased by 3 percent in 1992. This outstripped the normal population increase (about 93 million), and thus ensured that world stocks of food would be adequate on the whole, and would permit some replenishment of reserves drawn down during the previous year.

Distribution, too, continues to be a problem, because the growth in output took place, as usual, mainly in the industrialized exporting countries, on whose surpluses the less productive regions continue to depend. In the major exporting countries, stocks rose by 13 percent; in the developing countries, they declined by 3 percent. Prices declined slightly for most cereals, but remained higher than many people could afford.

Some regional shortages have been noted. Southern Africa's drought persisted through the year, continuing the food shortages that plague the continent.

There was acute distress in the Horn of Africa—in Somalia and the southern Sudan in particular. In other parts of Africa (e.g., Liberia) and elsewhere in the world where conflict rages (e.g., former Yugoslavia) or has not been fully resolved (e.g., Iraq), malnutrition remains a threat.

About 20 million people are directly affected by the kind of hunger that appears newsworthy; but the number suffering the chronic hunger that grows out of poverty and underdevelopment is more like 600 million. They are the victims of a global economic system that does not permit them to grow or purchase the food they need. It is generally acknowledged that without a sharp increase in food production in the countries in which they reside, their situation is unlikely to improve in the near term.

Hunger and malnutrition have not spared the industrialized countries. As a result of corporate downsizing, relocation of plants to low-wage regions and countries, declining wage levels, and a sharp increase in part-time employment, requests for emergency food assistance in major U.S. cities increased by 26 percent in 1991. In early 1992 the number of people relying on food stamps and AFDC (Aid to Families with Dependent Children) reached a record.

Martin M. McLaughlin

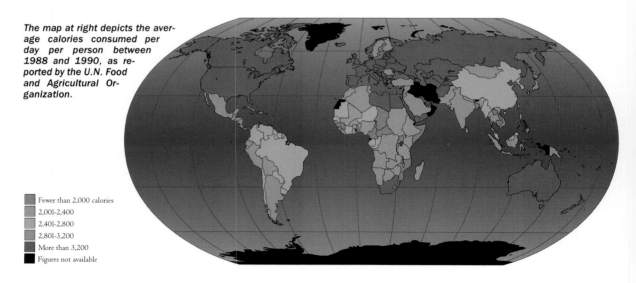

The map at right depicts the average calories consumed per day per person between 1988 and 1990, as reported by the U.N. Food and Agricultural Organization.

Fewer than 2,000 calories
2,001-2,400
2,401-2,800
2,801-3,200
More than 3,200
Figures not available

Genetics

GENE THERAPY

The ability to transfer genes from one organism to another, a process called *genetic engineering,* has led to a number of gene-therapy procedures aimed at combating various diseases. The first such procedure began on September 14, 1990, and involved adding a normally functioning gene for the enzyme adenosine deaminase (ADA) to the lymphocytes of a child whose cells could not produce the enzyme. The lack of this enzyme leaves an individual with virtually no natural immune defenses, and results in a situation in which even a minor infection becomes life-threatening. More than three years have passed, and this child, along with another who has undergone the same medical program, is alive and doing very well.

A Cold Virus to Carry a Missing Gene

Cystic fibrosis patients lack a gene for a protein to control salt flow in lung cells; mucus builds up and infections destroy tissue. Scientists will try a replacement method that worked in animals.

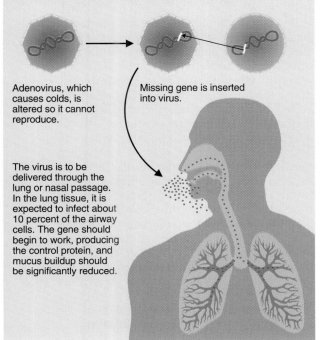

Adenovirus, which causes colds, is altered so it cannot reproduce.

Missing gene is inserted into virus.

The virus is to be delivered through the lung or nasal passage. In the lung tissue, it is expected to infect about 10 percent of the airway cells. The gene should begin to work, producing the control protein, and mucus buildup should be significantly reduced.

Since 1990, 37 gene-transfer procedures have been approved worldwide, involving a variety of diseases. One target disease is *melanoma,* a malignant skin cancer. Using gene therapy, a gene that produces a cell-membrane protein that attracts T cells will be introduced into the cancer cells. The subsequently produced cell-membrane protein will attract T cells from the patient's immune system, which will destroy the malignant melanoma cells.

Another disease that has been targeted for gene therapy is *cystic fibrosis,* which results from the lack of production of a specific cell protein. In this procedure a fully functional gene is introduced into the upper-respiratory-tract cells of the patient to produce the necessary protein.

The disease *non-small cell lung carcinoma* is a cancer often caused by a mutation in the K-*ras* gene of the patient. The mutation causes the gene to become overactive, resulting in an abnormal increase in cell growth and division of the lung tissue. In this case a gene is introduced into the cancerous cells to suppress the activity of the mutated K-*ras* gene, thereby stopping the growth of the tumor.

GENE-ACTIVATION THERAPY

Each red blood cell has about 200 million hemoglobin molecules. During embryonic development, each hemoglobin molecule consists of four polypeptide chains —two *alpha* and two *gamma* chains— and is referred to as *fetal hemoglobin.* Shortly before birth the gamma gene gradually ceases to function, and a different gene becomes active, producing a *beta*-type chain. Thereafter, each hemoglobin molecule consists of two alpha and two beta chains, and is referred to as *adult hemoglobin.*

There are two severe genetic diseases that are the result of alterations of the beta chains—*sickle-cell anemia* (prevalent in certain African populations) and *beta-thalassemia* (prevalent in certain Mediterranean populations). In both cases, doctors are hoping that a medical procedure that activates the gamma gene would lead to the production of fetal hemoglobin, resulting in a reduction in the severity of the particular disease.

In January 1993, a group of medical geneticists led by S. S. Perrine, M.D., of the Childrens Hospital Research Institute in Oakland, California, reported that three severely ill patients suffering from sickle-cell anemia and three suffering from beta-thalassemia underwent a hospital gene-activation treatment program during which the compound *arginine butyrate,* a gamma-gene-activating fatty acid, was continuously infused intravenously for periods of two to three weeks. All six patients experienced a significant production of fetal hemoglobin and a dramatic reduction in the severity of their diseases, without any noticeable deleterious side effects.

If these beneficial results are successfully repeated using larger groups of patients, it will mean that *gene-activation therapy* can be added to the growing repertoire of medical procedures available for genetic diseases.

PREIMPLANTATION GENETIC DIAGNOSIS

In September 1992, A. H. Handyside, M.D., and his colleagues at Hammersmith Hospital in England reported the birth of a normal baby girl to parents who were both carriers of the recessive gene for *cystic fibrosis.* This event followed a unique procedure that involved obtaining an egg from the mother and fertilizing it in the laboratory, using the father's sperm. The developing embryo was then cultured in the laboratory for three days to reach the eight-cell stage. One of the cells was then removed (removal of a single cell at this stage has no effect on an embryo's development). The removed cell was examined for the presence of the cystic fibrosis gene using *polymerase chain reaction* (PCR), a procedure that yields a large number of copies of any specified section of the cell's DNA. The embryo was found to be free of the cystic-fibrosis gene. The embryo was then transferred to the mother's uterus, with the subsequent birth of a normal child.

The birth of this normal baby girl demonstrated that it is possible to analyze an embryo for the presence of a deleterious gene before implantation, thereby avoiding the grief and hardship that follows the birth of a child doomed to chronic illness and early death.

PROGRAMMED CELL DEATH

During the normal embryonic development of all complex organisms, there is a pattern of cell death. The cells that die include those that served a transient purpose and those that are in excess of the number needed. This type of *programmed cell death,* or *apoptosis,* involves all systems of the body. While necessary for embryonic development, apoptosis has also become a focus of cancer research, where there is the possibility that some cancers may be the result of interference with the normal cell-death process.

Experimentation involving apoptosis has focused on a number of oncogenes associated with various types of tumors. One such gene is called *bcl-2,* discovered in lymphoma tumors composed of B cells. In December 1992, D. L. Vaux, M.D., and colleagues from Stanford University Medical Center in California published a study on how *bcl-2* operates through a suppression of apoptosis. The study centered on the roundworm *Caenorhabditis elegans,* which during its lifetime produces exactly 1,090 cells. Each one of the cells has been accurately traced from its formation to maturity or death. Of the 1,090 cells produced, 131 are known to die in a programmed way as the worm matures. The scientists transferred the human *bcl-2* gene to the roundworm, and found that the cells that normally would have died in fact continued to live.

This discovery opens the possibility that the control of some cancers may depend on therapeutic drugs or gene-therapy procedures that induce apoptosis. In the case of *bcl-2*-induced tumors, scientists hope to develop a drug or gene-therapy procedure that inhibits the action of the gene itself or its protein end product.

Louis Levine

Health and Disease

IMPEDING CANCER'S DEVELOPMENT

Evidence suggests that fruits and vegetables play important roles in impeding or preventing cancer. Now researchers are zeroing in on the mechanisms involved. Some plant-food compounds are believed to absorb or break down carcinogens; others may induce the body to produce protective enzymes.

Still another cancer-fighting property of plant foods was reported in 1993 by scientists at Children's University Hospital in Heidelberg, Germany. They isolated a compound called genistein from the urine of people who ate a traditional Japanese diet, which contains little meat, but large amounts of soybeans and other vegetables. Indeed, genistein levels in these individuals are at least 30 times greater than those of people who eat a typical Western diet. Experiments indicate that genistein inhibits the formation of new blood vessels. In order to grow and spread, a tumor requires new blood vessels, which can bring it food and oxygen. Without the vessels, scientists suggest, tumors may remain tiny and harmless.

TREATING CARDIOVASCULAR DISORDERS

A relatively new class of heart drugs called ACE inhibitors reduces the risk of death and improves the quality of life among people who have survived heart attacks. The drugs counteract angiotensin-converting enzyme, a protein that causes blood vessels to tighten and narrow. By keeping the blood vessels relaxed, the ACE inhibitors make it easier for the heart to pump blood throughout the body.

The Survival and Ventricular Enlargement Trial tracked 2,231 adults whose hearts had been permanently damaged by heart attacks. Half of the patients received the ACE inhibitor captopril; the others were given a placebo. The re-

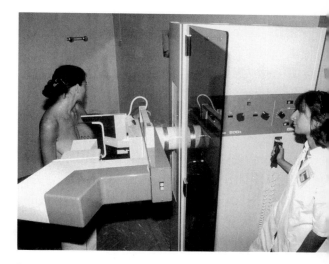

Despite a 1992 study that failed to show that women under 40 benefited from mammograms, the American Cancer Society refused to alter guidelines for younger women until further definitive studies are performed.

searchers found that the captopril reduced the incidence of new heart attacks by 25 percent, and the death rate from cardiovascular disease by 20 percent.

The American College of Chest Physicians urges people with atrial fibrillation to discuss with their physicians the advisability of drug therapy. Atrial fibrillation occurs when the upper chambers of the heart vibrate instead of pumping effectively. This allows the blood to form clots, which may travel to the brain and cause strokes. Researchers have found that low doses of blood-thinning drugs can reduce the risk of stroke by as much as 80 percent.

HEPATITIS A, B, C, AND E

One of today's most serious health problems is viral hepatitis, a sometimes fatal liver disease. Five forms of hepatitis are known, each caused by a different virus.

In 1992 a vaccine developed by Merck Sharp & Dohme was shown to be effective in protecting children against hepatitis A, a disease spread mainly through fecal contamination. The vaccine, made from inactivated hepatitis A viruses, was given to 519 children in a New York community that had been plagued by hepatitis A infections, another 518 children in the community were given placebos. None of the vaccinated children developed hepatitis A, but 25 cases occurred among those who received placebos.

It also was announced, by Genelabs Technologies and SmithKline Beecham, that a vaccine had been developed for the prevention of hepatitis E. The virus that causes hepatitis E was first identified in 1988 by Genelabs researchers.

Vaccines for hepatitis B became available in the 1980s, but they have not been widely used, even though 200,000 to 300,000 cases of hepatitis B occur in the United States each year. The U.S. Centers for Disease Control and Prevention (CDC) recommends that all infants be vaccinated against the disease, but many pediatricians disagree with this procedure. They cite the high cost of the vaccine, plus the fact that hepatitis B usually is acquired in the late teens or in early adulthood; it is not known if vaccines given to infants will remain effective for 15 years or more. CDC officials point out that 8 percent of hepatitis B infections are acquired before age 10, and children are three times as likely as adults to develop a fatal form of the disease.

No vaccine is yet available for hepatitis C, which is believed to be the most common form of viral hepatitis in the United States, and a major cause of cirrhosis and liver cancer. The only treatment for the disease is costly, has unpleasant side effects, and isn't always effective. Nor is it known whether the treatment actually cures a patient. Studies published in 1992 suggest that, in the words of one of the researchers, "once you get it, you can't get rid of it." The hepatitis C virus apparently can remain in a person's blood for years, causing no visible symptoms or damage, but making the person infectious.

NEW AIDS DEFINITION

The number of reported AIDS cases rose sharply in the United States in early 1993, reflecting a new definition of the disease. The CDC expanded the list of AIDS "indicator diseases" to include pulmonary tuberculosis, recurrent bacterial pneumonia, and invasive cancer of the cervix. These three illnesses often are fatal to women and intravenous drug users infected with the human immunodeficiency virus (HIV), which causes AIDS. The AIDS definition also was revised to include HIV-infected people with CD-4 cell counts lower than 200 per cubic millimeter of blood (healthy people have 800 to 1,200 CD-4 cells per cubic millimeter). Of the 12,243 new AIDS cases reported in

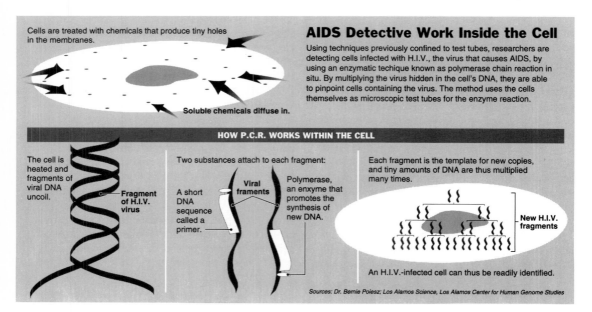

AIDS Detective Work Inside the Cell

Using techniques previously confined to test tubes, researchers are detecting cells infected with H.I.V., the virus that causes AIDS, by using an enzymatic techique known as polymerase chain reaction in situ. By multiplying the virus hidden in the cell's DNA, they are able to pinpoint cells containing the virus. The method uses the cells themselves as microscopic test tubes for the enzyme reaction.

Cells are treated with chemicals that produce tiny holes in the membranes.

Soluble chemicals diffuse in.

HOW P.C.R. WORKS WITHIN THE CELL

The cell is heated and fragments of viral DNA uncoil.

Fragment of H.I.V. virus

Two substances attach to each fragment:

A short DNA sequence called a primer.

Viral framents

Polymerase, an enxyme that promotes the synthesis of new DNA.

Each fragment is the template for new copies, and tiny amounts of DNA are thus multiplied many times.

New H.I.V. fragments

An H.I.V.-infected cell can thus be readily identified.

Sources: Dr. Bernie Poiesz; Los Alamos Science, Los Alamos Center for Human Genome Studies

the first two months of 1993, 41 percent fell under the new definition, almost all of them because of low CD-4 counts.

An HIV-infected person typically harbors the viruses within the body for about 10 years before developing symptoms of AIDS. Scientists at the National Institute of Allergy and Infectious Diseases and at the University of Minnesota reported in 1993 that the HIV hides in the person's lymph nodes during the long latency period. This discovery may enable researchers to develop methods to attack the HIV during this stage, before it appears in large numbers in the blood.

KILLING BACTERIA TO CURE ULCERS

Research at the Baylor College of Medicine provides strong support for the theory that most gastric and duodenal (stomach and upper intestinal) ulcers are caused by the bacterium *Helicobacter pylori*. Some ulcer patients infected with *H. pylori* were given the drug Zantac, a standard treatment that suppresses the production of stomach acid. Other patients were given a regimen of four drugs, including two antibiotics to kill the bacteria. Among patients with gastric ulcers who received the new therapy, 95 percent suffered no recurrence during the following two years; in contrast, only 12 percent of those receiving the standard treatment were ulcer-free in that period. Among patients with duodenal ulcers, 74 percent of those receiving the new therapy had no recurrence, versus 13 percent of those receiving the standard therapy.

THE EFFECTS OF LEAD

The CDC considers lead exposure to be the country's number one preventable pediatric-health problem. Exposure to even low levels of lead early in life impairs physical and mental development. Australian researchers report that lead poisoning was as likely to afflict middle-class children as poor children.

Major sources of lead include paints, ceramics, soil, and water. Lead in drinking water is estimated to contribute between 10 and 20 percent of total lead exposure in young children. The U.S. Environmental Protection Agency (EPA) reports that lead levels in 130 of the nation's 660 large public water systems exceed the federal guideline of 15 parts per billion (ppb) during 1992. The highest levels were reported in homes in Charleston, South Carolina (211 ppb); Pensacola, Florida (175 ppb); and Newton, Massachusetts (163 ppb).

A study of 154 children suggests that it may be possible to at least partially reverse the effects of lead. A research team headed by Holly A. Ruff, a psychologist at Albert Einstein College of Medicine in New York, reported in 1993 that over a six-month period, children with moderate lead poisoning improved their intelligence-test scores as the amount of lead in their blood was reduced.

EXPLAINING OSTEOPOROSIS

Osteoporosis is a bone-weakening disease that afflicts some 20 million people in the United States. Although the disease is most apparent among women age 60 and older, it actually begins much earlier, as the women begin to enter menopause and their production of the hormone estrogen begins to decline. Stavros Manolagas, a scientist at the Indiana University School of Medicine, reports that studies with mice indicate that falling estrogen levels result in a series of chemical changes that disturb the balance between bone formation and bone destruction.

Henry Lukaski, a physiologist at the Grand Forks Human Nutrition Research Center, finds that women who lose weight also lose bone mass, even if they exercise regularly and consume adequate amounts of calcium. Lukaski studied 14 overweight women between the ages of 20 and 40. Over a period of five months, the women lost an average of 18 pounds (8 kilograms), and 2 to 3 percent of their bone mass. This finding is particularly disturbing because many women who diet do not exercise regularly or consume sufficient calcium, and thus may lose even more bone mass.

Jenny Tesar

Physiology or Medicine

Two American biochemists received the 1992 Nobel Prize in Physiology or Medicine for their insights into a fundamental process that regulates the functions of cells. The Nobel committee honored Edwin G. Krebs, M.D. and Edmond H. Fischer, Ph.D., colleagues for nearly 40 years at the University of Washington School of Medicine, Seattle, for their discoveries concerning reversible protein phosphorylation. This process is an essential mechanism that controls the interactions of proteins within a cell, as in the response of the immune system to infection, as well as in the formation of certain kinds of cancer.

PROTEINS: THE "LIVING TOOLS" OF THE ORGANISM

The human body contains several trillion cells. Each cell contains about 10,000 different types of proteins, which are constructed of links of amino-acid residues shaped into three-dimensional structures. Proteins carry out most cellular functions, such as maintaining the cell's metabolism, dictating growth, releasing hormones, and mediating the work of muscles. In fact, it was through their experiments on muscle tissue that Drs. Krebs and Fischer made their important discoveries. In the late 1940s, Dr. Krebs joined the laboratory of Carl and Gerty Cori, a husband-and-wife team of researchers at the Washington University School of Medicine in St. Louis. The Coris had won the Nobel Prize in Physiology or Medicine in 1947 for their discovery of an enzyme known as phosphorylase. Enzymes are catalysts—proteins whose role is to make biological reactions possible. The enzyme phosphorylase plays a crucial role in breaking down glycogen—the body's principal energy-storage compound—into glucose, which

is the sugar that muscle cells use for energy when they contract. The Coris knew that phosphorylase existed in both active and inactive forms, but they were not sure how these two forms differed. The Coris decided to drop the problem into Dr. Krebs's lap and turn their research in a different direction.

Dr. Fischer, meanwhile, had been performing research in Switzerland, studying a plant version of phosphorylase. In 1953 Fischer and Krebs joined forces at the University of Washington to try to elucidate the different forms of phosphorylase and the factors controlling their activity. The two researchers soon determined the biochemical mechanism underlying the conversion of phosphorylase from an inactive to an active form. The change, they observed, is brought about by the addition of a phosphate group, which is transferred to the protein from an energy-rich compound known as adenosine triphosphate, or ATP. This phosphate transfer, or phosphorylation, alters the biochemical properties, as well as the function, of a protein.

Drs. Fischer and Krebs also discovered that the transfer could go in both directions. Removal of the phosphate group returns the protein to its inactive state. Thus, the two-way process—the revers-

Dr. Edwin G. Krebs (left) and Dr. Edmond H. Fischer shared the 1992 Nobel Prize in Physiology or Medicine for their discovery of a key cellular mechanism that affects virtually all cells in the human body.

ible protein phosphorylation—constitutes a kind of "on-off" switch that starts and stops a variety of cellular functions. In the case of muscle tissue, the phosphate transfer activates the protein and causes the muscle to contract.

The next step was to identify the enzymes responsible for carrying out this phosphate transfer. Fischer and Krebs subsequently isolated and characterized the first such enzyme, known as a protein kinase. The two researchers also identified the enzymes that remove the phosphate group and thereby inactivate proteins. These enzymes are called phosphatases. Subsequent to the key work by the two Nobel laureates in the 1950s and '60s, hundreds of other protein kinases have been identified.

APPLICATIONS

The mechanism of reversible protein phosphorylation has a wide range of effects on the cell. For example, the protein kinase that was first identified by Krebs and Fischer turned out to be essential in the cellular response to various hormones, such as epinephrine. This hormone releases glucose to provide energy for an animal's "fight-or-flight" reaction in times of stress. In addition to muscle contraction and hormone response, phosphorylation also influences such cellular functions as protein synthesis, gene regulation, and neurotransmitter release. These effects are manifest in blood pressure, in inflammatory reactions, and in the transfer of signals in the brain—to name just a small selection.

INSIGHTS INTO IMMUNOLOGY AND CANCER

The response of the human immune system is another process in which the work of Drs. Fischer and Krebs has provided key insights. When an organism invades the body, part of the immune system's response is to invoke a chain reaction involving a cascade of phosphorylating and dephosphorylating enzymes. As a result the body marshals the proper cellular defenses to neutralize the invader.

When the body is fighting a viral or bacterial intruder, this response is desirable. However, a newly transplanted organ will also rally the body's immune response, causing rejection of the transplanted tissue. In such cases, doctors use drugs to suppress this immune response, giving the new tissue time to be accepted. One such immunosuppressant drug that has been used with great success in transplants is cyclosporine. It works by interfering with the phosphorylation reaction in the body's immune response.

Cellular growth is another function in which protein phosphorylation is essential. Unfortunately, however, phosphorylation can also contribute to the abnormal growth of cells—the process known as cancer. The nuclear DNA within the cell contains approximately 100 oncogenes. These are genes that normally control cell growth, but, when altered or mutated, actually promote the growth of tumors. Roughly half of these cancer-causing oncogenes are now known to encode protein kinases. Some forms of cancer, such as chronic myelocytic leukemia, are apparently initiated by an abnormality in the regulation of protein-kinase activity. By understanding more about the role of protein kinases and phosphatases in cell growth, researchers hope to find new ways to inhibit the enzymes that give rise to cancer.

Dr. Edwin G. Krebs was born on June 6, 1918, in Lansing, Iowa, and earned his medical degree from the Washington University School of Medicine in St. Louis. After completing his residency in internal medicine at Barnes Hospital, St. Louis, Dr. Krebs decided on a career in science, returning to Washington University as a research fellow in biological chemistry. He joined the University of Washington faculty in Seattle in 1948.

Dr. Edmond H. Fischer was born on April 6, 1920, in Shanghai, China. He earned his doctoral degree in chemistry from the University of Geneva, Switzerland, and joined the faculty of the University of Washington in 1953.

Christopher King

Nutrition

THE NEW FOOD LABELS

Health-conscious Americans should find grocery shopping easier by mid-May 1994, when nearly all foods will carry the new user-friendly food labels. Food-labeling reform began in November 1990, when Congress passed the Nutrition Labeling and Education Act to encourage the Food and Drug Administration (FDA) to propose tighter regulations on the labels and health claims made by manufacturers.

The new regulations require that all packaged foods supply nutrition information, with the exception of small packages (no larger than a package of breath mints) and restaurant foods. The nutrition information required on the label will continue to include the serving size and the amount per serving of the following: total calories, total fat, total carbohydrates, and protein. In addition, the new label must now list saturated fat, cholesterol, complex carbohydrates, sugars, dietary fiber, and the number of calories from fat. The only vitamin and mineral information now required is for vitamins A and C, calcium, iron, and sodium. The FDA will set the standards for serving sizes to eliminate the problem of manufacturers manipulating the portion to make their product appear healthier.

The new labeling format also includes "Daily Reference Values" (DRVs), which are recommended levels of certain nutrients based on two daily calorie levels, 2,000 and 2,500. DRVs are provided for total fat, saturated fat, cholesterol, sodium, carbohydrate, and fiber. Percentages of Daily Values are included for these nutrients so that consumers may compare what the product contains to what they should be eating.

Consumers will no longer be fooled by such misleading terms as "light" or "cholesterol-free," since the FDA has set precise definitions for these and other descriptors. For example, a "low-fat" product cannot contain more than 3 grams of fat, and 50 grams of the food must have less than 3 grams of fat. Therefore, a product such as nondairy creamer with a small serving size could not be called "low-fat," even though the serving of 1 tablespoon has less than 3 grams of fat.

The new regulations restrict to seven the health claims that manufacturers may make about the relationships between diet and disease prevention on food labels. These associations are: calcium and osteoporosis; sodium and hypertension; fat and cancer; fat and heart disease; fiber and cancer; fiber and heart disease; and fruits and vegetables, a source of antioxidants such as vitamins C and beta carotene, and cancer.

Consumers can learn how to use the new food label through a comprehensive education program developed by the FDA.

PARTIALLY HYDROGENATED OILS

Further research has substantiated an earlier investigation that revealed that partially hydrogenated oils found in margarine, vegetable shortenings, and many commercial baked goods raise cholesterol. Owing to concern about saturated-fat intake in the American diet, hydrogenated oils from soybean and corn oil have replaced butter, lard, and coconut and palm oils in commercially made baked products.

Trans-fatty acids, formed when vegetable oils are converted to margarine or shortenings, are the cholesterol-raising culprits. This most recent study by the U.S. Department of Agriculture (USDA) showed that people who followed diets containing moderate to high levels of trans-fatty acids developed higher cholesterol levels than those who followed a diet high in oleic acid. Still, their cholesterol levels were not as high as those individuals placed on a diet high in saturated fat and low in trans-fatty acids. But the people who followed the high-trans-fatty-acid diet had lower levels of heart-protective HDL cholesterol than those whose diets were rich in oleic or saturated fats.

According to Mary Enig, Ph.D., who studied trans-fats for many years, marga-

Another Suspect in Heart Disease

New research implicating iron in heart disease may help explain the discrepancy in the heart attack rates between men and women. Because it is carried in the blood, menstruating women lose iron each month. Their iron levels, and their risk of heart attack, rise at menopause. Research suggests that iron can interact with oxygen to form highly destructive compounds.

The rise of stored iron, in milligrams

Men   Women

Age

CLOGGING THE ARTERIES

Iron interacts with oxygen in a process called oxidation. New research suggests that when one kind of cholesterol, low density lipoprotein, or LDL, is oxidized, it can be trapped in arterial cells, initiating the process that leads to atherosclerosis and clogged arteries.

Native low density lipoprotein
Free iron
Oxidized low density lipoprotein
Artery wall
Fatty streak lesion

TISSUE DAMAGE IN HEART ATTACKS

Iron is also implicated in tissue damage after the onset of a heart attack. When normal blood flow is disrupted, heart cells release iron stored in proteins there.

When blood flow and oxygen transport resume, the iron can react with oxygen and damage muscle fiber and cell walls.

Blood vessel
Iron is released
Block
Blood

Released iron
Oxygen in resumed blood flow

Sources: "Iron Balance" (St. Martin's Press); The New England Journal of Medicine.

rine accounts for only a small portion of the American intake of these fats. Her findings suggest the average American consumes 11 to 28 grams of trans-fatty acids per day, much of which comes from commercially baked and fried foods.

Health experts continue to recommend a diet with no more than 30 percent of the calories from fat, which will also limit intake of trans-fatty acids. For cooking or baking, it is wise to use a liquid oil such as canola or olive oil rather than margarine, vegetable shortenings, or butter. Since it appears there are no healthful solid fats, both margarine and butter should be used sparingly. Finally, Americans can reduce their trans-fat intake considerably by eating fewer commercially baked goods and fried foods.

IRON AND HEART DISEASE

A landmark Finnish study revealed that high levels of iron in the blood are a major risk factor for coronary heart disease, possibly second only to smoking. The researchers measured serum-ferritin (an iron-storing protein) levels in 1,900 healthy Finnish men and followed them for five years. During the course of the investigation, 51 men experienced heart attacks. The scientists concluded that men with ferritin levels greater than 200 micrograms per liter were twice as likely to suffer heart attacks as men with levels less than 200. In men with both high levels of low-density lipoprotein (LDL, often called the "bad" cholesterol) and ferritin, heart-attack risk was quadrupled.

This study has shed more light on the process of atherosclerosis (hardening of the arteries). One theory holds that iron combines with oxygen, triggering oxidation of LDLs in the arteries. White blood cells called macrophages, which scavenge for debris, envelop the oxidized LDLs and swell into "foam" cells. These foam cells collect along the arterial walls and create plaque, which narrows the blood vessel and can lead to a heart attack.

The iron theory also suggests a reason why women who lose iron monthly during menstruation have a lower incidence of heart disease. It also provides an explanation for why there is a higher rate of heart disease in countries where there is a high consumption of red meat (a rich source of iron), and why vegetarians rarely suffer from heart disease.

The study's results do not warrant drastic dietary changes, since there are still questions about the role of genetics in determining a person's iron level and how dietary iron affects the body's iron stores. It's wise for men and postmenopausal women to avoid multivitamin and mineral supplements that contain more than the suggested daily allowance for iron. For now the most prudent advice is to limit the intake of animal protein to 6 ounces per day, and to limit red meat to about three meals per week.

Maria Guglielmino, M.S., R.D.

Public Health

HEPATITIS B

The hepatitis B virus (HBV) is the leading cause of acute and chronic liver disease worldwide, according to an article in the March 1993 *American Family Physician.* The Immunization Practices Advisory Committee has proposed a comprehensive strategy to eliminate transmission of HBV in the U.S. Universal vaccination of all infants and the selected vaccination of high-risk adolescents and adults have already been recommended by the American Academy of Family Physicians.

Approximately one in 20 persons in the United States will become infected with HBV in his or her lifetime. Already an estimated 300,000 Americans contract the virus annually. Many cases, especially asymptomatic ones, go undetected; a large number of symptomatic cases are simply not reported.

Of the 20,000 to 25,000 cases of acute HBV infection that are reported to the Centers for Disease Control and Prevention (CDC) annually, 90 percent occur in young adults. Jaundice is present in about 25 percent of the persons with HBV. More than 100,000 HBV sufferers require hospitalization each year.

Between 18,000 and 30,000 Americans develop chronic HBV each year, adding to the up to 1 million chronic HBV cases already existing in the United States. Chronic HBV increases the risk of cirrhosis of the liver or primary hepatocellular carcinoma, the incidence of which is estimated to be 100 times greater in HBV-infected people than in the general population. HBV infection is responsible for approximately 5,000 deaths a year from cirrhosis of the liver, and 1,500 from liver cancer.

The CDC estimates that the direct health-care cost of HBV infections exceeds $200 million a year in the United States, while the cost of vaccinating an infant against the virus is as little as $20 to $40. Health-care workers are at great risk for contracting HBV, with more than 9,500 being infected each year. Other high-risk groups include homosexual males, intravenous drug users, and prison inmates.

VIOLENCE

Writing in the *Journal of the American Medical Association (JAMA)* in June 1992, Antonia C. Novello, the U.S. surgeon general, notes that there is a growing realization that violence constitutes a public-health emergency. The statistics are chilling.

Homicide is the tenth-leading cause of death in the United States, and suicide is the eighth. The U.S. has the highest homicide rate of any Western industrialized country. Each year, more than 1.5 million individuals are victims of assault; 650,000 women are victims of rape; 1.5 million to 2.5 million children are abused or neglected; and 700,000 to 1.1 million elderly people are abused or neglected.

Homicide is the second-leading cause of death among all Americans 15 to 24 years old, and the leading cause of death among 15- to 34-year-old black males. For blacks, the homicide rate is eight times greater than for their white counterparts of the same age. Overall, homicide rates for children and adolescents have more than doubled in the past 30 years.

Thirty thousand Americans commit suicide each year, according to an article

Health officials have grown increasingly concerned about the escalation of homicides and suicides in the U.S.

in the February 1993 *American Journal of Public Health*. Partly in response to this problem, the CDC recently established the National Center for Injury Prevention and Control. The previously noted *JAMA* article also points out that suicide is the third-leading cause of death for 15- to 34-year-old Americans, and the second-leading cause for 15- to 24-year-old Native Americans. Since 1950 the suicide rate among children and adolescents has nearly tripled. The surgeon general says that the public-health community must search for creative solutions to these problems.

THE COSTS OF HEALTH CARE

A 1993 American Medical Association study says that of the $666 billion that Americans spend on health care each year, nearly 25 percent goes to treat victims of drug abuse, violence, and other kinds of social behavior that can be changed. Crime and alcohol and tobacco use add $171 billion to the health-care bill. The report estimates that $22 billion in costs are due to cigarette smoking and other tobacco use, while alcohol use accounts for $85 billion. Street and domestic violence costs $5.3 billion. The study also points to other factors that add to health-care costs, including the failure to use lifesaving technologies such as seat belts and smoke detectors, failure to get routine medical checkups, and unprotected sex.

INFANT MORTALITY

Black children under the age of 1 year die at twice the rate of white infants, according to an article in the *New England Journal of Medicine*. Large numbers of African-American babies have very low birth weights—3.3 pounds—and therefore are at great risk. A closer look at maternal health is called for, since nearly all the excess mortality is directly related to four treatable pregnancy problems: infection or rupture of the amniotic membrane (38 percent); premature labor (21 percent); high blood pressure (12 percent); and uterine bleeding (10 percent).

While quitting smoking is undeniably beneficial to health, a 1993 study suggests that a former smoker faces a higher risk of lung cancer than does a person who has never smoked at all.

CIGARETTES

Two recent studies suggest more reasons to never begin smoking, or at least to quit as early as possible. The pooled results from 15 studies of 4.5 million people say that smoking cigarettes increases the risk of contracting leukemia by 30 percent, and causes up to 3,600 cases of adult leukemia per year in the United States. The 30 percent increase is relatively small, considering that longtime smokers run about a 1,000 percent higher risk of lung cancer than do nonsmokers; it is nonetheless one more reason to stop smoking. The American Cancer Society expects 26,700 new smoking-related cases of leukemia in 1993, up from 25,700 in 1992.

A University of Michigan study says that although giving up smoking at any age is beneficial, a former smoker's risk of dying of lung cancer is consistently greater than that for people who have never smoked. Moreover, the risk of dying from lung cancer is directly related to the age at which one stops smoking.

Previous studies had suggested that the cancer risk for a former smoker goes down to the risk of a lifelong nonsmoker after five or 10 years. The new study suggests otherwise, finding that the cancer risk for people who smoked for at least 15 years never returns to the levels of a person who never smoked. Smokers are 33 times more likely to die from lung cancer by age 75 than are people who had never smoked.

Neil Springer

PAST, PRESENT, and FUTURE

CONTENTS

The SAGA of the DEAD SEA SCROLLS

by Elizabeth McGowan

Professor Geza Vermes, D.Theol., a reader in Jewish studies at Oxford University, called it the "academic scandal par excellence." And that's probably not much of an exaggeration. Few issues have rocked the ivied towers of biblical scholarship like the controversy over the Dead Sea Scrolls, a saga with some of the cloak-and-dagger of an Indiana Jones movie—and all the duplicity of "As the World Turns."

What Are the Scrolls?

The Dead Sea Scrolls are nothing less than a treasure trove of ancient writings discovered in 1947 by a Bedouin shepherd who, no doubt, had no idea that his find would still be causing such a stir 46 years later. The shepherd wandered into a cave in the Qumran desert on the shores of the Dead Sea in search of a missing member of his flock. Instead of his sheep, he stumbled across a hidden cache of earthen jars containing sacred scrolls written between 200 B.C. and A.D. 50 by the Jews of Palestine.

When word got out about the discovery of these "Dead Sea Scrolls," an army of archaeologists, biblical scholars, and fortune seekers swooped down on Qumran to scour the area's collection of caves. They hit the jackpot in 11 of the caves: 800 manuscripts secreted in the Judean desert for approximately 2,000 years, arguably the greatest archaeological prize of the 20th century.

The scrolls included the earliest known copies of the complete Old Testament, with the exception of the Book of Esther, analysis of which proved how accurately Jewish scribes preserved the Bible over the centuries. Just how accurately? In 66 chapters of the Book of Isaiah, for instance, only 13 minor variations were found from the text used today in churches and synagogues. To put that into context, the version of the Bible that modern people read dates to the Middle Ages—transcribed approximately 1,000 years later than the cave version!

The documents also describe ancient Judaic laws, rules of behavior, ethics, and religious practices, illuminating the previously unknown beliefs and social mores of people who lived during the mysterious and turbulent period between the Old Testament and the time of Jesus Christ.

Written in Hebrew, Greek, and Aramaic on papyrus, leather, and copper, the scrolls varied in their stages of preservation. Many survived in one piece, amazing given their antiquity. To others, time was not so kind. Almost half the documents were blackened and fragmented, literary jigsaw puzzles with thousands of near-illegible pieces.

Who Shall Study?

Reconstructing and translating the scrolls, and figuring out who wrote them and how they got into the caves, was a challenge that any biblical academic would gladly assume. But the coveted task was awarded exclusively to a cadre of seven scholars by Jordan's King Hussein, who, at the time, reigned over the Qumran area.

In 1967, during the Six-Day War, Israel wrested control of Qumran from Jordan, winning possession of the scrolls, along with the stark desert terrain that is now called the West Bank. The original scrolls were stored, and can still be found, in the Rockefeller Museum and the Shrine of the Book Museum, both in Jerusalem.

Of surprise to many (especially given that all of the designated scroll scholars were Christian, a credential mandated by Hussein), Israel decided to permit the original team to finish the job they started, under the auspices of the Israeli Antiquities Authority.

Unfortunately, the team never seemed to quite get around to completing the project. Though they started their work at a gallop in the 1950s—the scrolls that were recovered in good shape were published almost immediately—their progress on the more difficult scrolls soon slowed to a crawl. As decade after decade slid by, and with access to the documents still limited to what some derisively refer to as the "cartel," scholars outside the chosen circle began to lose patience. Many were getting older and growing anxious about missing out on the opportunity to test lifelong-held theories about the scrolls by reading and interpreting the text themselves.

As the years passed, new computer technology and image-enhancement infrared techniques were developed, making reconstruction of the scrolls a less arduous task. But the Israeli Antiquities Authority consistently dismissed complaints about inept preservation and management of the translation and publication process, pointing to the poor condition of the fragments and the time-consuming methods needed to piece them together. The agency was equally cool

Only a select group of scholars has been allowed to study the Dead Sea Scrolls—a collection of sacred manuscripts not discovered until 1947. Oxford's Geza Vermes (right) and other biblical scholars are demanding greater access to the ancient writings.

to the suggestion that it would help speed things up to allow more scholars to study the scrolls. Authorities maintained that open access to the documents would be unfair to scholars who had already spent years on the project.

John Strugnell, a professor at the Harvard Divinity School who served as editor in chief of the scroll project from 1987 to 1990, further raised critics' hackles by telling *Maclean*'s magazine, "My problem is to get the scrolls published, not satisfy the vanities of particular scholars."

Needless to say, few of those "particular" scholars were particularly distressed to learn in 1990 that Strugnell was being dismissed as editor in chief. Still, Strugnell's separation from the scroll project did not stay the academic world's mounting anger about his team's apparent lack of progress.

"By 1985 people had started to say, 'Good grief. One generation of scholars has died, and another young generation is coming along, and none of us have seen the scrolls yet,' " explains Michael Wise, Ph.D., assistant professor of Aramaic at the University of Chicago.

Adding to the discord was the fact that as the original scholars retired or passed

away (only one member of the original group is still working on the scrolls), their positions were passed down to hand-chosen successors. "This brought up the question of why young graduate students should have access to the scrolls when masters of paleography and Semitic languages were denied it," adds Dr. Wise.

Conspiracy Monopoly
Herschel Shanks, editor of the *Biblical Archaeology Review (BAR),* who for eight years has been crusading for open access to the

The Dead Sea Scrolls were discovered in a cave (left) some 10 miles east of Jerusalem (see map). Photographs of the scrolls made soon after they were found (below left) have been used to circumvent the policy that restricts access to the originals.

scrolls, believes that the delay is due to "scholarly lethargy, greed, and hunger for power." Shanks has long used the Washington, D.C.-based *BAR* as a bully pulpit to fan the flames of anti-scroll-monopoly sentiment. By the 1990s that sentiment had built to the point of mutiny.

In September 1991, Professor Ben-Zion Wacholder, Ph.D., of Hebrew Union College in Cincinnati, who at age 67 had been waiting his entire career for a look at the scrolls, figured time was running out: If he couldn't join the cartel, he'd beat them.

Using desktop computers, Wacholder, with the help of doctoral student Martin Abess, input a concordance (a listing of the 52,000 Hebrew and Aramaic words that appear in the scrolls, as well as a portion of the sentence preceding and following the words) that had been prepared in the 1950s by the official scroll scholars. Stringing the phrases together on a Macintosh computer, the two were able to reconstruct the document with an accuracy they claimed to be 90 percent. *Voila!* Access to the scrolls without having access to the scrolls. Herschel Shanks, of course, was more than happy to publish the renegade reconstruction. The work met with acclaims of academic liberation by the anticartel crowd, and denunciations of criminality and questionable accuracy by the official researchers.

Improving Access

About two weeks later, the Huntington Library in San Marino, California, dealt the elite group of scholars another debilitating blow. Unbeknownst to academics, the public, and even to its new director until he took inventory of its archives, the Huntington possessed a complete photographic copy of the scrolls, one of several created to safeguard the contents of the documents against decay and natural disaster.

William A. Moffett, Ph.D., director of the Huntington, says he didn't realize the significance of the library's photo collection until he read about the scroll controversy in the *London Times* and received a letter from an editor in the cartel asking that the Huntington's archives be surrendered to the official cadre of researchers.

The library had to decide quickly what its official policy was to be. "We did our homework very carefully," Dr. Moffett explains. "Legally and ethically, we had one card to play: Our purpose as a library is to provide freedom of access to information. When we decided to allow free access to our files, we knew that there would be an attempt to stop us with legal actions and injunctions." To prevent that from happening, the library decided to get the public involved.

Luck was on the Huntington Library's side. Moffett quickly announced to the press that the library would make the scroll photos available to scholars without restriction. To his surprise, the announcement received front-page play in *The New York Times.*

According to Moffett, the response from the media and from institutions all over the country was overwhelmingly positive. Recog-

The decision by William Moffett (far left), director of the Huntington Library in San Marino, California, to make a photographic copy of the scrolls accessible to scholars caused an uproar among biblical archaeologists. Scholars are now trying to piece together fragments of the scrolls, many of which survive only in tatters (near left).

nizing a losing battle, the Israeli Antiquities Authority backed down, dropping its threats of a lawsuit against the library.

The academic community was still abuzz about the Huntington decision, when yet another photographic copy of the scrolls turned up—delivered by an anonymous donor to the doorstep of Robert Eisenman, Ph.D., chairman of religious studies at the University of California at Long Beach. Eisenman quickly edited the photographs with James M. Robinson, Ph.D., of California's Claremont Graduate School, and a two-edition facsimile of the collection was published by Herschel Shanks. The scrolls were now available to anyone who could afford $195 and had the ability to read them!

Access to the scrolls was still, for all intents and purposes, limited to scholars fluent in Hebrew, Greek, and Aramaic. But not for long. An English translation and interpretation of the photographs, the *Dead Sea Scrolls Uncovered,* by Drs. Eisenman and Wise, was rushed to press, adding new impetus to the official team's argument that unlimited access to the scrolls invited shoddy scholarship. The official scholars also claimed that proper credit for the past work conducted by other scholars had not been given.

Emotions in the scroll sector were running high. At an academic conference held in New York City in December 1992, the storm over the *Dead Sea Scrolls Uncovered,* published a couple of weeks earlier, threatened to eclipse the more important debate over the meaning and origins of the scrolls.

Dr. Wise, who apologized at the conference for not giving credit where credit was due, said, "We wanted to produce the texts for the popular reader, the nonspecialist. This was never meant to be the be-all, end-all. It was just meant to be the first version of the texts. . . . The ones who were most vocal were those who benefited by keeping the scrolls under wraps."

Norman Golb, Ph.D., professor of Near East languages at the University of Chicago, also defends the rapid publication of the book. "Because the scrolls were kept secret for so long, it's natural and legitimate for scholars to want to get their ideas across before they are pushed aside. Scholars will go back to [make revisions] to their work after the original publication, and later offer their own more scholarly editions."

Ghostwriters

Traditional scroll scholars claim that the scrolls were written by a small, ascetic, separatist Jewish sect known as the Essenes, who hid the scrolls in the caves to protect them from Roman invaders. Those who subscribe to this theory believe that ruins found in the vicinity of the caves are those of a monasterylike structure, used as a communal center for the Essenes.

Other scholars argue that the remains of women found buried at the Qumran site negate the theory that the occupants were celibate monks.

Still other scholars, including Dr. Golb, theorize that the ruins are actually those of a military site occupied by Jewish troops, and that the scrolls are a library of the work of many sects, representing a wide spectrum of Jewish thought from the time. "At every turn of the page, more scribe handwritings are revealed," he says. "I've counted at least 500 handwritings so far, which destroys the theory of the small sect of Qumran monks."

While his colleagues, despite their differences, emphasize the significance of the scrolls as ancient Judaic texts, Eisenman stresses what he sees as their strong connection to early Christianity. Though he has retreated from a much-publicized earlier view that the "Righteous Teacher" mentioned in the scrolls is James, Jesus' apostle and brother from the New Testament, Eisenman maintains that "the last stages of the messianic movement represented by the literature at Qumran are all but indistinguishable from that of the early church of the Palestinian Jerusalem community of James the Just [the apostle]." In simpler terms, Christianity was an outgrowth of a messianic movement that included the scroll sect.

Naysayers claim that carbon dating proves that the scrolls are too old for this link to be possible, going so far as to claim that the Christian connection is a figment of Dr. Eisenman's imagination. Eisenman retorts that the carbon-dating tests were not secure enough to be definitive. He claims that his theories are being rejected because they turn traditional scholarship on end.

As Dr. Eisenman debates scroll scholarship with his critics, yet another character in the continuing saga is finishing preparations for a trip to Qumran to put one of the scrolls to more-practical use.

Vendyl Jones, a verbose biblical archaeologist given to the type of folksy sound bites for which his fellow Texan, Ross Perot, is famous, claims that the Copper Scroll contains geologic and geographic clues to the location of the Lost Ark of the Covenant, said to hold the stone tablets of the Ten Commandments given to Moses. On a prior visit to Israel, Jones claims to have come close to unearthing the Ark; this time, he says, he'll be coming home with the prize.

Like Eisenman, Jones, who is the president of the Institute of Judaic-Christian Research in Arlington, Texas, makes no bones

about what he thinks about the scroll monopoly: "I abhor people like those academic snobs sitting on the scrolls who stand in the way of the truth. . . . My position is that I don't kowtow to anything."

Copyrighting History

But even Jones may find that he has to kowtow to the courts. In March 1993, a lawsuit filed against Herschel Shanks by Elisha Quimron, Ph.D., an Israeli scholar, was decided in Quimron's favor, a victory for the official-scholar contingent. Dr. Quimron, a member of the sanctioned team, had charged that Shanks published his decipherment of one of the scrolls in the foreword to the facsimile-photo editions without permission or credit. The judge rejected Shanks' argument that Quimron could not copyright an ancient text he did not write. She ruled that the reconstruction was a creative endeavor involving arranging fragments and filling in gaps in the text, and, as such, belonged to Quimron. Shanks is considering an appeal, and has already filed a countersuit in Philadelphia.

If, as the court ruling implies, scroll reconstructions can be copyrighted, the implications for scroll scholarship are enormous. The translator of a particular scroll would be able to control interpretations of that scroll by controlling access to the reconstruction.

William A. Moffett is among the many who disagree with the verdict, and find Shanks' argument persuasive.

He offers an illustration: "Say you have an ancient text written on a piece of stone, and you hand it to me, and I drop it and it shatters. Can someone then put that text back together and say that he can copyright it? If you don't own it before it breaks, how can you own it after it's put back together? . . . This is just another case of the official scholars trying to make it impossible for those outside the cartel to function."

Will Shanks win his appeal? Will biblical scholars ever resolve their differences, or at least agree to disagree? That remains to be seen. But at the rate things are going, it might be another 2,000 years before the rest of us find out what the original authors of the scrolls were trying to tell us.

The STATUE of ZEUS
at OLYMPIA

The TEMPLE of ARTEMIS
at EPHESUS

The COLOSSUS
of RHODES

The MAUSOLEUM at
HALICARNASSUS

The PHAROS at
ALEXANDRIA

The PYRAMIDS
of EGYPT

MEDITERRANEAN SEA

AEGEAN SEA

BLACK

NILE RIVER

The Seven Wonders
of the
Ancient World

BY THEODORE A. REES CHENEY

Since the dawn of civilization, people through their skill and artistry have created marvels that challenge the imagination. In ancient times, travelers began to list the awesome structures they saw. In the 2nd century B.C., two such lists appeared, one written by the ancient engineer Philon of Byzantium, and the other in a poem composed by Antipater of Sidon. Both lists refer to the seven wonders of the world: architectural accomplishments of such enormity and splendor that, as much as they amazed the Roman travelers of yore, they truly astound us, especially since no modern technology was used in their construction. Of the seven wonders of the ancient world, only the pyramids remain intact. All that is left of the other six wonders are the descriptions in ancient writings and legends. The following discussions and illustrations of the seven wonders are derived from these magnificent tales.

The Pyramids of Egypt
Some 80 pyramids have been located in Egypt, 42 of them built along the Nile during one intensive period of pyramid building

The Hanging Gardens of Babylon

EUPHRATES RIVER

NORTH

each and total 6 million tons, covering 13 acres (5.3 hectares). Cheops organized his entire nation to provide 50,000 men and women to work for years building this one monument. He may have built it not so much to memorialize himself as to give his nation full employment on one giant, unifying cause.

The Great Pyramid's base measures 756 feet (230 meters) on a side. Somehow these ancient builders had the engineering sophistication to make the sides agree in length to within a hand's width, a minuscule one-tenth of 1 percent of 756 feet. These sides run exactly north, east, south, and west, and meet precisely at right angles. The sides incline at 52 degrees to meet with precision at the peak 481.4 feet (146.8 meters) above the ground, making the pyramid only 70 feet (21 meters) shorter than the Washington Monument. Brilliant white Tura limestone originally sheathed the pyramids. The Giza Group of three, consisting of Cheops's pyramid and the smaller pyramids of two other Egyptian kings, must have been a spectacular sight in the flaming desert Sun.

Pi in the Pyramid. Scholars have been fascinated by the many mathematical surprises found in the pyramid's architecture. For example, Egyptians had known since about 1650 B.C. that the ratio of a circle's perimeter to its diameter approximates the transcendental number 3.1416. It wasn't until 200 B.C. that the Greek mathematician Archimedes rediscovered that ratio as somewhere between $3^{10}/_{17}$ and $3^1/_7$, which approximates our decimal expression of *pi* as 3.1416. Take the Great Pyramid's height (485.5 feet, or 148 meters) as the radius of a circle (and hence with a diameter of twice that), and multiply that diameter by 3.1416. You'll find that the circumference equals 3,050.49 feet (929.65 meters). Amazingly, the perimeter of the four sides around the square base also equals 3,048.04 feet.

Building the Pyramid. Modern Egyptologists calculate that to complete the Great Pyramid during his 23-year reign, Cheops's workers must have labored 10-hour days all year every year to set in precise position one multi-ton block every two minutes, day after day, year after year—for 23 years.

(2750 B.C. to 2275 B.C). The grandest of all was The Great Pyramid at Giza.

The Great Pyramid at Giza. Cheops, an arrogant, tyrannical man, erected the Great Pyramid during his 23-year reign as king of Egypt. Ironically, only a 3-inch (7.6-centimeter) ivory statuette remains to give us the likeness of this king responsible for the largest monument ever built. Cheops's pyramid held the height record for man-made objects for over 4,000 years.

The Great Pyramid's 2.3 million shaped blocks of limestone average about 2½ tons

Egypt's pyramids are the oldest of the ancient wonders and the only wonder that still stands.

Speculative drawings show the workers hauling rocks up ramps, but many scholars now feel that typical ramps would not have worked. A ramp of suitable slope would have extended a mile into the desert, yet the main quarries were but 500 yards (460 meters) away. A solid ramp up to the peak of the pyramid would, itself, have used as much stone as all three pyramids combined.

Herodotus, father of all historians, reported around 500 B.C. that Egyptians (2,000 years before) had used the pyramid's blocks themselves as steps. Then, using short wooden levers, the workers lifted each block step by step, tier by tier, to the top. He made no mention of ramps. Scholars have been systematically studying "how they did it" for 200 years. Much has been learned, yet much mystery remains.

The Hanging Gardens of Babylon

The City of Learning, known as Babylon, located not far south of today's Baghdad, Iraq, reigned as the regional cultural center for 3,000 years. The powerful Babylonian armies routinely kidnapped and brought back royal hostages, military leaders, scientists, and talented craftsmen. The Old Testament (II Kings 24:14) describes such an event: "And he [Nebuchadnezzar] carried away all Jerusalem, and all the princes, and all the men of mighty valor, even ten thousand captives, and all the craftsmen and smiths: none remained, save the poorest sort of people of the land."

Babylon treated these captives well, training and educating them in Babylonian ways and wisdoms. They would then return to their homes as governors and scientific leaders, teaching and supporting the advanced Babylonian cultural ways. The hostages also carried home with them their new Babylonian names; for example: Shadrack, Meshack, and Abednego.

Although no Babylonian texts refer directly to any spectacularly different "Hanging Gardens," earlier kings did prize their royal gardens, so they probably did exist somewhere. Even Berossus, writing around 280 B.C., was already referring to "the *so-called* hanging gardens" of King Nebuchadnezzar II. Berossus wrote of the king that, within his palace, Nebuchadnezzar "erected lofty stone terraces, in which he closely reproduced mountain scenery, completing the resemblance by planting them with all manner of trees and constructing the so-called Hanging Garden; because his wife, having been brought up in Median, had a passion for mountain surroundings."

Indeed, Nebuchadnezzar loved to build, a passion evidenced by the multiple palaces, temples, and great gates he erected in Babylon and the double walls he constructed around the city. Scholars speculate that he built the spectacular hanging gardens between his palace and the high city wall, providing himself with access from his royal chambers into the garden, and from the garden through a pair of 22-foot (6.7-meter)-thick walls into the surrounding parklands where he kept wild beasts.

Another writer of the 1st century B.C. wrote that the garden's trees stood 50 feet (15 meters) tall and 12 feet (3.6 meters) around. The garden, he wrote, was supported by 20 thick walls 10 feet (3 meters) apart and carved columns in between. Large Archimedes'-screw devices lifted water to the topmost terraces 80 feet (24 meters) above the Euphrates River, where it then flowed down through the gardens in beautiful streams, cataracts, and falls.

The Temple of Artemis at Ephesus

In ancient times Artemis, the Goddess of Fertility, was worshiped all over the Middle East. Statues of her have been dated to as early as 7000 B.C. One statue the people worshiped was actually a meteorite, said in the Bible to have fallen from Jupiter, but most were of wood. Some showed her with numerous pendulous protuberances on her chest. A marble statuette from the 1st century A.D. has 21 such protuberances, apparently depicting breasts to symbolize fertility. Another, dating from the 2nd century, has protuberances shaped more like another symbol of fertility, eggs. The earlier one has a definitely Asiatic face, while the latter looks more Grecian, typical of this area, the crossroads between East and West.

When the Crusaders marched through Ephesus in Asia Minor, the villagers didn't know that a temple to this goddess lay right under their muddy village—in fact, seven

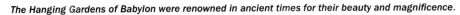

The Hanging Gardens of Babylon were renowned in ancient times for their beauty and magnificence.

The Temple of Artemis at Ephesus purportedly covered an area four times that of the Parthenon in Athens.

temples glorifying Artemis lay beneath them. Invading armies had destroyed the temple seven times during several thousand years. The one the Crusaders may have heard most about was the fifth temple from the bottom. According to Pliny, the Roman writer, this fifth temple was not yet complete when Alexander the Great came through in 333 B.C. Even that long ago, Pliny was calling it "one of the seven wonders of the world." To the non-expert eye, this temple greatly resembled the Parthenon in Athens, Greece.

The Parthenon stands 230 by 100 feet (70 by 30 meters), and its 58 stately columns reach up 34 feet (10.3 meters). Pliny wrote that the Temple of Artemis at Ephesus, in dramatic contrast, stood 425 by 225 feet (130 by 70 meters), and had "a forest of 127 columns" soaring to 60 feet (18 meters). Unlike the Parthenon's, many of these columns had sculptured figures—and, most unusual, many of the sculptures and temple features were in splendid colors.

The first list of seven world wonders reported in the poem by Antipater of Sidon around 200 B.C. made this comparison of the seven: "When I saw the sacred house of Artemis that towers to the clouds, the others were placed in the shade, for the Sun himself has never looked upon its equal outside Olympus."

The Statue of Zeus at Olympia

Zeus, father of the Gods, sat on a golden throne in a magnificent temple on the grounds of the Olympic Games in northern Greece. The temple, built by the architect Libon, resembled the Parthenon in Athens. Phidias, who had carved Athena's statue in the Parthenon, was selected to create one of Zeus for this new temple. A wooden statue of Zeus had been worshiped there for a few hundred years, but the people of the modern 5th century B.C. felt they needed something far better.

Phidias had found a new way to build colossal statues of ivory and gold. He first built a wooden figure similar to what he envisioned for the final statue. Thin sheets of ivory were sliced from innumerable elephant tusks and carved to represent the exposed flesh parts of the statue. He fastened these thin ivory sheets to the wooden underfigure, and then carefully fitted and camouflaged each joining line, so that the final figure looked like a solid ivory statue.

Since the Christians have long since destroyed this wonder of the Golden Age of Pericles, we have to rely on words written by people like Pausanias of the 1st century A.D., who had seen the awesome statue: "On his head lies a sculpted wreath of olive sprays . . . on his right hand he holds a figure of victory made of ivory and gold [about 10 feet, or 3 meters, tall] . . . in his left hand . . . a sceptre inlaid with every kind of metal, and the bird perched on the sceptre is an eagle . . . sandals of gold . . . and the throne decorated in gold, precious stones, and ebony with ivory."

When a worshiper, with eyes at about the level of Zeus's golden sandals, looked up at the magnificent head 43 feet (13 meters) above him, he must have felt incredibly insignificant, the feeling enhanced further by the unusual visual effect of the god's head practically scraping the ceiling.

Whether Phidias deliberately designed the figure to fit so uncomfortably within his own temple, we'll never know, but Zeus's overwhelming presence made itself felt throughout the temple. A mere person could not escape its presence in any corner.

The magnificent Statue of Zeus at Olympia had skin made of ivory and robes made of gold.

The Mausoleum at Halicarnassus

When Mausolus, king of Caria (the area of southwest Asia Minor, now Turkey), died in 353 B.C., his sister Artemisia (to whom he was married) completed a tomb in his honor at the capital, Halicarnassus. Since the royal couple admired the Greek civilization more than they did the Persian one to which they owed their allegiance, they retained the Greek architect Pythios to build the tomb.

Artemisia, unimpressed by the gigantic Egyptian pyramidal tombs, wanted the world to know her brother-husband's tomb, not for its great size, but for its beauty, richness, and grace. The structure she created (probably begun long before Mausolus's death) was so lavishly beautiful it immediately became known as one of the seven wonders of the world. It so impressed the Romans that they coined the word *mausoleum* to describe any large and stately tomb. Because Queen Artemisia died only two years after Mausolus, she lies with him in a sculpted marble sarcophagus below the grandiose structure.

The building remained essentially intact until the earthquakes of the 15th century. What didn't come tumbling down was soon pilfered by the Christian Crusaders who had settled there after expulsion from Palestine by the Turks. The Christians broke up many of the remaining statues to make lime for building their own fortresses and churches.

Since nothing but fragments of this wonder of the world remain, we depend largely on what observers at the time, like Pliny the Elder, wrote. The massive mausoleum resembled no other building of its time. No one is positive about its shape, but it did seem to be about 63 feet (19 meters) on its longest sides and somewhat shorter on the other two sides. About 140 feet (43 meters) above its marble base stood a mammoth marble sculpture of four horses harnessed in bronze pulling a giant chariot carrying Mausolus and Artemisia. The wheels of the chariot were over 7 feet (2 meters) in diameter.

Sloping down from this sculpture on the summit was a 24-step pyramid. Below this stood an open colonnade of 36 Ionic columns each 30 feet (9 meters) tall, possibly arranged nine to a side. Below and around the colonnade ran friezes depicting battles between Greek men and Amazon women warriors. Below the friezes the walls may have dropped vertically or sloped again in pyramid fashion down to a level where a surrounding wall enclosed an open terrace about 100 feet (30 meters) square. A fierce warrior on a prancing horse guarded the wall's four corners.

Along the wall between those vigilant marble guards stood 10-foot (3-meter) figures, each a marvel of sculptural genius. Artemisia had assigned the design of each side of the mausoleum to a different Greek sculptor. Twenty huge lion figures guarded the broad stairway leading up to the platform on which the memorial itself stood.

The Colossus of Rhodes

Shakespeare may have fallen for the wildest of wild popular legends about the Colossus of Rhodes when he wrote: "Why, man he doth bestride the narrow world like a Colossus; and we petty men walk under his huge legs, and peep about. . . ." (*Julius Caesar,* act 1, scene 2, lines 134–37). This legend purported that a statue of the sun god Helios straddled the entrance to a harbor on the island of Rhodes, all ship traffic into that eastern Mediterranean harbor passing between his legs.

Helios was indeed far taller than any other colossal sculpture of the time. Even

Atop the Mausoleum at Halicarnassus stood an immense sculpture of a chariot.

The towering Colossus of Rhodes stood for only 56 years before an earthquake toppled it.

Zeus at Olympia stood only 43 feet (13 meters) tall, while Helios stood 110 feet (33 meters) above its base, putting his torch (if he held one) 160 feet (49 meters) above the waters of the harbor. By comparison the torch of America's Statue of Liberty towers a somewhat less imposing 150 feet (46 meters) above New York harbor.

Scholars disagree on whether Helios held a torch; whether rays of the Sun came from his head, as they do on the Statue of Liberty; whether his face looked like the one on a contemporary coin; and even on where he stood, and his stance. We know little, because this magnificent work of art and engineering stood for only 56 years before an earthquake cut him off at the knees, toppling him to the ground.

Ancient writers told of a colossus by Chares of Lindos at a harbor on Rhodes, but they gave few details. Not one of them mentioned an unusual stance. Engineers today say that erecting a figure astride the entrance would have disrupted sea traffic for at least 12 years, a most unlikely undertaking for this maritime crossroads. Scholars calculate that a figure with legs spread enough for traffic to pass comfortably between his massive thighs would have to have stood a half-mile (0.8 kilometer) tall. In any event, so huge a statue of bronze and iron could not have stood very long on slanted legs. Most modern experts do agree that the Colossus of Rhodes stood at attention, naked, and carried a long spear.

After this wonder crashed to the ground in about 300 B.C., it lay there for 900 years. Pliny wrote in the 1st century A.D., "Even as it lies, it excites our wonder and admiration. Few men can clasp the thumb in their arms, and its fingers are larger than most statues."

Finally, someone in A.D. 658 cut it into scrap and shipped it to Syria, where a dealer cut it into still smaller pieces. Then they lugged away the magnificent Colossus of Rhodes on the humped and hairy backs of 900 camels.

The Pharos (Lighthouse) at Alexandria

Alexander the Great renamed many cities after himself, but just west of the Nile delta, he built an entirely new city he hoped would become the world's cultural center. For the island of Pharos in Alexandria's harbor, he planned a lighthouse, the likes of which had never been seen.

The Pharos at Alexandria served as a lighthouse until A.D. *1303, when it collapsed in an earthquake.*

After Alexander the Great died, King Ptolemy I Soter began construction of the lighthouse, his son, Ptolemy II Philadelphus, completing the project in 270 B.C. Mariners could see this magnificent light from 100 miles (160 kilometers) out to sea, about a day's sail.

A plume of black smoke arose from the fire that was kept continually burning in its colonnaded summit. Large burnished bronze plates reflected the firelight out to ships at sea. Sometimes these "mirrors" reflected the Sun as heliograms to city or ship.

Lighthouse keepers fueled the fire by wood, charcoal, or even dried animal dung in this all-but-treeless land. Some ancient sources claimed that in those days one could see in the curved mirrors ships unseen by the naked eye—perhaps the first use of a telescopic lens?

Atop the beacon chamber stood a colossus of Poseidon, God of the Sea. Depending on how one measures, the tip of Poseidon's trident stood anywhere from 413 to 600 feet (126 to 183 meters) above the sea.

Around the base of the 60-foot (18-meter)-tall cylindrical beacon chamber, which held the fire, smoke, and mirrors, architect Sostrates of Knidos built a balcony for the lookouts. Inside the building was a dumbwaiter lift that raised the fuel from several hundred feet below.

The second tier of this three-tiered building was octagonal, standing 118 feet (36 meters) tall. Where it joined the bottom tier ran a much wider balcony where sightseers could view the sea. Kiosks sold these onlookers souvenirs, tea, and lamb-on-a-stick snacks.

Those two tiers surmounted a wide, square building of marble that housed many government offices, a military barracks, and enough space to stable 300 horses. And below this was a platform that began below sea level.

The Pharos continued to serve as a lighthouse, and remained for many centuries the tallest building in the world. But then, on August 8, A.D. 1303, a severe quake toppled this wonder of the world. Its name lives on to this day, some mariners referring to any bright light on the horizon as a pharos.

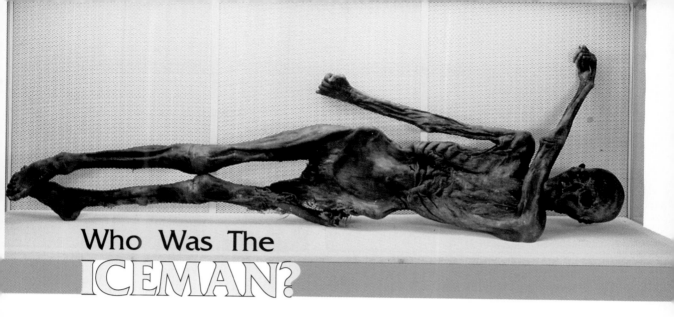

Who Was The

ICEMAN?

by Sandy Fritz

After forensic expert Rainer Henn found a flint-tipped knife in the slush, he told his assistants to stop working on the body. "When I saw this knife, I had the idea that this man was very old," Henn later reported. "From that moment, I ordered all the people to be most careful while getting the body out of the ice."

Henn and his team had no idea that the man they were freeing from the ice could be the most important discovery in modern archaeology. The man had been sealed in the Similaun Glacier in the Tyrolean Oetztaler Alps for 5,300 years. He died wearing his buckskins and grass cape. His bow and arrows, a copper ax, and other tools were recovered nearby. His skin, internal organs, and even his eyes are still in place. Dubbed the Iceman by the public, but referred to in scientific literature as *Homo tyrolensis,* the Similaun man, and the Hauslabjoch mummy, he is the oldest and best-preserved human body ever found.

Freak of Nature?

When the corpse and its bunch of artifacts arrived via helicopter at the University of Innsbruck's Institute of Forensic Medicine in October 1991, the university's dean of prehistory, Konrad Spindler, was on hand. "I felt like Howard Carter staring onto the likeness of King Tut," he recalls.

In fact, the find will probably prove to be far more important than the discovery of King Tutankhamen because Tut's 3,344-year-old tomb only served to further illustrate the opulent lives of Egypt's well-known pharaohs. The Iceman is nearly 2,000 years older, and his discovery illuminates a far more mysterious time period. Scant traces remain of the people who farmed and hunted in the forests of Europe during the late Neolithic Age. Now a single individual from that shadowy time has emerged from a retreating alpine glacier, completely outfitted with clothes, tools, and weapons.

The preservation of the body ranks among the true freaks of nature. In late summer or autumn around 3300 B.C., a 25-to 35-year-old man wandering above the tree line at about 10,500 feet (3,200 meters) took shelter in a natural trench nearly 6 feet (2 meters) deep and 20 feet (6 meters) wide. He died there. Exposure to several weeks of cold winds mummified the body. Snowfall froze the mummy, and centuries of snowfall became a glacier. Sheltered in the trench, the frozen body was spared most of the shearing forces of the glacier. Then, during the unusually hot European summer of 1991, a pair of German hikers spotted a leathery skull and a shoulder poking out of the glacier, and contacted the police.

Nobody rushed to the scene. The melting alpine glaciers had already released six

Scientists rejoiced in the discovery of the frozen remains of the "Iceman"—the oldest and best-preserved human body ever found.

Who Owns The Iceman?

When the Iceman was first found in the Oetztaler Alps, his 5,300-year-old body and equipment were presumed to be the property of Austria. But a closer inspection by surveyors revealed that the corpse was found about 100 yards (90 meters) inside the border of Italy's South Tyrol. He is therefore the property of Italy. The body will remain in Innsbruck, Austria, and the equipment in Mainz, Germany, until September 19, 1994—three years from the date the Iceman was discovered. If the researchers need more time to investigate the remains, they may petition for an extension. Meanwhile, it costs about $10,000 a month to maintain the body in its chilly preservation chamber.

Very few people have seen or examined the finds recovered from the site last summer. These artifacts are being held in South Tyrol, tangled in red tape. Provincial officials there plan to build an Iceman museum; they would like to display the body in a refrigerated case.

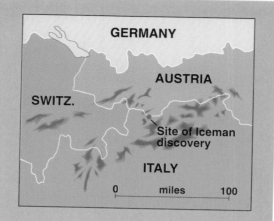

The Iceman was discovered about 100 yards inside Italy's border by a pair of German hikers. The body is being studied in Innsbruck, Austria.

Werner Platzer, head of the University of Innsbruck's Anatomical Institute and leader of the team researching the body, says ownership of the body is meaningless. Where he ends up is not important, Platzer says, "so long as he is well taken care of and receives the respect that is his due."

corpses that summer. The others were 20th-century climbers whose bodies, full of moisture, had turned waxy and had been partially pulverized by the glacier's slow, ponderous movement. The Iceman's body, however, was virtually undamaged. That is, until his rescue. Unfortunately, the man who died fully clothed arrived at Innsbruck totally naked, except for a shoe.

Rescue Damage

"Thirty men with picks and compressors worked on him," groans Werner Platzer, dean of physiology at the University of Innsbruck and leader of the team studying the body. "The body froze at night and thawed during the day while the rescue was under way. They had no idea how old he was."

The rescue caused the most prominent disfiguration of the body, tearing a sizable chunk from the left hip. The penis is miss-

ing, probably ripped off with his pants. What was most likely a backpack was largely destroyed, and what later turned out to be a longbow was broken. In their haste to ferry the find back to Innsbruck, the recovery team left some material behind. And by the time everyone had collected their wits after the first wave of excitement passed, 10 feet (3 meters) of snow covered the site, sealing it for the winter.

The Iceman has spent most of the past 17 months swaddled in an icy cocoon at the University of Innsbruck's forensic laboratory. In life, he stood 5 feet 3 inches (160 centimeters) tall, weighed about 110 pounds (50 kilograms), and sported dark hair and a beard. Today he weighs 44 pounds (20 kilograms), and lies frozen in his death position, naked, shriveled, and bald. To slice off a section of skin for testing, researchers would need a saw, not a scalpel.

A blanket of surgical gauze protects the Iceman's skin; on top of that is a layer of crushed ice rendered from sterilized water. A layer of plastic covers the ice, more sterilized ice is heaped on top, and a final plastic wrap encloses the body, guaranteeing humidity of 96 to 98 percent next to the skin. The cocoon sits inside a high-tech freezer maintained at a glacial temperature of 21.2° F (−6° C) and monitored by an elaborate system of sensors. Six temperature sensors are rigged to set off alarms and por-table pagers should the temperature rise by a few degrees.

No Touching Allowed

Removing the body from its cocoon for inspection is somewhat traumatic. The uncovered mummy reclines inside a chilled glass box filled with filtered, sterilized air. Researchers have but 30 minutes for their work before the mummy is hurried back into its icy chamber. It takes 48 hours for the body to return to its storage temperature.

Because physical examination of the body endangers it, researchers are developing a comprehensive database that allows them to study the body in detail without touching it. Using computerized axial tomography (CAT) scans—three-dimensional images constructed by a computer from a series of cross-sectional X-ray pictures—researchers can view the Iceman's bones and organs on a computer screen. The same CAT-scan data have allowed researchers to create a three-dimensional plastic skull that is an exact replica of the original.

Platzer has made a few tentative conclusions based on preliminary investiga-

The scientific world remains shocked at the recovery team's rough handling of the Iceman. What had been a perfectly preserved and completely clothed corpse arrived at the lab naked (except for a shoe) and missing various body parts. Much of the Iceman's hunting equipment was damaged or destroyed during the recovery.

tions: The man died lying on his left side, with his left arm extended across his rump. His current posture—facedown with his left arm cushioning his forehead—is the result of the glacier dragging his arm upward.

A CAT scan revealed that the Iceman's left arm is broken above the elbow. Initially, Platzer suspected the break occurred when the brittle body was rescued. "But his dry skin was not torn," Platzer explains. "It may have been an old break that healed, or it may have happened just prior to his death."

Platzer also says the man has a deep groove in his right earlobe, suggesting that he wore an earring. But a 2-inch (5-centimeter) disk of white stone with a spray of leather fringe was the only jewelry found, and it was probably worn on a leather thong around his neck.

Enigmatic Tattoos

Other adornment includes a series of enigmatic tattoos: four 3-inch (7.5-centimeter)-long stripes mark the top of his left foot, a cross dots the left kneecap, and 14 bars run down the small of the back. "His clothing would have covered the tattoos," observes Konrad Spindler, who is overseeing the entire investigation of the body and the equipment. "They were not meant to be seen publicly. In this sense, they are not like modern tattoos at all. They must have had significance to him alone."

A team of Dutch researchers will be studying the tattoos to ascertain how they were made. Some primitive peoples tattoo their bodies by pricking the skin with a needle and then rubbing ash or pigment into the wound. The Iceman could have used a similar technique to make the tattoos on his own foot and knee, but the bars on his back are clearly the work of another person.

Another visible clue about the man is his worn teeth, suggesting that his diet included gritty bread. Two grains of primitive wheat were recovered from the fur of his coat; the grains are evidence that he may have lived in or near a lowland farming community at the foot of the Alps.

Farming communities spread through the virgin lands of Neolithic Europe some 7,000 years ago. The first farmers slashed and burned hardwood trees, sowed wheat in the clearings, and pastured their sheep and oxen in the woods. The area was already peopled by a seminomadic folk skilled in hunting, tracking, and fishing. The two cultures eventually merged, as the idea and practice of farming spread throughout Europe.

While the Iceman's body lies preserved in Innsbruck, his belongings are being housed and studied at the Roman-Germanic Central Museum in Mainz, Germany. Under the watchful eye of archaeologist Markus Egg, leather objects have been meticulously cleaned, greased, and dehydrated, and can now be handled. Artifacts of grass have been freeze-dried to dispel moisture. Wooden finds have been cleaned and bathed in vats of thin, warm wax, which replaces moisture in the wood to prevent rotting.

The Bow

One of the most remarkable wooden objects found was a longbow, hewn from the strong but flexible heartwood of a Tyrolean yew tree. Yews still grow in the valley below the Similaun Glacier where the Iceman was found, and in the past, this valley was farmed for the high-quality bow staves it produced. This, and the fact that the Iceman's bow was unfinished at the time of his death, suggests he had just cut his stave in the valley.

The sheer size of the bow tells us something about the man. Broken in half during the rescue, the bow originally spanned 5 feet, 10 inches (178 centimeters)—7 inches (18 centimeters) taller than its owner. "You can't make them much bigger than that," says Christopher Bergman, principal archaeologist with 3D-Environmental Services in Cincinnati.

Bergman is an experiential archaeologist who studies prehistoric bows from around the world. He crafts replicas in the same manner as the ancients, and then uses them in real-world settings to see how they perform. Bergman once made a 5-foot, 8-inch (172-centimeter) yew bow, and says it drew about 90 pounds (40 kilograms): to pull the bowstring back for firing required the same strength needed to lift 90 pounds with one hand. Bergman tested this weapon on a hunt. His shot blew a hole through a

HIGH TECHNOLOGY FROM THE LATE STONE AGE

The Iceman's grass cape served as a water-repellent, insulating topcoat. At night, the Iceman may have used the cape as a bedroll. Considerable skill was used in weaving, knotting, and splicing the grass. According to experts, the Iceman himself almost certainly constructed the cape.

The Iceman may have worn a hat. A ball of fur, still frozen in the ice that surrounded it, is among the objects being stored in Italy; it may well be a hat. This has not been officially determined yet, however, so this painting depicts the Iceman hatless.

The Iceman's clothing was made almost entirely of deerskin that was tanned in a vegetable-based solution. This painting depicts the coat as if it were worn with the fur side next to the skin, but it may have been worn with the fur side out.

The Iceman carried an assortment of tools that allowed him to exploit the natural resources around him. A stone knife, a drill, some retouchers, and an antler — together with intimate knowledge of the local plants and animals — were all that the Iceman needed to survive. The Iceman's powerful longbow was incomplete at the time of his death.

The Iceman's shoes were very cleverly designed. In warm weather, the Iceman could wear the shoes snugly around his bare feet. On cold days, he could relace the strings, allowing the shoes to accommodate wads of insulating hay.

Prehistoric leather work can now be studied by examining the tattered remains of the Iceman's clothing. More details about the Iceman will be learned when artifacts being stored in Italy are officially released for study.

deer's chest cavity from 30 yards (27.5 meters) away, "and right out the other side, and still going," he recalls of the arrow's flight. "I wouldn't want to stand in front of it."

The bow's underbelly bears marks from the discovery's most striking find, a copper ax. The 4-inch (10-centimeter) blade appears to have been cast from molten metal poured into a mold; when cooled, it was worked with a hammer. The blade was hafted in an L-shaped crook of yew wood and lashed in place with leather thongs soaked in glue, probably derived from birch-tree sap. The ax marks the end of one age and the beginning of another. By 3300 B.C., stone was giving way to metal as the material of choice for tools. The technological revolutions arising from the practice of metallurgy would change the world forever.

The Iceman was an expert at exploiting the natural resources around him—the gear he carried tells us that. Along with his ax, he carried an ash-handled knife with a tiny flint blade, just 1.5 inches long. "Had we found the blade without the handle attached," says Konrad Spindler, "we would have assumed it was an arrowhead."

The man also carried a small tool bag that held a bone awl, two flint blades, and what appears to be a flint drill. One of the

Scientists studying the Iceman's clothing remnants and hunting gear have gained an entirely new appreciation of prehistoric technology.

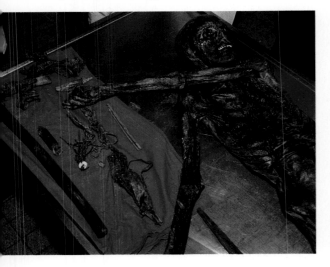

blades bears pollen traces from some of the 46 species of grass he used to make strings, lashes, and other gear. Unique grass artifacts, including a woven sheath for his knife and a voluminous cape, were manufactured as needed. The cape seems to be made from strands of plaited grass that hung vertically from the collar, as if a poncho were cut into thin ribbons. "Running down to his knees, it looked like a grass skirt you would bind around your shoulders," says Egg.

World's Oldest Quiver

The Iceman's equipment includes a U-shaped backpack frame made from hazelwood, and the oldest quiver ever found. The quiver contained 12 dogwood arrow shafts and two finished arrows, tipped in expertly chipped flint held in place with gum derived from boiled birch-tree roots. X rays of the quiver's interior show a ball of string, a piece of deer antler, and a couple of raw flints. A puzzling tool—a willow dowel with horn or antler running through the center—may have been used to shape flints. Broad calfskin straps recovered from the glacier may be a belt for the quiver. Somewhere along his trek, the Iceman harvested *Piptoporus betulinus,* a birch-tree fungus that is used in folk medicine as an antibiotic.

All the equipment was found carefully stashed about 15 feet (4.5 meters) from the body, just beneath the lip of the trench. Near the body were charcoal remnants from six different types of wood. They could mean the spot was a hideaway that had been frequented in the past, or that the Iceman carried embers for starting a fire.

The Iceman was prepared for cold weather. His shoes, oval shapes cut from what is possibly cowhide, cunningly fold up around the foot and leave plenty of space for insulating hay. The hay, from grass that grows only at 10,000 feet (3,050 meters) in the Alps, was anchored with grass laces threaded though numerous eyelets. A leather flap on each shoe shielded the laces from moisture.

People at the scene of the find described leather pants that "wound around the legs," but only uncertain fragments survived. The Iceman's jacket, worn under the

BOGMAN: Peat Moss Of Britain

In 1984 a peat cutter spotted a foot protruding from a clod of peat near Manchester, England. But it wasn't until 1990 that the full story behind the body —preserved in an acidic bog rather than in ice—finally came to light.

Only the upper body and one foot were found; the other parts were probably chopped up with the surrounding peat and sold as fuel many years ago. The victim, estimated to be a 33-year-old man, was murdered in a particularly gruesome manner around 50 A.D.

Peat Moss, as the British public calls him, was brained so hard with an ax that the force of the blow sheared off the top of his molars. A thin leather noose was then twisted tightly around his neck with the aid of a stick, crushing his windpipe. At the same moment, an expertly placed jab lanced his jugular vein, apparently to drain his blood. Finally Peat Moss was dumped into a pool of black water.

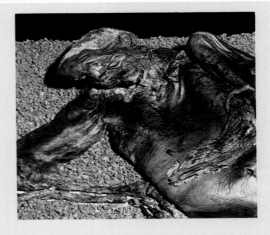

The overkill has been linked to a human-sacrifice ritual that was held during the Celtic celebration of Beltane. Three bloodthirsty deities needed attention at this time of year, so it appears that Peat Moss was triple-killed in order to satisfy all of them. Foniore says that whoever selected the burned bread at the feast would be sacrificed. And indeed, Peat Moss had the remains of a burned grain pancake in his stomach.

---

grass cape, is also in tatters. Small patches of finely stitched deerhide with the fur facing out seem to form the torso section, while broader strips probably served as long sleeves. The remaining fragments show many careful repairs over the jacket's lifetime—and one coarse repair. Egg thinks that the man may have been away from home for quite a while, damaged his coat during his trek, and quickly repaired it before he died.

A Shepherd or Hunter?

When a team of researchers returned last summer to continue the investigation, the site was blanketed in 20 feet (6 meters) of snow. The Sun melted 12 feet (3.6 meters) of it away, and volunteers shoveled through the remaining 600 tons of snow to reach rock. They found what may be a fur hat, or a cap for the quiver, along with the other half of the longbow and numerous scraps of leather and wood. Konrad Spindler collected many pelletlike animal droppings, tentatively identified as belonging to sheep. The Iceman may have been grazing a flock when he died.

An alternate scenario is that of a hunter, who hiked to the valley to cut a replacement for a broken bow, and was forced to seek shelter when a storm struck. Experts studying the body speculate that the Iceman, perhaps exhausted as a consequence of adverse weather conditions, fell asleep in the trench and froze to death.

Without more study, researchers are reluctant to draw further conclusions about how the Iceman lived and died. It will be many years before the frozen body reveals all of its secrets. Microscopic pieces of the Iceman's organs and equipment are spread among 120 researchers in Europe and the United States, and their work slowly progresses. A man who lost his life 53 centuries ago has become a messenger from the past, emerging from the ice to bring tidings of a world that has long since vanished.

TIME IN A CAPSULE

by Joseph DeVito

I f you could send a letter to the future, what would you say? Better yet, if you could mail a package to people generations from now, what would it tell them about your life? Thousands of such time-traveling presents have been sent throughout history. They're called time capsules.

A time capsule can be as grand as a scientifically constructed underground chamber, or as simple as a cardboard box kept in a special place. All you need is a sealed container that holds items important in your life. In the future, those who open the capsule will know what life was like in the past.

Instant Archaeology

Throughout the ages, pits, burial chambers, and sacred wells have been dedicated to the preservation of important objects. Many of these inadvertent time capsules, such as the pyramids of ancient Egypt, have outlasted their creators by thousands of years. "It's a very human desire to reach people beyond our own lifetimes. It's a way of carrying on our legacy by letting those in the future know what was important in our lives," says Paul Hudson, one of the founders of the International Time Capsule Society (ITCS), an organization dedicated to promoting and studying time capsules throughout the world. Established in 1990, the ITCS is headquartered at Oglethorpe University in Atlanta, Georgia. "Americans are especially interested in time capsules because our history is still relatively new," Hudson adds.

The ITCS has established a registry of time capsules. Of the approximately 10,000 time capsules worldwide, most of them are lost due to poor planning or secrecy. The ITCS database will help monitor these projects so they are not forgotten, stolen, or lost.

Treasure Trove

It's no coincidence that Oglethorpe University is home of the ITCS. The university is the site of the famous Crypt of Civilization,

Children in Florida (above) buried a time capsule filled with memorabilia from 1986, the last time that Halley's comet crossed the sky. The capsule will be opened in 2061, when the comet again appears.

considered the first successful attempt to encapsulate a record of Western culture for examination by future generations.

The swimming-pool-sized chamber contains over 640,000 pages of microfilmed material, hundreds of newsreels and recordings, inventions, toys, and thousands of other items from everyday life, past and present. "The Crypt is the granddaddy of them all, a museum covering everything from the Egyptians to this century," Hudson explains. "If the scheduled retrieval works, it will convey 6,000 years of civilization."

Included are instructions on how to perform surgery, a Kodak camera, and film clips of famous personalities from Adolf Hitler to Popeye the Sailor. Some of the pop culture items from the 1930s seem like curious relics even today: an electric "Toastolator" and Artie Shaw records.

The Crypt is set within the granite foundation of the university's administration building. The vault is airtight for better preservation of the objects, and has had its steel doors welded shut. The Crypt was sealed on May 28, 1940, and is not to be opened until A.D. 8113, 6,177 years from the start of its construction. That date is as far in the future as 1936 is from 4241 B.C., the date experts calculate is the first recorded date in history.

The U.S. Government has a record of the longitude and latitude of the Crypt's location, although it's unlikely there will be a U.S. Government in the year 8113. In fact, there is also a device to teach the English language to the Crypt's finders. The Language Integrator is a hand-cranked multimedia device that displays pictures of objects with their names written in English underneath. The device includes a phonograph with the recorded pronunciation for each word.

The Time Torpedo

The Crypt of Civilization sparked the modern interest in time capsules that has escalated over the years. A similar project was developed by the Westinghouse Electric Corporation for their pavilion at the 1938–39 New York World's Fair. G. Edward Pedray, the program's director, was the first to coin the phrase "time capsule."

The torpedo-shaped capsule was about 7 feet (2 meters) long, and made of Cupaloy, a metal of almost-pure copper that could resist the long-term effects of underground burial. The capsule was buried 50 feet (15 meters) beneath the Earth, and is not scheduled to be opened until A.D. 6939.

Inside were a variety of objects from the world up until 1938. Samples of fabrics, metals, rubber, asbestos, and food grains were included with human-made objects such as an alarm clock, a safety pin, a toothbrush, a pair of bifocals, and a Bible. A

newsreel and a microfilm essay with more than 10 million words describe the culture, science, and society of the world of 1938.

When another world's fair came to New York in 1964–65, a second capsule was interred with examples of the changes that had occurred in the 27-year gap. This capsule contained the new 50-star U.S. flag, a bikini bathing suit, contact lenses, and a piece of a U.S. spacecraft. Also included were records of the new atomic age, including a sample of radioactive carbon and a film strip about the USS Nautilus, the world's first nuclear-powered submarine. Replicas of the capsules and their contents are on display at the George Westinghouse Museum in Pittsburgh, Pennsylvania.

Time capsules were also part of two later world's fairs: Expo 67 held in Montreal, Canada; and Expo '70, in Osaka, Japan. At the Japanese fair, two capsules were created containing thousands of unusual items, ranging from artwork to a typical Japanese

Time capsules need not be buried. At the Museum of Discovery and Science (below) in Fort Lauderdale, Florida, the capsule was entombed in a vault.

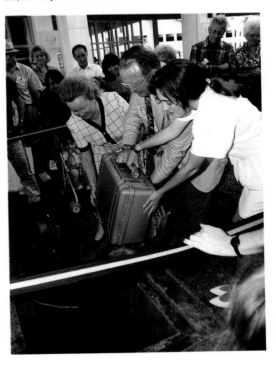

The Crypt of Civilization (above) in Atlanta contains items dating from the ancient Egyptians through 1940, when it was sealed. The granite shaft (above right) marks the burial site of a time capsule from the 1939 World's Fair. At the 1964–65 Fair, officials (right) reviewed the capsule's contents prior to the burial ceremony.

business-appointment book. Stranger still, the capsules contain the blackened fingernails of an atomic-bomb survivor and a recording of a dog barking at night!

The Keepers of the Capsule

Officials planning a centennial celebration for Olympia, Washington, wanted something of lasting significance to mark the city's birthday. The result? With the help of Knute Berger, a Seattle-based independent newspaper editor and ITCS co-founder, the Washington Centennial Time Capsule became the world's first renewable time capsule.

"Time capsules are a great way to commemorate ceremonial occasions," Berger says. But several hundred capsules are lost for every one that makes it to a future generation, Berger estimates. "Most time capsules are lost because people overestimate memory," he claims. "How do you communicate when the capsule is to be opened, when the language will be so different? What kind of sign do you leave?" He believes that many planners make the "arrogant presumption that future people will do your bidding."

To guard against the capsule being forgotten, Berger appointed hundreds of children as "keepers of the capsule."

In 1989, 300 10-year-old Washington youngsters born within a week of the state's birthday were registered for the program. New keepers will be appointed every 25 years as the capsule is updated, until the opening ceremony on the state's 500th birthday in 2399. "The generations will serve as keepers, and later as recruiters, as part of a civic ritual," says Berger.

The time capsule is contained in a safe with separate compartments that keepers will update to create "layers of history." Each container is filled with microfilmed memorabilia and messages. Situated in plain sight, the 7-foot (2-meter)-tall, 3,500-pound (1,600-kilogram) steel vault sits in the capitol building rotunda, where it can't easily be moved or misplaced.

What Will We Send Beyond?

Why do humans want to communicate with their descendants? "I'm not sure why people do it," Berger admits. "As we've realized that we are transitory, time capsules may be our only chance to be remembered."

The written messages of Washington State's Capsule Keepers were snapshots of their concerns—many asked for forgiveness for the state of the environment, while others

A time-capsule ceremony makes a fitting climax for important community events. In Ossining, New York, citizens (right) had the opportunity to contribute items of their choice to the local time capsule.

A few hints on organizing your time capsule

• Set a date for reopening the capsule. The longer the amount of time, the more difficult the job becomes. Be sure to leave notes or other documentation so you won't forget its existence or its location.

• Select a container. As long as the interior is cool, dark, and dry, the contents can be preserved. This can mean a safe or airtight box for long-term projects, or a sturdy box that can be kept in an attic or closet.

• Find a secure indoor location. Although many time capsules are placed underground, this romantic notion of a "buried treasure" is often impractical and hostile to the capsule. Those that are buried tend to be easily forgotten or fall victim to the elements.

• Make a thorough list of the items you will include. While examples of technological advances and the best culture of the time are good, it's important to keep a balance of funny and everyday items.

These can often speak volumes about how the people who used them lived their daily lives. "Encapsulate pop culture and what you think history is like today," advises Paul Hudson, co-founder of the International Time Capsule Society (ITCS). "Who knows? In three centuries, dental floss may be a great artifact!"

• Have a sealing ceremony, and keep a photographic record of the event. This will not only help those present remember their mission, but it will also provide support for those who will someday have to open the capsule. Marking the location with a sign helps ensure recovery.

• The ITCS can help in managing your time-capsule project. It will add your name to its time-capsule database so it won't be forgotten.

• **Contact: International Time Capsule Society, c/o Registrar's Office, Oglethorpe University, Atlanta, GA 30319–1441 (404–261–1441)** expressed hope for cures for AIDS and other diseases. "Time capsules can also be a way of making peace with the future," says Berger.

Time capsules represent a serious effort to manufacture a possible archaeological cache of a culture that will be poorly understood by the time the capsule is opened. "An encyclopedia on CD-ROM represents a great stride in technology," Paul Hudson observes, "but will those who uncover our civilization have any means of recovering it?"

As an example, Knute Berger mentions a time capsule in Idaho that contained recordings made prior to the advent of phonograph technology. The wire recordings were intact, but no one could locate an apparatus on which to play them.

The Good, the Bad, and the Simply Misplaced

"As we approach the year 2000, people are more interested in time capsules," says Hudson. "Time capsules are replacing ribbon-cutting ceremonies." The ITCS receives inquiries every day about time capsules, especially about family capsules, because they are so popular with children.

Time capsules come in many shapes and sizes. The world's largest time capsule is the Tropico Time Tunnel, located outside Los Angeles, California. Scheduled to be opened in 2866, the former gold mine contains entire rooms of furniture and a baseball autographed by Willie Mays.

A capsule begun at the 1992 World's Fair in Seville, Spain, has an unusual twist —it's a tar pit! Art objects of all types are thrown in to be preserved the way ancient animals were at the La Brea Tar Pits in Los Angeles, California. The anarchic format of this project shows the humorous side of time capsules. "They've even offered to take Lenin's body from the former Soviet Union," remarks Hudson.

Alas, many interesting time capsules are missing in action. For example:

• In 1983 the cast of the hit television series "M*A*S*H" buried a time capsule beneath a Hollywood studio parking lot in a secret ceremony. Too secret, as it turns out—no one remembers exactly where it's buried!

• In 1941 workers lowered a copper capsule into the Kingsley Dam in Nebraska. A plaque describing the capsule's whereabouts was sent to the state capitol building for safe-keeping, where it was promptly lost.

Not all capsules are victims of poor planning. Officials in Oklahoma buried a 1957 Plymouth to commemorate the state's 50th anniversary. The capsule containing the car will be opened in 2007, with the car being awarded to the person or heirs of the person who came closest in 1957 to predicting Tulsa's population 50 years later. In 1976 the city of Braintree, Massachusetts, sealed a capsule that contained a U.S. savings bond. The bond should mature enough to pay for the scheduled 2076 opening celebration. The capsule is checked at an anniversary ceremony held every Fourth of July.

Transcending Space and Time

Time capsules have been sent into space with the Voyager 1 and Voyager 2 space probes. Each capsule carries a gold-plated copper phonograph record containing visual and audio images. The sounds range from the calls of humpback whales to classical compositions and a song from American rock-'n'-roll singer Chuck Berry. Also included is information on the Earth's position and the genetic makeup of human beings. The capsules should last for 1 billion years.

In 1992 a space time capsule sponsored by the Rochester Museum & Science Center in New York was prepared to commemorate the International Space Year. Called Space Arc: The Archives of Mankind, the capsule contains digitally scanned messages, drawings, photos, essays, and poems provided by thousands of schoolchildren. The master optical disk will be aboard a satellite headed for Earth orbit.

Not all time capsules need to be closed for a very long time. Smaller capsules for families and children can yield a great deal of information after only a few years. Photos, a newspaper from the day your capsule was closed, and a handwritten note describing your daily life and what you think you will be like when you open the capsule are all important artifacts. Imagine: the future person who reads that message could be you!

RESTORING MOUNT
RUSHMORE *by Neil Springer*

"... let us place there, carved high, as close to heaven as we can, the words of our leaders, their faces, to show posterity what manners of men they were. Then breathe a prayer that these records will endure until the wind and the rain alone shall wear them away."
—Mount Rushmore sculptor Gutzon Borglum, at the dedication of the Washington head, July 4, 1930

Mount Rushmore has begun to show its age after 50 years of facing the elements. The imposing granite structures are undergoing a face-lift using state-of-the-art technology.

It is a mark of Borglum's faith, foresight, tenacity, and, perhaps, conceit that the sculptor should have spoken those words only three years after work had officially begun on the monument. Indeed, it would be another 11 years before the sounds of drills and air hammers echoed their last around the faces of the four great presidents—George Washington, Thomas Jefferson, Abraham Lincoln, and Theodore Roosevelt—carved into the smooth-grained granite in the Black Hills of South Dakota.

Borglum's vision spoke to the eternal, but his life, like the lives of all mortals, was circumscribed. At 4:00 P.M. on October 31, 1941, when, in the words of author Rex Alan Smith, the mountain "returned to the timeless silence from which Gutzon Borglum so many years before had roused it," the great sculptor had already been dead for seven months.

No carving on Mount Rushmore has been done since then, and Borglum's project was never completed as he had planned. Now, however, more than 50 years later, innovations in science, technology, and chemistry are being used to fulfill Borglum's wish that the memorial endure for centuries, until the wind and rain slowly wear it away.

Shifting Goals

It's a wonder that Mount Rushmore was ever carved at all. In 1923, when Doane Robinson, the South Dakota state historian, first conceived the idea of carving giant statues in rock, he was thinking more on the lines of such western heroes as Kit Carson and Buffalo Bill, not presidents. Moreover, his site choice was the granite formations in the Black Hills known as the Needles. Indeed, Borglum was not even his first choice as a sculptor. Fortunately, public reaction to Robinson's original idea was tepid at best.

However, Robinson persisted in his dream of a giant carving, gaining the influential support of men like U.S. Senator Peter Norbeck, U.S. Congressman and attorney William Williamson, and John Boland, a successful businessman.

In 1924 Gutzon Borglum was invited to the Black Hills to study Robinson's proposal. Due to financial difficulties, Borglum had just been dismissed from a site in Stone Mountain, Georgia, where he was to have carved a bas-relief of Confederate generals and their troops. Borglum's misfortune would be the answer to Robinson's dreams.

A renowned sculptor, Borglum had begun his art studies in San Francisco and completed them in Paris. Initially successful there as a muralist and illustrator, Borglum developed a friendship with Auguste Rodin, who ultimately influenced Borglum to shift his interest to sculpture.

Upon his return to the United States, Borglum became popular as a sculptor of portraits and public monuments, such as the 12 stone apostles for New York's Cathedral of St. John the Divine, and the 6-ton marble head of Lincoln that stands in the Capitol building's rotunda in Washington, D.C.

After much searching around South Dakota's Black Hills, Borglum chose for the colossal monument the 5,724-foot (1,745-meter)-tall Mount Rushmore, which dominates its surroundings and is bathed in sunlight most of the day.

Private donations were gathered for the project, and work officially commenced on August 10, 1927. The project was eventually authorized as a national memorial by Congress in 1929, and the federal government contributed $836,000 of its $990,000 cost. But appropriations never came easily—it was a constant struggle, and work was delayed repeatedly, either due to lack of funds or bad weather. Although 14 grueling years would pass before the project ended, the National Park Service estimates that only six and a half of those years were spent actually carving the monument.

The four presidents—Washington, Jefferson, Roosevelt, and Lincoln—were chosen to represent, respectively, the nation's founding, philosophy, expansion, and unity. Borglum had to make changes in his design nine times because of cracks and fissures in the rock. For example, Jefferson's head was originally carved on the right side of Washington in 1931, but faulty rock was encountered, and the unfinished head had to be blasted down.

The carving of the 60- to 70-foot (18- to 21-meter)-tall faces was an impressive engineering feat. The drillers, who climbed 506 stairs to get to work each day, and spent their days suspended over the face of the mountain in "swing seats," were, in a sense, the real heroes of the project. They removed some 450,000 tons of rock by blasting with dynamite. At each stage when the work required more finesse, drillers would exchange jackhammers for small drills, then wedging tools and hammers, and finally air hammers, to smooth the surface.

Today the memorial remains unfinished, at least according to Borglum's original plans. Nevertheless, by 1941 the "Shrine of Democracy" monument stood whole and complete in its majestic splendor.

Wrinkles of Time

In 1989, as the 50th anniversary of the completion of Mount Rushmore approached, the monument was beginning to show signs of

Working from a small model, sculptor Gutzon Borglum (left) transferred crude dimensions for the monument's features using artistic, rather than engineering, techniques.

its age. Indeed, cracks and fissures have always been present in the rock, both on the faces of the presidents and behind their heads. Each year, the Park Service practices a kind of gigantic cosmetic surgery, filling cracks in the faces with a compound made of granite dust, linseed oil, and white lead, a mixture that Borglum himself had used.

Still, water would seep into the joints, freeze in the winter, and then undergo daily freeze/thaw cycles in the spring. Borglum's brittle mixture could not bear up to the strains of this contraction and expansion. Worse, the cracks behind the heads, which had gone untreated, were horizontal water canals also subject to the freeze/thaw cycle. Due to the cracks' strategic location, the failure to repair them could lead to intrinsic damage to the monument.

It became clear to the National Park Service and the Mount Rushmore National Memorial Society that the monument would need a face-lift. Moreover, the associated visitor facilities, originally designed for 1 million visitors annually, would need to be modernized and expanded in order to better accommodate the throng of visitors, which has grown to more than 2.5 million.

The National Park Service developed an ambitious plan that called for work on the visitor information center, the museum, the amphitheater, and the parking areas, as well as a historic restoration of Borglum's sculpture studio and the preservation of original artifacts and equipment. The expected cost: around $28 million.

It was obvious that the National Park Service, with some 350 parks to manage and other priorities to be met, could not fund all the needs for Mount Rushmore in this century. So the venerable society made an offer the Park Service couldn't refuse.

In an unusual example of private and public cooperation, the society proposed to run a national campaign to raise money, setting its sights on $21 million. Any money raised beyond that goal would be appropriated by the Park Service, explained Gordon Brownlee, director of the nonprofit Mount Rushmore Preservation Fund.

The Park Service agreed to this plan, which began with a structural analysis of Mount Rushmore to assess the long-term stability of the monument.

Enter Tim Vogt, a geologist with Re/Spec Inc., a geologic-research company in Rapid City, South Dakota, who had heard through a chance conversation that the Preservation Fund needed to make a structural survey. Vogt, who was interested in creating realistic three-dimensional computer images of structures, offered his company's unique services.

"The main objective of the study was to find out if the carving is solid and stable. Then, what is the [stability] projection for the next lifetime or two," Vogt says. In order to look at the cracks, project their path into the granite, and see how they fit together

Mount Rushmore is one of the most popular tourist spots in America. The visitor facilities, originally designed to accomodate 1 million tourists a year, now host over 2.5 million visitors annually.

to make blocks of rock, Vogt's team used a technique called photogrammetry.

Traditionally, this technique calls for the use of two aerial photographs, taken from slightly different positions in a plane that is directly over the target. These photos produce a "stereo" image that can be manipulated to create three-dimensional measurements.

"We did that, but we also took pictures out of the side of a helicopter and on the ground. We took more than 300 photos in order to make a number of different stereo combinations, plot them on a computer, and come up with a 3-D model," Vogt explains.

3-D Imaging

The results were quite amazing, says Brownlee. For the first time, accurate measurements, within centimeters, were available of the relationships between the faces, like the distance from Washington's nose to Jefferson's nose.

During its construction, Borglum used a model of the project from which he transferred crude dimensions up to the mountain. But he made all his corrections visually, not like an engineer, but rather, like the true artist that he was, Vogt says.

In addition to providing exact measurements, Vogt was able to look at each crack and fracture—he cataloged 144 significant cracks—in 3-D space, and determine that the monument is composed of 22 separate and defined blocks.

From this data the Re/Spec team was able to focus on which blocks would most likely move if there were no friction at all, and whether they would likely move in toward the mountain or out from the face.

According to Vogt, even in the absence of friction, the planes in the large majority of blocks would tend to push inward. Since the granite is so solid, this doesn't present a problem. However, it was determined that two "key blocks," one located in the forehead and forelock of Roosevelt, and the other in the center of Lincoln's face, needed monitoring. In addition, a tabular-shaped block in the right temple area of Washington, while not a key block, also bears watching.

"There is no imminent danger," cautions Vogt, "but if humanity were suddenly to disappear and there was no maintenance done, the action of the water freezing and thawing in and out of the cracks would likely cause those blocks to fall out."

Early results made it clear to Vogt that something more had to be done to seal the cracks, particularly those in the back of the carving that were less visible but far more

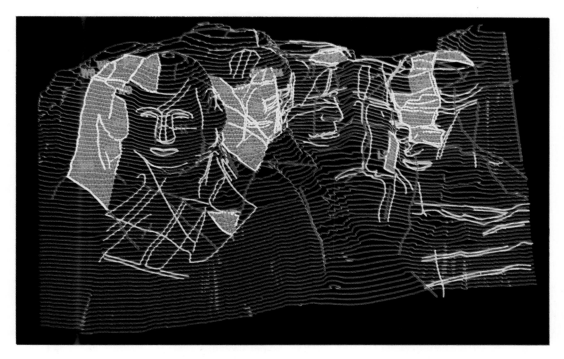

Restoration specialists used computer imaging to analyze the monument's underlying structure and stability.

important to structural stability. The sealant needed to be an off-the-shelf type of product that was easy to apply.

High-tech Sealant

Coincidentally, a Dow Corning employee, visiting Mount Rushmore as a tourist, heard about the project and the need for a new sealant. When he returned to work, he triggered a reaction at Dow. Jerry Kehrer, a construction specialist, was called on to be the point man.

He contacted Vogt and arranged to bring a sample of a silicone construction sealant for any in-house testing Re/Spec wanted to perform. Kehrer knew his product would be the perfect answer for the monument's woes.

"It was an established product on which we had done extensive testing. We had tested for durability, adhesion, flexibility, and resistance to staining, among other things. It's been used on two or three high-rise buildings in Minneapolis, on the Admiral Perry memorial at Lake Erie, and a formulation for metal was used on the Statue of Liberty," he says.

Most important, he thinks, is the sealant's ability to retain its flexibility through the wide temperature swings on Mount Rushmore that range from −30° F (−34° C) in the winter (with wind chills as low as −70° F, or −56° C) to 120° F (48° C) in the summer. The sealant is also resistant to ultraviolet light.

The sealant was compared with similar products and was found to excel in all performance testing by Re/Spec. The first application started in late 1991, and sealing all the monument's cracks is expected to take five years, says Kehrer.

Vogt has submitted a proposal to install monitoring instruments on Mount Rushmore to keep track of any rock movement over the long term. And he points out that the sealant is being applied just on the top part of the rock joints, not shot all the way into the cracks. This allows for easy removal in case there is a major technological advance in the next few decades, he says.

Vogt is optimistic that Borglum's wish will be fulfilled. "The rock is solid, the climate is favorable for preservation, there is low rainfall, and no acid rain. The monument will erode, but very slowly. I read that Borglum, knowing about the erosion factor, left extra rock on the upwind side of Washington's nose. In 10,000 to 50,000 years, Washington may look just a little bit better."

HORSE POWER

by John H. White

During the second half of the 19th century, low-tech, horse-drawn street railways were the preferred means of urban mass transit. In New York City (right), millions of passengers used the service.

Public transit before the age of the trolley was a decidedly low-tech operation. Horsecars moved through city streets at speeds scarcely faster than a walk. Their tracks, embedded in the pavement, were primitive assemblies of wood and strap rail that hearkened back to the earliest days of railway engineering. The cars were tiny four-wheel affairs with hard seats, minimal lighting, and no heat. Braking power depended on the strength of the driver's arm. The motive power was nature's own beast of burden, which had been in haulage service since the beginnings of civilization.

The horse was a serviceable, if unenthusiastic, street motor, but horse railways were costly and inefficient. The beasts were expensive, subject to illness, capable of working only short hours, and always hungry. Their presence on city streets presented a health hazard, as well as an aesthetic one, in the form of urine and feces. Because thousands of horses were in regular use in the larger urban areas, the sanitary problem was inescapable. Human labor was also expensive. The cars required a two-man crew: a driver to manage the horses and brakes, and a conductor to look after the passengers and collect the fares. Since the cars were so small, a great many were needed, thus swelling the demand for men, horses, and rolling stock.

This rudimentary form of city railway was clearly hard on both man and beast. With so many shortcomings, why did it last for nearly 50 years? Why were thousands of miles of track laid down in every major city in the nation for these hopelessly inefficient little vehicles? Why were millions of dollars invested each year? The answer is simple: Horsecars worked. For all their failings, they could move passengers and earn their owners a decent profit. Reliability was their virtue. They did a

job that nothing else could do until the advent of cable and electric traction. (Steam power was in very wide use for long-distance travel, but the need for compactness, rapid changes of speed, and frequent stops and starts made it inconvenient for most urban transportation.) When practical cable lines came into use during the 1870s, they proved economical only for heavily traveled lines or unusual topographical situations, such as hilly San Francisco. Electric cars were not perfected until the late 1880s. And so, in an age devoted to science and industrial progress, city railways remained loyal to the most antiquated form of power.

The Beginnings

It all started quite by accident. The New York and Harlem Railroad (NY&H) was incorporated in 1831 to build an ordinary steam railroad north through New York City to the Harlem River. It was never intended as a

street railway—the term was not even in use at the time—but intended or not, the Harlem line quickly evolved into the classic model for future street railways.

Service on the NY&H began late in 1832, running along the Bowery between Prince Street and 14th Street. At a festive City Hall dinner celebrating the opening, New York's mayor declared, "This event will go down in the history of our country as the greatest achievement of man." The NY&H used mules, then horses, then steam in its first few years as it expanded northward. The noisy, smoky, pine-burning locomotives drew considerable public opposition, however, and after an engine blew up at Union Square in 1839, killing the engineer and injuring 20 passengers, mobs tore up tracks on the Bowery. The city responded by banning locomotives from the densely populated southern part of town, and from then on, horses pulled the cars to a safe rural area —eventually as far north as 42nd Street— where steam engines took over.

New York was ripe for some form of cheap public transit. Omnibus—horse-drawn bus—service was just getting under way, offering the right compromise between walking and paying cab fare. But the buses were very small—a dozen passengers were a load—the streets were rough, which made for an uncomfortable ride, and climbing up the steep, high back steps of a bus was too much for many potential passengers.

The Harlem's cars could handle 20 passengers with ease. The rails offered a fairly smooth ride, and getting on and off was no problem. As the NY&H extended its tracks northward to Harlem, city folk found it a convenient way to go uptown or downtown, particularly when the cars started running on a six-minute headway by about 1845. In 1863 the Vanderbilt family assumed control of the NY&H. The city end was operated separately, but remained under Vanderbilt control. The horsecar operation grew and prospered until by 1893 the Fourth Avenue line, though only 8.5 miles (13.6 kilometers) long, carried 21.8 million passengers a year and was valued at $24.9 million. Even this late in the electric era, it was a horse-powered line. Behind the times and ineffi-

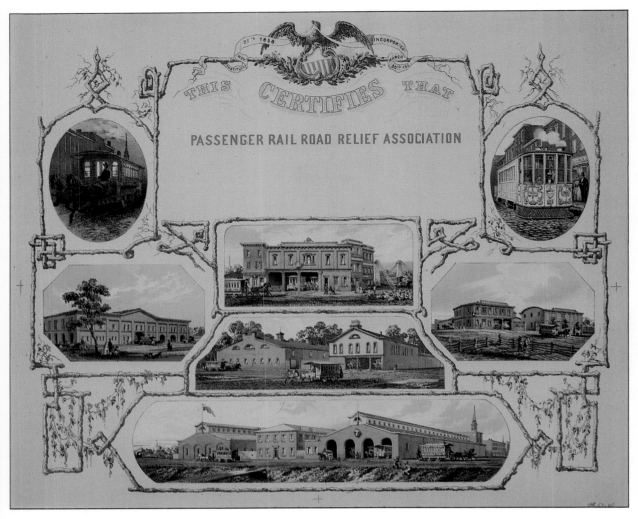

THIS CERTIFIES THAT

PASSENGER RAIL ROAD RELIEF ASSOCIATION

A variety of horse-drawn streetcar scenes from Philadelphia appear on a certificate from an 1860s benefit society offering insurance to railroad employees.

cient though it was, there could be no question about its profitability.

The early success of the Harlem experiment did not, surprisingly, bring on a rush of imitators. In 1852 a second horsecar line, the Sixth Avenue, was opened in Manhattan. And then, as if some magic curtain had been pulled aside, other investors belatedly recognized the potential of city railroads. The Second and Third Avenue lines opened in 1853, as did a line in Brooklyn. A Boston-to-Cambridge line, opened in April 1856, was the first outside the New York area (except for a railroad in New Orleans that some have called a streetcar line). Within four years, Boston had 57 miles (92 kilometers) of horsecar trackage, and was carrying 13.6 million passengers per year. At the same

time, New York had a dozen lines, and was adding track and equipment rapidly.

A trend had started. Now everyone wanted a street railway. They were fashionable, they were convenient, and they promised to be very profitable. Philadelphia scurried to catch up with its rivals to the north by opening its first line in early 1858. During the next two years, just about every city with any pretension of being major joined the ranks. Lines opened in Baltimore, Cincinnati, Chicago, St. Louis, Milwaukee, and even distant San Francisco. Washington, D.C., was a little slow, with its first horsecar appearing in 1862, and Denver, Colorado, a very young city, did not put tracks down until 1871.

The industry had come a long way in a short time. In 1860 there were just over 400 miles (645 kilometers) of street-railway tracks in this country. The industry's investment value was $14.8 million. Within a decade, that value went up to $65 million. Horsecar operations were spreading into the suburbs from the center city. By the early 1880s, America's horsecar business was a major operation.

During the 1880s city railways grew faster than a well-fertilized pumpkin vine. Mileage almost tripled, and while horse-powered lines prevailed, other forms of power began to gain a foothold.

Investors were attracted by the faddish street railways, though the prospects for a large return were somewhat oversold. Market analysts of the day could point to rising traffic: the passenger count on the New York City lines had doubled between 1855 and 1864. Yet few lines had net earnings any better than 8 percent, a so-so return in those days. The inflation caused by the Civil War doubled the price of feed, while wages went up by half. Most pioneer horsecar lines suffered from the inflation; Cincinnati operators reported losses; the St. Louis horse-railway system, in particular, was said to be a very poor investment.

Four-legged Engines

Rich and poor city railways agreed that the problem with the bottom line was horses. A hundred thousand of them were literally eating up the profits. The feed bill exceeded wages by nearly half, and made track repairs and building maintenance look trifling. The care and feeding of horses made up 40 to 50 percent of operating costs, and about 40 percent of total investment.

Horses needed more than just food. They needed shelter, grooming, medical care, long rest periods, bedding, stable clearing, and—sometimes as often as every three days—new shoes. The purchase and repair of harness was a costly extra requirement. In 1877 the Market Street Railway in Philadelphia registered daily costs of $9.31 for keeping nine horses, the number required to dependably power a single two-horse car.

Add to this depreciation ($1.15) and interest on the cost of the horses ($0.20), and the daily bill comes to $10.66. Why the con-

The ASPCA owes its creation to public concern over the health of streetcar horses. Officials staged surprise inspections to guarantee the health of the horses (left). The advent of electrically powered trains (on elevated tracks, below) led to the demise of the horse-drawn system.

ductor's wages were omitted from this list is not clear; if included, they would add another $2.00 per day.

Why nine horses per car? The cars rarely traveled more than 50 miles (80 kilometers) in a 12- to 14-hour day, but a horse could never be worked more than six hours, and three was preferable. Hence, teams had to be replaced regularly. Men and cars ran the full day, but the horses worked short shifts and were pampered and treated with care, like the valuable assets they were. Their numbers were swollen by the need to keep 10 percent extra on hand for illness, and another 10 percent for traffic emergencies. If normal needs called for 1,000 horses, an experienced street-railway manager would insist on stabling 1,200.

Even more horses were needed by lines that had steep grades. Tow boys drove the double-headed teams required to pull a loaded car uphill, then rode back down to catch the next car waiting for a boost.

Illness and Excrement
Horses not only spent 20 hours or more of their workday eating and resting, but got a full day off each week, and active horses were kept separated from resting ones in the stables, so as not to disturb their sleep. Stables had to be huge—two or three stories high—because so much space was needed for hay and grain storage, shoeing, harness repair, and storage and manure bins. The North Chicago Railway's barn at Belden Avenue and Jay Street measured 125 by 238 feet (38 by 72.5 meters) and housed just 260 horses. Large systems had car barns and stables scattered around the city, and of course the price of land and structures further inflated the capital costs of the typical horsecar line.

Yet for all this good care, horses failed their masters when the need was greatest. Their energy tended to flag in the very weather when people used the cars most. When a heat wave hit, horses lay down in the street. When a blizzard blew in, they lost all interest in motion except toward the stable. Even in the best of weather, horses were given many days of sick leave because of minor illnesses, sprains, and sore feet.

If there was concern over the health of the horses, there was also concern over the health of people required to live around them. Horses spent many hours a day on the streets, and paraded through the most densely populated areas of the city, spraying their excrement everywhere, splashing it against the dashboards of cars or dropping it to the street to be run over and picked up by each passing vehicle. An average 1,000-pound (450-kilogram) horse could yield nearly 50 pounds (23 kilograms) of dung a day, much of it while working.

If horse manure had simply been smelly and annoying to step around, it might have been tolerable, but it also presented serious health problems, especially in the spreading of tetanus. Systematic efforts to sweep up helped, but not enough, considering how rapidly the horse droppings accumulated. Curiously, there was a good side to this noisome by-product of horse traction. The manure shoveled out of the stables produced $2 to $6 per horse per year when sold for fertilizer. A big operation such as New York's Third Avenue line could thus boost its annual income by $14,000.

Mules were thought to offer relief to street-sanitation problems because, amazingly, they could be toilet-trained. In 1880 a street railway in Louisville demonstrated this miracle. At the end of their run, the mules first were walked to a manure station to defecate, then to a drainage area to urinate, and finally back to their stalls. No one could ever do that with a horse, except perhaps a well-trained trick horse. Mules had many other advantages. Known for their ability to withstand heat, they were favored in the South. They rarely suffered from hoof or leg ailments, and could work 20-mile (32-kilometer) days even in scalding San Antonio, Texas. Plucky, enduring, strong, and swift, mules were good for 10 years' work. But then they would grow fat and lazy—and therein lay the problem. There was a poor market for secondhand mules. Horses, by contrast, were readily resellable, and at a good price, because few street railways worked them for more than four years.

Street railways bought about 25,000 new horses each year. Purchasing agents

Passengers expected streetcar punctuality no matter what the weather. A misstep by a horse on a muddy surface (top) could easily lead to a broken leg. Some horses objected to working in snow (inset).

were advised to acquire an 1,100-pound (500-kilogram) animal, about 15½ to 16 hands (1 hand equals 4 inches, or about 10 centimeters) high, that was not too leggy and did not have too much daylight beneath him. Good legs and feet, of course, were very important, and so was a good disposition, because the horse would be commingling with other vehicles and crowds of people. Some horses adapted very well to railway service, and would start and stop at the sound of the conductor's signal bell without the driver's even having to move the reins. Whenever possible the same drivers and horses were paired, because the animals responded better when handled by someone they knew.

City railways were a rough-and-tumble business, and nothing in their service was gently used. Yet curiously, the component that received the best treatment became the subject of a zealous reform movement. Henry Bergh, a wealthy New Yorker, was horrified by the cruel treatment of animals he witnessed while serving as a diplomat in Russia in the 1860s. At about the same time, he became aware of the work of the Royal Society for the Prevention of Cruelty to Animals in England, and so founded a sister organization soon after his return to New York in 1865. The focus of the American Society for the Prevention of Cruelty to Animals (ASPCA) became horses, particularly those in heavy dragging service.

Car Designs

Horsecars came in a number of sizes and configurations. The most basic division was between open (or summer) cars and closed (or winter) cars. Only the larger systems had summer cars; smaller lines could not afford to maintain a double set of vehicles. If any one style of closed car was typical, it was the 23-foot (7-meter)-long, 22-passenger double-horse car, which cost around $1,000. Its short wheelbase of 6 feet (1.8 meters), made necessary by sharp curves common on city railways, gave it a bouncy, unstable ride.

Ventilation was achieved in several ways: opening the end doors and windows, perhaps while drawing louvered wooden shades, and opening the small glass windows mounted in the roof eaves. Most horsecar roofs featured a low clerestory built into the curved sides of an ogee-shaped roof, its ornamental effect enhanced by ruby-red or royal-blue glass in the clerestory windows, sometimes cut or etched decoratively.

The body was set low on the wheels to make getting on and off as easy as possible, and a broad step helped all but the most feeble climb aboard without too much trouble. Small children and the very elderly might need assistance, as did fashionably dressed ladies whose frocks made all but the most delicate steps nearly impossible to mount.

John B. Slawson, a street-railway manager in New Orleans, invented the compact "fare box" or "bobtail" car in 1860 for lightly traveled lines where patronage was limited. It seated 10 passengers, weighed just over a ton, was pulled by a single horse, and operated without a conductor. The little bobtails were wonderfully economical, but they were never popular with the public, which objected to their cramped quarters and a ride even bumpier than in their larger cousins. But the bobtails prevailed, despite all grumbles, and were found in service everywhere in the United States.

City railways kept trying to imitate their big-brother steam roads. In 1871, with the palace-car craze running at full fever, the Third Avenue Railway began running a sumptuous parlor horsecar that featured revolving chairs, sofas at either end, carpets, large plate-glass windows, silver spittoons, and fancy interior woodwork. The West Philadelphia Railway began operating a similar car a few years later. Apparently, too few riders were ready to pay a 15-cent fare, and parlor-car service was never much expanded.

Most cars had plain, bent-veneer seats with holes punched for ventilation, and had little light and no protection against the cold —except for a few pitchforks of hay to insulate the passengers' feet from the car floor. Those who complained about drafts were advised to buy warmer overcoats. As for lighting, tiny oil lamps, mounted in opposite corners at either end, cast a feeble glow just bright enough to let someone with good eyesight enter and leave the car. A few lines operated cars fitted with a center ceiling or dome light, which did much to brighten the interior. But the elegant lamps added $100 to the cost of a new car, and most city-railway managers felt they were too dear.

End of the Line

Electricity had been the energy source of choice from the beginning, but not until around 1880 were there generators that could supply the power needs of a large street railway. By the 1880s a host of inventors and mechanics had finally devised a truly practical system. Cable and horsecar lines were converted to trolley operations, and by 1890 one-sixth of all U.S. street railways were electric. Ten years later the conversion was virtually complete; only 259 miles (417 kilometers)—1 percent of the total system—were still horse-powered.

The horsecar became a dinosaur in a single decade. The clomping hoofs and jingle of the harness bells were soon a memory. The new electrics were an enormous improvement: they were powerful, relatively cheap to operate, fairly quiet, smokeless, dependable, and prone to few ailments. Best of all, they did not eat when off duty, and they never soiled the streets.

Technical progress rarely sweeps clean, and horsecar holdouts endured well into the electrical age. Most were marginal short lines—"dinky operations"—where traffic was too modest to warrant the investment of bonded rails, poles, overhead wires, and electrical feeder lines. Horsecar operations stayed on until heavier traffic developed or some way was found to abandon the line entirely. Not all such operations were in remote rural settings. New York City held on to a number of lines after 1900 because of unusually high installation costs, laws against overhead wires, and the expense of third-rail systems. Motor buses finally offered a way out, and the last New York horsecar ran on Bleecker Street in July 1917. Pittsburgh kept a small horsecar operation going until 1923. A little line in Sulphur Rock, Arkansas, which did not shut down until 1926, is believed to have been the last one in the country. Right to the end, the old-fashioned hayburners seemed determined to demonstrate that sophisticated technology is not required in the public-transit business.

Anthropology

ALPINE ICEMAN

The well-preserved, clothed body of a man, found with his tools and ornaments high in the Alps in September of 1991, continues to be a subject of intensive study. For more information, see the feature article on page 197.

NAMIBIAN SKULL

A fragmentary jaw found in Namibia in southern Africa provides important new evidence for the widespread distribution of creatures about 13 million years ago that were probably ancestors of modern humans, chimpanzees, and gorillas. Until now, fossils of these particular animals had been discovered only to the north, in East Africa. The new specimen may provide scientists with a better understanding of the evolutionary changes that led to a split in development of hominoids, with one branch leading to modern chimps and gorillas, the other to modern humans. But anthropologists emphasize that they will need more specimens in order to assess confidently the evolutionary position of this creature.

The new find includes the right half of a jaw, with five teeth intact—two premolars and three molars. The shapes of the teeth indicate that the animal ate soft plant foods such as fruits and leaves and that it probably lived in a damper, more heavily wooded environment than that which exists in the region today. This fossil joins the growing body of evidence indicating that hominoids of the Miocene Period (from 24 to 5 million years ago) were more widespread and more varied anatomically than scientists had believed. The Miocene is a critical time for human evolution because, at its end (about five million years ago), creatures emerged that are identified as human.

Anthropologists estimate that the individual represented by this fragment weighed between 30 and 45 pounds (13 and 20 kilograms).

CHINESE FOSSILS

Two fossil skulls found in China that may be over 350,000 years old could yield important new evidence about how anatomically modern humans evolved. The skulls were excavated in 1989–1990 by a team from the Hubei Institute of Archaeology. Preliminary dating of the specimens was based on fossils of other animals in the deposit; further studies to determine the age of the skulls for full assessment of their significance have yet to be carried out. Some anthropologists believe that the skulls show features transitional between *Homo erectus* and modern *Homo sapiens*. They argue that if additional evidence supports the dates suggested so far, these new finds will lend support to the theory that modern humans evolved not only in Africa, but in other parts of the world as well, including East Asia. Most anthropologists believe

Two human skulls discovered in China cast doubt on the popularly held belief that modern humans originated in Africa.

that modern humans evolved in only one place, and that was in Africa.

Other scientists who have examined photographs of these specimens suggest that the skulls may be too fragmentary and distorted to allow definitive comparisons with other fossil human skulls. The uncertainty of the dating evidence represents another major concern.

Peter S. Wells

Archaeology

NEW CAVE ART

A spectacular cave with wall paintings about 18,000 years old has been discovered at Cap Morgiou near Marseilles in southern France. The cave's entrance, about 120 feet (37 meters) beneath the surface of the Mediterranean Sea, was discovered by divers. This new find is far to the east of northern Spain and southwestern France, where the majority of the caves with elaborate paintings have been found. The creatures portrayed on the walls include bison, chamois, deer, horses, and ibex, as well as seals and great auks, two animals not represented in other European cave art.

Archaeologists believe that wall paintings in a cave that now lies deep beneath the Mediterranean Sea were drawn some 18,000 years ago.

The cave was used during the very cold period near the end of the Ice Age. At that time the sea level was several hundred feet lower than it is today, and the cave was probably several miles from the water. Besides the painted animals, scientists discovered representations of human hands, made by outlining hands pressed against the cave wall with charcoal. Remains of two fireplaces were found, and fragments of charcoal were recovered along the cave walls, perhaps remains of wood torches used for illumination.

EQUINE DOMESTICATION

The earliest known sculpture of a domesticated horse was discovered recently at the site of Tell es-Sweyhat in Syria. The figurine, made of baked clay, is about 5 inches (13 centimeters) long and is believed to date to about 2300 B.C. Two pieces of evidence indicate that the figurine is meant to portray a domesticated horse rather than a wild one: The mane lies flat against the animal's neck, and a hole at the mouth shows where the bit of a bridled horse would fit. This object joins a growing body of evidence showing that horseback riding began earlier than previously thought, and that it played an important role in the cultural development of western Asia and eastern Europe.

BACCHANALIA HISTORY

The earliest direct physical evidence of wine and beer consumption has been discovered at a Sumerian settlement named Godin Tepe in western Iran. Chemical analysis of residues on the inside surfaces of pieces of pottery dating to around 3500 B.C. shows that they once contained these beverages. Beer was apparently a popular drink among the Sumerians, who lived mainly in southern Mesopotamia (now southern Iraq). Before this new chemical evidence was discovered, archaeologists had identified representations in Sumerian art showing people drinking beverages, and barley grains had been found at settlements of the period. Some archaeologists think that beer was produced soon after barley and other cereals were domesticated at the beginning of the Neolithic period in the Near East, around 8000 B.C.

PHILISTINE ORIGINS

Ongoing archaeological research is creating a new view of the Philistines, a people who lived on the east coast of the Mediterranean Sea. Excavations at Ashkelon, Ekron, and other sites in Israel are yielding evidence of a highly developed culture that existed between about 1200 and 600 B.C. Archaeologists are comparing Philistine pottery, loom weights, and hearths to those from Mycenaean Greece. The similarities of these artifacts suggest that the Philistines may have originated in Greece. Archaeologists hope that they might find Philistine inscriptions in the course of future excavations, because the language represented and the script used would provide important additional information about the question of Philistine origins, one of the greatest mysteries of the ancient Near East.

BYZANTINE CAPITAL

A large and extraordinarily well-preserved city from Roman and Byzantine times is being investigated at Beit Shean in the Jordan Valley in Israel. The city was destroyed by an earthquake in A.D. 749, and archaeologists now have the unusual opportunity of excavating a large portion of the 30-acre (12-hectare) site. It was an important trade center between Mesopotamia and Egypt, and served as a provincial capital during Byzantine times. Streets, mosaics, public buildings, and statues are in exceptionally good condition, and provide an unusually complete view into the physical character of an ancient city.

SPANISH CARAVEL

A remarkably well-preserved Spanish ship discovered on the ocean floor off the Bahamas may be the oldest sunken vessel known in the New World. The wooden ship, thought to date to the early 16th century on the basis of the character of the cannon and swords found, may be the only known surviving caravel, a type of ship that played a vital role in the overseas explorations by European nations during the 16th century. Archaeologists have mapped the wreck and inventoried the more than 5,000 artifacts on board.

Numerous weapons have been identified, including cannon, swords, chain mail, and an early type of rifle. Ceramic vessels and plates and small glass containers represent some of the objects used by the sailors. As study of the ship and its artifacts progresses, archaeologists will be able to develop a detailed picture of shipbuilding technology, and of everyday life aboard a vessel during that period.

EARLIEST VILLAGE IN THE AMERICAS

In northern Alaska, archaeologists have reported the earliest known site of human activity in the Americas. Chipped-stone spearheads were found together with remains of campfires left by hunters. Michael Kunz of the Bureau of Land Management (BLM), who identified the site in the Brooks Range of northern Alaska, believes that early hunters used the location to seek out game animals, including mammoth. Pieces of charcoal submitted for radiocarbon testing yielded a date of around 11,700 years ago.

Scientists believe that humans migrated from eastern Asia to North America during the latter part of the last Ice Age. Back then, sea levels were much lower than they are today, exposing an extensive land bridge between what are now eastern Siberia and western Alaska. This site provides added support to the current thinking that pushes back the date for the beginnings of human occupation of the Americas. It also provides valuable new information about the way of life of some of the first Americans.

EARLY LANGUAGE DISCOVERED

Specialists have deciphered a language that preceded Mayan. A carved stone was discovered in 1986 in Vera Cruz, Mexico, bearing a portrait of a ruler and an inscription recounting his ascent to power. Archaeologist John Justeson, from the University of New York, Albany, and linguist Terrence Kaufman, from the University of Pittsburgh, suggest a date for the inscription of A.D. 159, making this the earliest New World text yet to be deciphered.

The length of the inscription allowed scholars to identify repetitions of specific words in different forms. The inscription itself bore enough resemblance to later Mayan inscriptions to provide clues from that recently deciphered language. Dr. Kaufman was also aided by his study of a modern version of the same language, which is still spoken by some groups of people in rural parts of southern Mexico. As the languages of ancient Mesoamerica are increasingly well understood, we are learning much about specific early rulers and their personal histories, as well as about the culture and the history of the Olmec, the Maya, and other peoples who lived in the region.

ROYAL MAYAN TOMB

A royal Mayan tomb dating to the 6th century A.D. has been discovered at Copan in Honduras by a research team from the University of Pennsylvania working with teams from Northern Illinois University and from Honduras. In a chamber underneath a staircase, archaeologists found a skeleton outfitted with shell ornaments, jade beads, and other jewelry. Over 20 ceramic vessels were also in the tomb, some containing beads, shells, and bones. The walls of the chamber were coated with red plaster. The investigators think the grave contained one of the chief personages of this rich ancient metropolis.

New evidence is revealing that the Maya of Central America had a much more complex society than once thought, and that they may have included an ancient equivalent to a middle class. Arlen and Diane Chase of the University of Central Florida interpret their findings at the site of Caracol in Belize as signifying that a large portion of the population at this Mayan center comprised a social stratum made up of persons in particular occupations, such as specialized crafts workers, administrators, and warriors. Evidence for the existence of this social group consists of burials in which non-elite individuals were treated in ways believed to have been reserved for the elite. Other evidence was identified in crafts areas on the site, where quantities of ornaments were manufactured of shell and jade. These production areas were located not just in the elite precincts of the settlement, but elsewhere as well, suggesting that the ornaments of these special materials were made for a sizable portion of the population. Similar evidence has been identified at other Mayan sites as well.

Peter S. Wells

Paleontology

MOST PRIMITIVE DINOSAUR

While prospecting in the badlands of northwestern Argentina, paleontologists from the University of Chicago and several Argentine colleagues discovered a new genus of dinosaur called *Eoraptor*, the most primitive dinosaur yet found. About the size of a dog, this tiny animal lived 230 million years ago, during the late Triassic period, when dinosaurs were a relatively minor member of the reptile world. From this meager beginning, dinosaurs soon rose to power, ruling as the dominant land animals for 150 million years until their sudden extinction. The discovery of *Eoraptor*, along with other early dinosaurs, sheds new light on how these animals inherited the Earth. Standard theories have held that dinosaurs took over because they had special characteristics, such as an erect posture or a fast metabolism, that made them better hunters or fiercer fighters. But the new evidence hints that dinosaurs may have risen to power more by accident after some tragedy befell the reptiles ruling before them.

SAVAGE UTAH DINOSAUR

Paleontologists in Utah discovered a new species of dinosaur called *Utahraptor* that was a vicious killing machine. Living 125 million years ago, this 20-foot (6-meter)-long carnivore sported 15-inch (38-centimeter)-long, sickle-shaped claws on its hind feet. *Utahraptor* belonged to a family of dinosaurs called dromaeosaurids—agile beasts that may have hunted in packs and attacked the large plant-eating dinosaurs alive at the time.

PROTEIN FROM DINOSAURS

A group of molecular biologists and paleontologists from the Netherlands and the United States identified a protein preserved in dinosaur bone. The protein, called osteocalcin, was discovered in several dinosaur bones dating from 75

The 15-inch-long claws of the Utah-raptor _helped make this dinosaur one of the fiercest predators ever to walk the Earth._ Utahraptor _hunted in packs, preying on plant-eating dinosaurs, and may have been more of a terror than the notorious_ Tyrannosaurus rex.

million to 150 million years ago. Many molecular biologists had thought that proteins could not survive such long periods because they are typically rather fragile molecules. But if future work can confirm this find, it will open the door for using ancient proteins to trace the evolutionary relationships between dinosaurs and other animals, a topic that sparks fierce debate among paleontologists.

ANCIENT DNA

In an unusual case of fact following fiction, scientists succeeded in extracting 30 million-year-old DNA from an ancient insect preserved within amber, a fossilized form of tree resin. This is the oldest known segment of DNA, which is the molecular building block of genes. A similar search for prehistoric DNA appeared in the novel _Jurassic Park_ by Michael Crichton, which chronicles researchers who clone a living dinosaur from its DNA. In the novel the scientists find the dinosaur DNA inside an ancient insect that had bitten a dinosaur and had later been encased in amber. Real-life scientists don't think such a cloning feat is currently feasible, but they do hope to use ancient DNA to help study evolution in a way not previously possible.

A TELLING TOOTH

The discovery of a tiny tooth in Australia has caused paleontologists to question an age-old explanation for the kangaroos, koalas, and other strange animals of the land down under. These oddballs are marsupials — a group of mammals that carry their undeveloped newborn in external pouches. Marsupials are known to have lived on most continents long ago, but they were eventually replaced by the more familiar placental mammals that carry their young to term internally. Because most marsupials went extinct around the world, paleontologists have presumed that they were in some way inferior to the placental mammals. Marsupials supposedly survived on Australia only because this continent split away from all others before placental mammals could migrate there.

The newly discovered tooth, however, belonged to a placental mammal that lived in Australia 55 million years ago. This suggests that marsupials and placental mammals coexisted for millions of years on Australia. While they lost out on other continents, marsupials appear to have dominated over placental mammals in Australia, suggesting that marsupials were not inferior animals.

TALES FROM A WHALE'S EAR

An ancient whale ancestor may have spent much of its life out of water. Paleontologists working in Pakistan uncovered the jaw- and ear bones of a species called _Pakicetus_, a carnivorous mammal that lived 50 million years ago. Ever since discovering the first specimens of this animal, researchers have recognized it as an evolutionary link between modern whales and the land animals from which they evolved. _Pakicetus_ is believed to have lived part of its life in the water. The animal did not have ears specialized for hearing underwater, suggesting the whale forebear led an amphibious life-style.

Richard Monastersky

PHYSICAL SCIENCES

CONTENTS

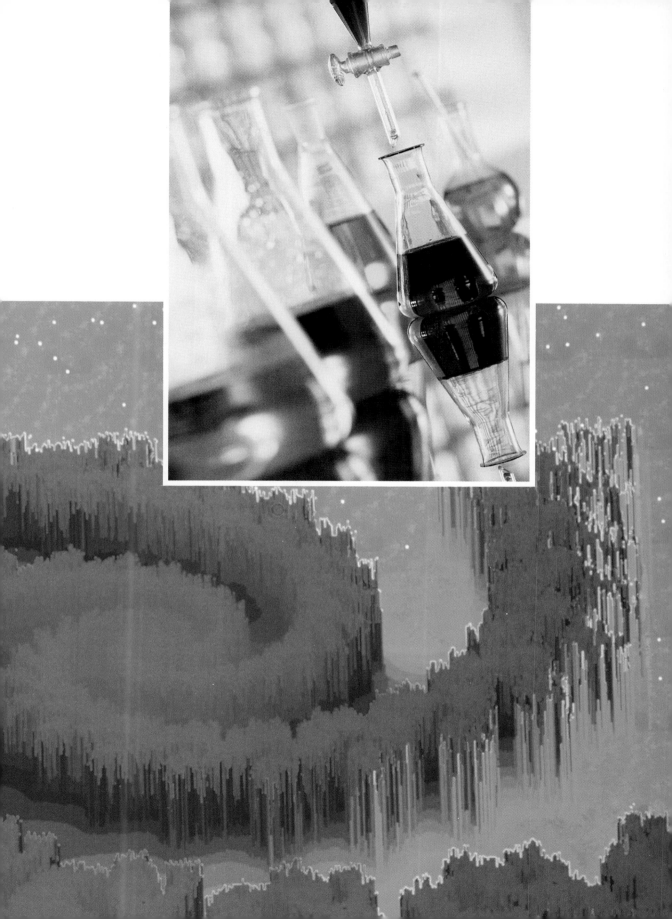

DESIGN BY NATURE

by Delta Willis

A hefty piece of tree limb is passed around and through the hands of architects, engineers, biologists, and the occasional philosopher. The speaker is a physicist. Claus Mattheck wears shaded wire-rims and a dark-brown leather jacket, and works at the Nuclear Research Center in Karlsruhe, just east of the German-French border. He is known to take long walks in the nearby forest, delving into what he calls the body language of trees. Mattheck uses this data to refine industrial designs, including those of car parts.

Mattheck's audience is gathered for the Second International Conference on Natural Structures, a rare multidisciplinary dialogue begun by German architect Frei Otto. Otto is best known for the undulating rooftops over the 1972 Munich Olympic Stadia. He was inspired by the work of spiders.

Industrial designers continually look to the natural environment for blueprints of efficiency and beauty. The graceful, aerodynamic Concorde was patterned after the streamlined flight of the swan.

The participants investigate the blueprints of nature. The dome of a sea urchin, for example, is used as a model by no fewer than five speakers. There is a debate about the architecture of sand dollars. The finesse of a butterfly's proboscis is compared to the design of an oil drill. The pleat of a poppy demonstrates how folding patterns strengthen even the most delicate material. A barnacle is seen as the epitome of efficient design. "These are lovely things that need nothing extra," says paleontologist Adolf Seilacher. "They have only the perimeters that need to be there."

Frugality is key, or, as Charles Darwin noted, "Natural selection is continually trying to economize every part of the organization." Castles of clay built by African termites are better thermoregulated than any of our skyscrapers; a single chamber is maintained at a constant 86° F (30° C), no matter the temperature outside; the hotter the climate, the taller the chimney. The shell of a pearly nautilus encases what amounts to a jet-propulsion engine with soft parts, less likely to break than rigid ones. The shell itself is capable of enduring the pressure at depths of 1,800 feet (550 meters), a model of maximum strength employing minimum material.

Learning from Trees
Trees invite particular envy. Their leaves are highly efficient solar panels, yet they have many other functions. Leaves fold to avoid drag when the winds blow, and by transpiration move small rivers up trunks as tall as 360 feet (110 meters). As the oldest living things on Earth, trees also win the prize for fatigue resistance. The oldest in North America are bristlecone pines in the White Mountains near the California-Nevada border. Now 4,600 years old, they were drawing sap via complex hydrologic systems before aqueducts were built in Rome. Trees even recycle their wastes: Forest duff is buried, then converted into fossil fuels like charcoal.

Trees also have the ability to repair themselves. Mattheck points out that a healthy tree is centered against gravity. A sapling born on a mountain slope has an inherent inclination to grow away from the center of the Earth, rather than perpendicular to the slope. Should the main trunk of any tree break, a side branch will fold in as a replacement. Should a storm throw a trunk off-balance, the tree works to get its center of gravity back over the rootstock again. Softwoods (like pine) grow new wood on the outside, compressing the trunk toward the center of gravity. Hardwoods (like oak) favor the inside, creating a tension that pulls the trunk back over the center.

Mattheck began to trace the history of trees by studying their form. He found trunks

embracing rocks, road signs, fences, garden statues, and a crucifix; he discerned the entanglement of a conifer that had overtaken its neighbor, parallel trunks that had married, and a beech that persevered in a horizontal position. (Friends learned of his interest, and sent him photographs of kinky trees; he was led down paths in England and France, in British Columbia, in Namibia, and in the forests of Maine and Idaho.)

The Biomechanics of Bones
In his search for optimum design, Mattheck also considered the biomechanics of bones. With his colleague Andreas Baumgartner, he

The shell of a sea urchin (below) looks deceivingly delicate, and yet its structure can withstand the force of pounding surf. The urchin's ingenious architecture was the prototype for the Epcot Center's geodesic dome.

devised the Soft Kill Option, a computer-simulation program that can be applied to industrial designs. "We get rid of waste," Mattheck says simply. "Most waste of energy is in oversized, much-too-heavy structures." The process eliminates "soft," unnecessary parts of a design and "hardens," or reinforces, areas vulnerable to fracture—exactly what a bone does.

Few of us think of our bones as plastic, but their malleability is evident to any orthopedic surgeon who monitors their repair. A badly set bone can actually grow to straighten itself, filling in the weak spot by increased mineralization, shrinking the unnecessary bulge through lack of minerals. The mineral is calcium, the same material that strengthens a nautilus shell, makes marble hard, and dissipates in the bones of astronauts. The last is an example of the Soft Kill Option as employed by nature: Bones designed to carry a human body against gravitational forces on Earth quickly adapt to weightless conditions.

Mattheck explains that trees and bones "always try to grow into a state with constant stress on their surface." Constant stress sounds like an unhappy situation, but, in engineering terms, it means relatively homogeneous loads. Loads for trees can come in the form of ice, wind, or snow. Loads for bones come in the form of the weight they must carry. "This constant-stress axiom seems to be one of the most general design rules in biology," Mattheck says, adding that any industrial applications must also strive for equal distribution. Areas with too much stress are subject to fatigue or fractures. Areas with too little stress denote a waste of material. "The optimum form has maximum strength with a minimum of material."

Blueprints from Nature
Drawing blueprints from nature is nothing new. During the 16th century, Leonardo da Vinci designed sleek ship hulls based on the movements of fish in water. His notebooks are rich in comparative anatomy and corresponding machines, several inspired by birds in flight. The Wright brothers devised stabilizers after the way a turkey vulture employs its primary feathers to reduce turbulence at low speeds. The cockpit of the supersonic Concorde is lowered on approach, like the head of a swan. The interior braces and struts of an eagle's wing have inspired bridge designs. Latticework for London's Crystal Palace was inspired by the veins and ribs of the water lily *Victoria regia*.

Bones offer obvious blueprints for engineers. Such was the case in 1866 in Zurich, when an engineer named Cullmann wandered into the dissecting room of an anatomist named Meyer.

Cullmann had been busy designing a crane. Meyer was studying a femur. Along the neck of the bone were visible, intersecting lines, known as trabeculae, a result of the stress and load imposed on the bone, and a record of its responsive growth. The engineer is said to have taken one look and cried out, "That's my crane!"

The anecdote comes from D'Arcy Thompson's remarkable 1917 book, *On Growth and Form*. Thompson suggested that the shape of all things depends on physical forces. He looked at nautilus shells and kudu horns and compared their logarithmic spirals. He dropped ink into water and saw the tentacles of a jellyfish. He wrote beautifully about everyday matters, from the surface tension that keeps a drop of water dangling from a faucet, to raindrops that marry in a free-fall. Stephen Jay Gould notes that Thompson's ideas have gained new influence in a science that only now has the technology to deal with his insights. Thompson, says Claus Mattheck, was a "true pioneer."

Mattheck's own work is breaking new ground. His computer programs are being used to refine automobile parts, as well as spare parts for the human body. He describes a screw used to position a steel plate in the spine. "These screws often broke under stress," he says, "but now they are 105 times more durable in laboratory tests." Areas subject to high stress were strengthened; in areas of low stress, excess material was trimmed. Mattheck describes the result as "more vital," as if the screw had developed a life of its own: "When a screw grows like a tree, it really grows."

In 1961, about the same time the slide rule gave way to computer calculations, Frei

Otto founded the Institute for the Study of Lightweight Structures, originally headquartered in Berlin. Otto, now 68, served in the German Air Force during World War II. He was shot down and captured, and during his time working in a POW camp near Chartres, he led a construction team that repaired bridges. Wartime circumstances forced him to work with very little material.

His quest for lightweight strength led him to favor a lattice girder known as the fish-belly design, and, eventually, the designs of spiders. In engineering terms, the strength of a structure is measured by the load that will break it. The tensile strength of wood is 15,000 psi (pounds of force per square inch); the tensile strength of muscle tendon is equal to that of hemp rope—12,000 psi. The tensile strength of a spider's web is 35,000 psi.

Also in 1961 Otto had an encounter that recalled Thompson's engineer and his crane nearly a century earlier. On a visit to the laboratory of Berlin biologist J. G. Helmecke, Otto noticed Helmecke was studying the silica shell of a diatom, a tiny aquatic animal. The two men began to trade observations, which resulted in a series of lectures on biology and architecture. It was the genesis of what became known as SFB 230. SFB is a German abbreviation for Special Research Projects, and 230 is the code for the research project.

Discerning how evolution designed such enviable forms became the focus of SFB 53,

The intricate chambered shells of the Nautilus permit movement via a unique jet propulsion system. The Nautilus' minimalist structure is mimicked by the engines of wide-bodied aircraft.

The clever suspension structures of spiderwebs were modified by German architect Frei Otto to produce tentlike roof structures over several of the sports stadiums at the 1972 Olympic Park in Munich.

led by Adolf Seilacher. Seilacher describes innovative designs in terms of internal construction and external constraints. In the case of a sand dollar, for instance, Seilacher notes its tiny plates are held together by collagen, the same material that gives elasticity to our skin.

These slender sea urchins prefer to feed on a sandy beach, their sleek design less buffeted by waves than that of fully vaulted cousins that feed in deeper waters. Sand dollars even take on sand as ballast, to keep them in place. Canadian biologist Malcolm Telford describes the sand dollar's finesse as "the art of standing still."

As a former geologist, Seilacher emphasizes how critical external forces can be to the changing form of a species. "At first glance the environment appears to be an extrinsic factor clearly separated from the organism," he notes. "In biological reality, however, it is not an independent entity. . . . An effective environment becomes an intrinsic part of an organism."

Changes in Design

What triggers changes in design? Some species develop weird forms (like peacock feathers and gross antlers) for reproduction. So function and behavior are factors, along with environmental changes. Diet is a paramount force in evolving new forms. By becoming the tallest plants on Earth, trees won the competition for sunlight, the source of their photosynthetic diet.

Change can even come to things that remain still. Seilacher found that sea urchins in the polluted waters of the Red Sea near Eilat were grossly distorted, with "bizarre morphologies . . . dramatically deviating from the evenly vaulted domes." These sea urchins had lost their symmetry in response to the toxic conditions of their environment.

Transformation occurs so readily in biology because structures can have more than one function, Seilacher notes. The parts can experiment and explore and change according to changing needs. "That's the way we have to use technology—in an evolutionary sense," Seilacher tells the gathering in Stuttgart. This transformation in technology includes changing some old ideas, such as that bulk denotes strength or power, which created some problems in Detroit. (Getting rid of waste in design has dominated postwar industry in Japan, and this precision goes all the way down the production line to the bottom line. The United States creates five times the waste that Japan does for every dollar of goods sold, and twice that of Germany.) As anatomist Alan Walker once remarked, "Imagine if we could design a car to work like the human body. We'd save billions in oil."

With the advantage of millions of years of evolution, natural forms offer us blueprints in efficiency. It's yet another argument for preserving the planet's biological diversity—within every form is a potential blueprint. This is exactly what Claus Mattheck has in mind. "Trees," he says, "are the design teachers of us."

The similarity between the head of an oil drill and a butterfly's proboscis is enhanced by an artist's humorous depiction.

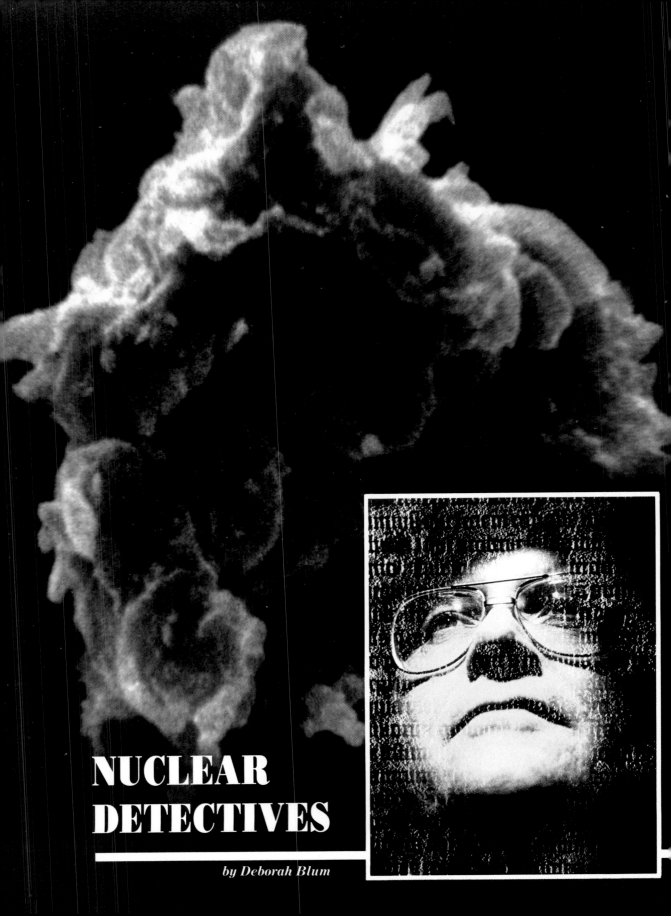

NUCLEAR
DETECTIVES

by Deborah Blum

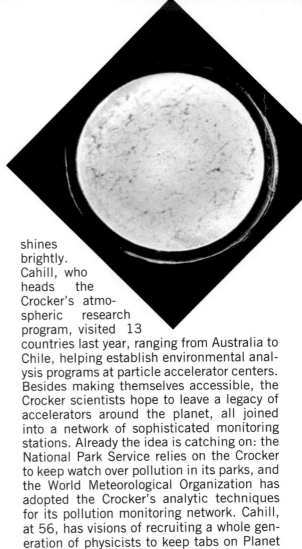

For a nuclear physicist to include a stroll through an almond grove as part of his research may seem a bit peculiar. Unless, of course, the physicist works at the Crocker Nuclear Laboratory at the University of California at Davis, where unusual is the norm. In the name of physics—and in the pursuit of paying customers—Crocker scientists have done everything from trudging through dusty lake beds to poring over ancient manuscripts. They have studied the ink in the famed Gutenberg Bible, tracked the source of air pollution in the Grand Canyon, verified the handwriting of Johann Sebastian Bach, and analyzed the smoke from burning Kuwaiti oil wells. For two decades now, as long as the project has called for good, solid everyday science, the Crocker group has been happy to fire up its small particle accelerator for a cost that's now just $384 an hour (volume discounts available). "We're the McDonald's hamburger of elemental analysis," jokes Davis nuclear physicist Thomas Cahill.

Cahill and colleagues have made the Crocker lab internationally recognized as a place where a standard cyclotron is used in nonstandard ways. "We attract people who are bored by doing repetitive, traditional work," says Crocker director Robert Flocchini, the physicist who's been analyzing the blowing dust clouds raised by California almond growers; he hopes to settle a furious dispute over the extent to which agricultural dust contributes to the often gritty air of the state's heavily farmed Central Valley.

A Legacy of Accelerators

While its choices have made the Crocker somewhat low-profile in the high-powered community of traditional particle physicists ("Do they *have* a cyclotron at U.C. Davis?" asked one surprised physicist from Stanford's powerful linear accelerator program), outside that inner circle the laboratory

Physicist Thomas Cahill (inset) and his colleagues at the Crocker Nuclear Laboratory at the University of California at Davis use their small particle accelerator for a number of unconventional investigations. Cahill determined from a sample of pollution particles (above) created by the Kuwaiti oil fires (left) that the atmospheric contamination caused by the fires did not present a serious health threat to people living in the Persian Gulf region.

shines brightly. Cahill, who heads the Crocker's atmospheric research program, visited 13 countries last year, ranging from Australia to Chile, helping establish environmental analysis programs at particle accelerator centers. Besides making themselves accessible, the Crocker scientists hope to leave a legacy of accelerators around the planet, all joined into a network of sophisticated monitoring stations. Already the idea is catching on: the National Park Service relies on the Crocker to keep watch over pollution in its parks, and the World Meteorological Organization has adopted the Crocker's analytic techniques for its pollution monitoring network. Cahill, at 56, has visions of recruiting a whole generation of physicists to keep tabs on Planet Earth: "What do I want out of this?" he asks slowly and then grins. "I want to be the Pied Piper of environmental physics."

PIXE Analysis

Circular accelerators, like the Davis cyclotron, use the force of their magnetic field—which tugs at right angles to the particles' motion—to bend the path of the moving particles into an orbit between the electrodes. On each orbit the particles cross the charged gap twice, as they gain a rush of acceleration each time. With each boost the radius of their orbit increases a bit. Eventually, when the ions are really sizzling, they are slung through a vacuum pipe. They emerge in a tightly focused beam that is aimed at a target. The Crocker, at full power, can whip particles up to about one-third the speed of light. Such highly charged beams can produce heady consequences in target materials—they can change the form of elements, make them radioactive, or—significantly, for Cahill's purposes—reveal the structure and composition of the target material.

In the early 1970s, just as Cahill was becoming interested in air pollution analysis, he learned that Swedish researchers had used cyclotrons to induce certain target materials, including air pollution samples, to emit X rays. The technique is called proton-induced X-ray emission, or PIXE. Cahill realized it would allow him to take a black fleck of pollution and read the elements out of it like words on a page.

To do a PIXE analysis, the Davis cyclotron is run at low power, with the proton beam held near its low end of 4.5 million electron volts. Operators aim the beam at a paper-thin Teflon air filter, held in the frame of a photographic slide, that has been used to collect pollution samples. The beam whispers through the smoggy grit and filter without doing any damage; it's just strong enough to nudge the electrons of the various elements in the target. A big charge beam would blast the electrons out of place, spraying them about like bird shot. But this light push merely elbows the electrons out of their customary atomic orbits for a moment. When the electrons then fall back into place, they release energy in the form of X rays.

What's important here is that the energy necessary to push an electron from its orbit —and the energy released when that electron returns—is different for each element. Thus, when the target material is hit by the proton beam, the X rays emitted can be used to identify the elements from which they came. The identification is done with the help of a silicon diode, which converts each discrete packet, or photon of X-ray energy into an electrical pulse. (The trick is to get just one photon to hit the diode at a time. The physicists accomplish this by lowering the number of protons in the beam until the odds of more than one proton hitting an electron in the target at any one instant are infinitesimal.) The resulting signal is fed into a computer and analyzed; the specific charge of the pulse is like a signature for the origin of the X rays.

Kuwaiti Armageddon?

The Crocker's expertise with this revolutionary technique was first acknowledged in 1977 when the Environmental Protection Agency (EPA) asked Cahill to do air pollution monitoring in Zion, Bryce Canyon, and Canyonlands national parks. Cahill installed air-filtering devices in each one, collected the dirty filters, then took them back to the cyclotron for analysis. Today the laboratory earns over $2 million a year through its PIXE work, largely from environmental researchers. Atmospheric scientists say the extreme sensitivity of the analysis, the ability to analyze a few specks of dust, has made even the barest shimmer of pollution identifiable.

Not everything the lab looks at is as subtle as a shimmer, though. In 1991, for example, the Crocker was asked to develop an elemental "fingerprint" of the smoke spiraling up from Kuwaiti oil wells that had been torched by departing Iraqi soldiers in the Gulf War. Atmospheric scientists knew from experience that oil fires often release a bewildering mixture of chemicals, which vary according to local geology. But, says Cahill, there were elements about these fires that were unique. "First, there was the sheer size of the disaster, which was unprecedented. Second, these wells were under tremendous geologic pressure, so the oil was squirting into the air as if from a syringe, burning in a plume. Third, some of the plumes were white instead of black, and there was much speculation about what they were."

Cahill notes that at the time there was also much speculation about the fires' possible environmental impact—speculation, he says, that bordered on the hysterical, with phrases like "ecological holocaust" and "Armageddon" being used by scientists who perhaps should have known better. Still, the situation was potentially threatening, and facts were needed to understand what was pouring into the air and whether it could become a regional or global problem.

When the National Oceanic and Atmospheric Administration (NOAA) extended an invitation, the Crocker team was eager to go. By obtaining samples of Kuwaiti crude oil, testing them in the laboratory, then comparing them with the atmospheric contamina-

Cahill's cyclotron proved that rising dust clouds from the bed of the now dried-up Owens Lake in California are laced with arsenic. The arsenic is a natural component of the volcanic bedrock that once formed the floor of the lake, but is now eroding. The border above shows the dust particles used in the study.

tion, they were able to identify the toxic metals coming from the burning wells: showering the region was as much as 1,000 tons of fine vanadium particles and 500 tons of nickel. They also found that the white plumes were ancient salt water that was being ejected with the oil. "It turns out that sodium chloride from seawater was being put into a spray and burned and then forming very fine salt particles," says Cahill.

Furthermore, he found, natural conditions in Kuwait were keeping the situation from becoming truly disastrous. Kuwait City has some of the "dirtiest" air in the world—the air is filled with fine particles of alkaline dust blowing up from the desert floor. "We found that most of the mass of the smoke plumes was desert soil," Cahill says. "And this fine desert dust was neutralizing the acids released by the burning oil. Basically you had a scrubber forming in the sky." The result was that the amount of sulfates in the air was almost exactly the same as it is during a typical summer over Los Angeles—worrisome, but hardly Armageddon.

"With all the talk about Kuwait being a disaster, these were not really toxic levels," Cahill says. "Our analysis showed too that as the acids combined with the dust they became heavier, so they didn't travel as far as they might have or go as high in the atmosphere; they tended to settle locally, making it more of a local problem than a regional one, and not at all a global problem."

What's Polluting the Grand Canyon

The situation had been much the same—speculative and murky—several years earlier, when the Crocker team was asked to take a look at the Grand Canyon. And that time the politics was as complex as the chemistry. Cahill was part of a Park Service study of pollution in the canyon. Park administrators had become increasingly dismayed about dirty air blurring the canyon views. They suspected that the culprit was the massive coal-burning Navajo Generating Station, located just 18 miles (29 kilometers) from the northeastern portion of the canyon, which had been fired up in the late 1970s. The plant, operating without scrubbers, was estimated to be releasing some 300 tons of sulfur dioxide daily, more than twice the sulfur pollution load of the entire Los Angeles basin. Nevertheless, the operators of the plant, which included the U.S. Bureau of Reclamation, blamed the park's clouded visibility on either pollution blown in from western urban areas or from copper smelters located south of the park.

The problem, of course, was that no one could actually see pollution coming from any of these sources. "The plume, consisting largely of sulfur dioxide, is largely invisible," says Cahill. "The SO_2 has to do two things before it can be seen: it must convert from a gas to a particle, and it must pick up water. That allows it to scatter light." Even with nonvisual techniques, though, the pollution was hard to pick up. Both the Park Service and the utility had been carefully checking for pollution drift, perching their air-sampling equipment like sentries along the canyon's 7,000-foot (2,135-meter) rim. But park rangers had been noticing that even on days when the view was clear at the rim, it was murky below, within the Grand Canyon itself. They suggested the canyon might be acting as a sink that collected and concentrated the dirty air.

With a colleague, John Molenar of Air Resource Specialists, an environmental consulting company, Cahill began monitoring deep in the canyon itself. In the winter of 1990–91 the work paid off. First of all, Cahill and Molenar determined that the plume from the Navajo generating station drifted over nearby Lake Powell, where it picked up water and cooled; when the colder air reached the canyon, it sloshed over the rim. They were even able to show part of this sequence of events on videotape. During an exceptionally cold winter night Molenar set up a video camera and recorded a flow of gritty air, like a tumble of water, falling over the rim of the canyon and almost splashing into the bottom.

How did they know the air was from the generating plant? "It was an indirect process," Cahill says. They couldn't follow the plume from the plant on video, but with PIXE analysis they could track tiny amounts of telltale elements. "The plume included trace metals from the burning of coal, such as selenium," Cahill explains. "The plume would mix with other pollutants in the air, but the selenium came along as a trace element. And during the times we saw a lot of haze in the canyon, we also saw high levels of selenium. It took the cyclotron to pick it up—these 'high' levels were still very low to anybody else."

In March 1991 hearings were scheduled on the Grand Canyon pollution. But the Park Service was barred from testifying. The decision had come from then secretary of the interior Manuel Lujan. Lujan was presiding over two agencies with competing interests —the Park Service, which wanted the plant cleaned up, and the Bureau of Reclamation, which had a financial interest in leaving it alone. He didn't want them slugging it out in public, and so he ordered them to submit written statements only.

Cahill was extremely frustrated. In his view, the whole point of having good data is to spread the word, He had the cyclotron date, and Molenar had his videotape. Both men decided to attend the hearing on their own.

The two flew to Phoenix, where the hearing was set to begin, and met with environmental groups that were scheduled to speak. The Grand Canyon Trust, a regional conservation organization, was impressed enough to give up an hour of its slotted time to the scientists. Cahill made his presentation as a private citizen. After a second hearing, representatives from the utility and the environmental groups decided to negotiate a cleanup plan: "No one wanted to be the one who trashed the Grand Canyon," Cahill says. The final agreement calls for the plant to be equipped by 1999 with $430 million worth of scrubbers, which will reduce the pollution load by some 94 percent.

Ink Analysis
The delicacy of the Crocker's analytic techniques has moved the lab into other surprising areas for nuclear physics, most notably the analysis of old documents. That venture really began some 15 years ago in a conversation between Cahill and Davis historian Richard Schwab. Schwab is an expert on the eighteenth-century *Encyclopédie* of Diderot, the world's first great encyclopedia and a work so popular that it was immediately followed by a wild number of forgeries. Schwab had learned to identify the fakes by various peculiarities in the text, which he admits did not endear him to universities that owned forged editions. He was cheerfully complaining about this to Cahill at a dinner party.

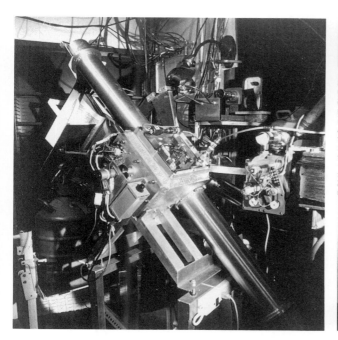

The Crocker physicists used their cyclotron (left) to analyze the pages of a priceless edition of the Gutenberg Bible (right). The scientists found that Gutenberg had used an ink that was essentially a slurry of copper and lead, a fact that explains why the ink has retained its blackness over the years.

"Then, right in the middle of dessert, we both had this brilliant idea," says Schwab. "We realized that a piece of paper would look just like an air filter to the cyclotron beam, and the ink on it would look like the pollution particles."

Schwab promptly donated a not very valuable 18th-century book from his library so the idea could be tested. Cahill and fellow physicist Bruce Kusko set about slicing small rectangles from the edges of several pages to fit them into the cyclotron's beam-analysis chamber. After the first run, Cahill called Schwab in puzzlement to report that the paper which the beam had also struck, seemed to change every eight pages. Schwab was struck with a shiver of recognition. In the 18th century, books were made by putting together quires of paper (a quire is one large sheet folded to make smaller pages.) The cyclotron was picking up the slight elemental changes from one quire to another.

But the scientists realized that slicing up valuable books was unlikely to appeal to the owners of rare old documents. Like every conventional cyclotron the Crocker's proton beam was enclosed in a vacuum pipe. In open air the beams fall apart as the protons collide with air molecules and begin losing energy. About 4 inches (10 centimeters) out of the pipeline the beam simply wears out. The Crocker physicists started wondering whether those 4 inches could be used. By the time an opportunity came along in 1982 to analyze a Gutenberg Bible, the Crocker group had put in place a "proton milli-probe." The milliprobe is a proton beam left open at one end, with an X-ray detector attached about 1 inch (2.5 centimeters) from the opening. The uncut pages of a book, a letter, or a map can easily be fitted between the beam line and the detector and analyzed in serious detail; the proton beam focuses to a tenth the size of a single period on a printed page.

Analyzing hundreds of pages of the Gutenberg Bible was a major production, taking 42 hours of cyclotron time. The analysis was further slowed by the security anxieties of the Bible's lender, St. John's Seminary in California, which was terrified that someone would steal the valuable object. They insisted that only the scientists and technicians directly involved know that the book was in Davis. Cahill, a devoted fan of spy novelist John Le Carré, was frankly thrilled by both the secrecy and the security. (The researchers promptly dubbed their experiment the ''fishing expedition'' and code-named the Bible the ''whale.'') But the scientists were even more excited by their results. One of the mysteries that still remained about Gutenberg's printing had been the ink he used—after some 540 years, it is still as black and glossy as a newly printed page. The X-ray analysis showed that Gutenberg had mixed an ink that was nearly a slurry of metal, almost all copper and lead. ''We know more about Gutenberg's ink than he did,'' Schwab says. ''If we had a few million dollars, we could probably do a Gutenberg forgery.''

Since that time the scientists have looked through a variety of artifacts. They may be proudest of their work with the personal Bible of Johann Sebastian Bach. The composer's signature was clearly scrawled across the flyleaf. But scholars were baffled by underlinings and exclamation points throughout the book—were they Bach's or someone else's? If they were Bach's, they might offer new insights into the religious influences on his music. After analysis, the Crocker scientists are convinced that Bach himself marked up the book. The ink was the same, exactly, as that in the signature.

Arsenic Clouds

For all his fascination with history, however, Cahill's first love has remained environmental studies, and it is here that the Crocker influence has been, and will continue to be, felt most strongly. As an example, Cahill cites the Crocker's role in a court battle in the early 1980s over water from Mono Lake —an ancient body of water set in dense volcanic rocks along the east side of the Sierra Nevada. Since World War II the city of Los Angeles had been diverting tributaries from the lake to provide water for its growing population. But as the lake dried, and as its wildlife began to suffocate and great clouds of dust began to blow off the now uncovered lake bed, Mono Lake became the focus of a furious environmental battle.

Los Angeles sought to portray it as a tug-of-war between helping people and helping wildlife. That changed after the state air resources board asked Cahill to analyze the dust clouds. He discovered they were full of arsenic, a known carcinogen. The arsenic was natural, part of the volcanic bedrock, and was being released as the lake bed dried and blew away. With the sensitivity of the cyclotron Cahill could track the dust, even as the clouds broke apart, into a near-invisible, hazardous haze drifting east toward Nevada. In 1989, when California state courts restricted water diversions from lake tributaries, they cited the arsenic risk as a major factor.

It's an issue that Cahill's group has not let go of; they are now deeply involved in an investigation of the dust blowing off the bed of Owens Lake, about 125 miles (200 kilometers) south of Mono and a body of water also eliminated by L.A. water diversions. At the turn of the century Owens Lake covered 110 square miles (285 square kilometers) in the Owens Valley, a spread of water so deep in waterfowl that residents 10 miles (16 kilometers) away claimed they could hear the thunder of beating wings as ducks rose off the lake. Now only a pool of salty brine, unfit to drink, remains. In a high wind the dust storms off the lake bed blow like the bitterest northern blizzard. The NOAA has calculated that the lake contributes 6 percent of the breathable dust particles in the air over the continental United States. ''It's a serious problem,'' says Cahill, whose group has, not surprisingly, found arsenic in the Owens Lake dust. Again, they painstakingly tracked its creeping spread 80 miles (128 kilometers); this time winds being funneled through the Owens Valley were pushing the dust south toward Los Angeles.

Crocker scientists are now studying the polluting effects of power plants in the Pacific Northwest and of agricultural burning in California's Central Valley and near Florida's Everglades. Cahill dreams of a whole new generation of physicists that will tackle such problems: ''I want to get all these young physicists who are frustrated and bored and want to do something that can make a difference and say, 'Look at what you can do.' ''

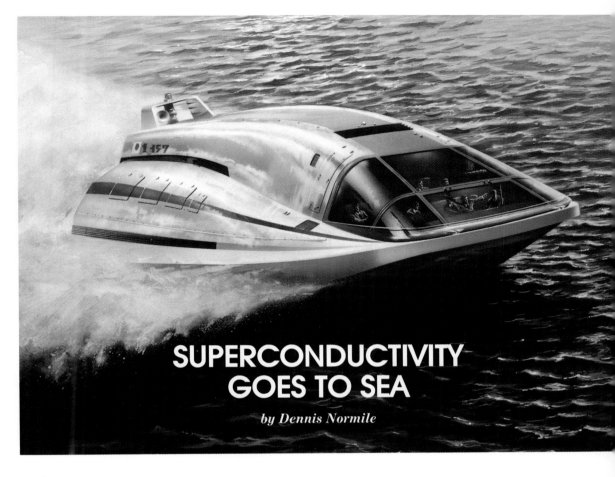

SUPERCONDUCTIVITY GOES TO SEA

by Dennis Normile

Sitting dead in the water, *Yamato 1* looks like a rocket. It doesn't take off like one, though. During its first sea trial in June 1992, the streamlined ship didn't even move fast enough to throw a wake.

Yet that maiden voyage could become as important in marine history as the 1807 launching of Robert Fulton's steam-powered *Clermont* or the *Nautilus* submarine's first trip under nuclear power in 1955. *Yamato*'s propellerless technology may eventually move ships faster and with less noise than any propulsion system yet developed. But with *Yamato* achieving a speed of only slightly more than 6 knots (about 7 miles—11 kilometers —per hour) as it glided through the harbor in Kobe, Japan, the technology has a long way to go. Before magnet-driven ships can become competitive with conventional watercraft, re-

searchers will need to develop superconducting magnets that are more powerful and lightweight than any known today.

High-tech Thrust

In place of a propeller or paddle wheel, *Yamato* uses jets of water produced by a magnetohydrodynamic (MHD) propulsion system. MHD technology is based on a fundamental law of electromagnetism: When a magnetic field and an electric current intersect in a liquid, their repulsive interaction propels the liquid in a direction perpendicular to both the field and the current.

In *Yamato* the liquid is seawater, which conducts electricity because of the salt it contains. The boat's futuristic shape channels the seawater into two MHD thrusters on the bottom of the hull—one on each side.

MAGNETOHYDRODYNAMIC PROPULSION AND THE "LEFT-HAND RULE"

Magnetic flux

Electric current

The "left-hand rule" is a simple method for calculating the interaction between a magnetic field and an electric current in a liquid. If you hold your left hand in the position shown at right, you can use three of your fingers to represent the three forces at work. With your thumb pointed in the direction of magnetic flux and your index finger pointed in the direction that the electric current flows, your middle finger will show the direction in which the liquid will move as a result of the repulsive interaction between the magnetic field and electric current. This rule can be tested on the drawing of one of the thruster ducts (below). Six of these ducts are arranged in a circle within each of two thruster pods (right). Diesel generators and cooling systems fill most of the remaining space inside the boat (bottom).

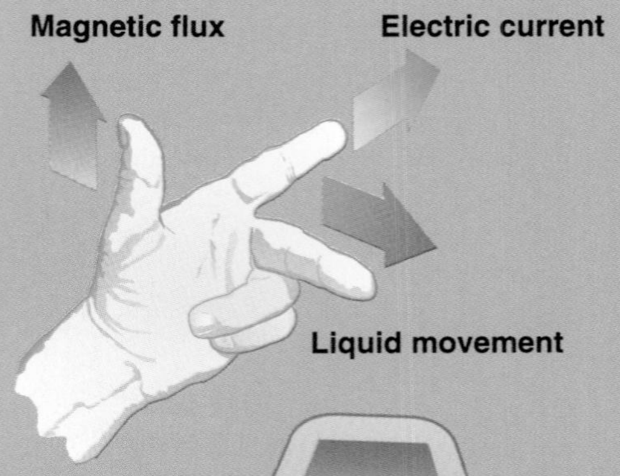

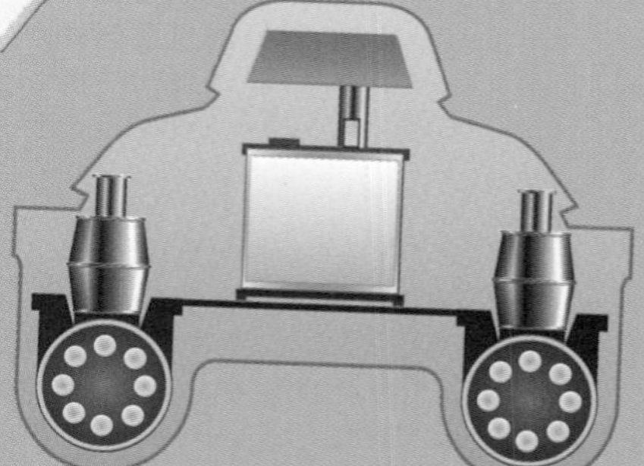

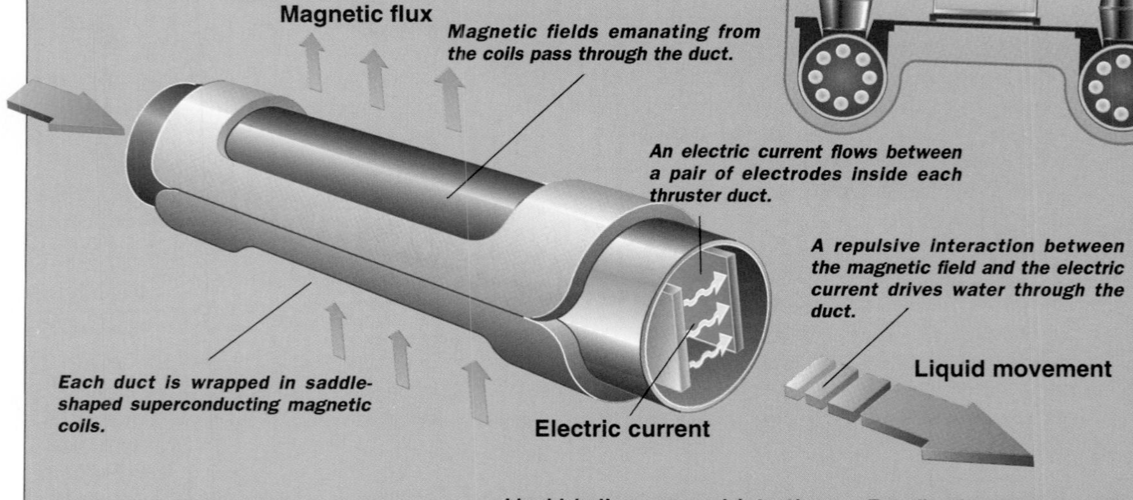

Liquid movement

Magnetic flux

Magnetic fields emanating from the coils pass through the duct.

An electric current flows between a pair of electrodes inside each thruster duct.

A repulsive interaction between the magnetic field and the electric current drives water through the duct.

Liquid movement

Each duct is wrapped in saddle-shaped superconducting magnetic coils.

Electric current

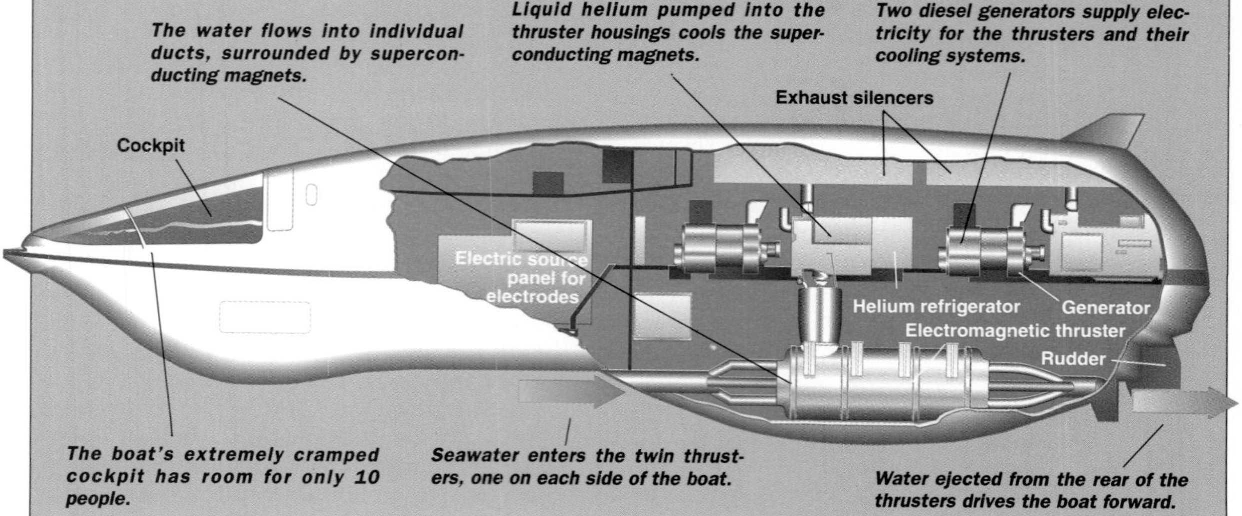

The water flows into individual ducts, surrounded by superconducting magnets.

Liquid helium pumped into the thruster housings cools the superconducting magnets.

Two diesel generators supply electricity for the thrusters and their cooling systems.

Exhaust silencers

Cockpit

Electric source panel for electrodes

Helium refrigerator

Generator

Electromagnetic thruster

Rudder

The boat's extremely cramped cockpit has room for only 10 people.

Seawater enters the twin thrusters, one on each side of the boat.

Water ejected from the rear of the thrusters drives the boat forward.

Inside each thruster the seawater flows into six identical tubes, arranged in a circle like a cluster of rocket engines. The 10-inch (25-centimeter)-diameter tubes are individually wrapped in saddle-shaped superconducting magnetic coils made of niobium-titanium alloy filaments packed into wires with copper cores and shells. Liquid helium cools the coils to −452.13° F (−268.96° C), just a few degrees above absolute zero, keeping them in a superconducting state in which they have almost no resistance to electricity. Electricity flowing through the coils generates powerful magnetic fields within the thruster tubes. When an electric current is passed between a pair of electrodes inside each tube, seawater is forcefully ejected from the tubes, jetting the boat forward.

The MHD thrusters have several advantages over conventional propulsion systems. Most important, they will enable ships and submarines to travel at high speed. Visionaries anticipate speeds of up to 100 knots (about 115 miles—185 kilometers—per hour), although researchers associated with the *Yamato* project regard that goal as extremely optimistic.

Kensaku Imaichi, professor emeritus at Osaka University in Japan and a key figure in the design of *Yamato 1*, believes the technology may have commercial ships cruising at 40 to 50 knots (46 to 58 miles—74 to 93 kilometers—per hour) sometime in the next century. Speeds could go even higher if breakthroughs can be made in hull materials and ship stability.

Silent Sailing

The speed of propeller-driven vessels is limited by a phenomenon known as cavitation. If the propeller turns too fast, an area of low pressure forms in front of the churning blades, causing the water to vaporize. This not only reduces a ship's efficiency, but can even destroy its propeller.

"By doing away with the propellers, we can avoid the phenomenon," says Seizo Motora, professor emeritus at the University of Tokyo and head of a scientific committee that designed *Yamato*'s hull.

The second major advantage of MHD propulsion is silence. "There is no noise from a propeller, there is no noise from cavitation," Motora says.

Silence is a central part of the plot of Tom Clancy's novel *The Hunt for Red October,* in which a Soviet submarine's noiseless propulsion system renders the vessel virtually undetectable by sonar. Clancy says that his fictional submarine propulsion system, dubbed the caterpillar drive, is not based on MHD technology. But the quest for stealthy submarines is what fuels interest in MHD propulsion in most countries, including the United States.

Because MHD thrusters have no moving parts, they are not only quiet and vibration-free, but are also expected to have lower maintenance requirements than conventional propulsion systems. And with no need for a drive shaft to link the power source to the propulsion system, shipbuilders could experiment with new ship designs. Among the possibilities: cargo submarines shaped like jet airplanes, and catamaran-style ocean liners.

A primary goal of the consortium that built *Yamato 1* was to revitalize the stagnant Japanese shipbuilding industry. The Ship and Ocean Foundation, a private organization that sponsored the $40 million project, received research and manufacturing support from Japanese giants such as Toshiba, Kobe Steel, and Sumitomo Electric. Mitsubishi Heavy Industries supervised the construction of *Yamato 1* at its shipyard in Kobe.

The Japanese researchers aim to develop fast, fuel-efficient commercial ships that could eventually ferry California oranges to Japan in less than a week, a trip that currently takes about two weeks. The technology might even be used to construct large cargo submarines, which could dive below the ocean surface to avoid storms. MHD propulsion might also be used in submarine ferries that would transport passengers between Japan's numerous islands.

If the Ship and Ocean Foundation has its way, MHD technology will never be used for military gain. "We will restrict the use [of the technology] to peaceful purposes," vows Yohei Sasakawa, director of the foundation and chairman of the steering committee for the *Yamato* project.

Yamato 1, *the first full-scale ship to use superconducting electromagnets for propulsion, made its maiden voyage in June 1992. Crew members monitored the ship's performance from the cockpit (right).*

An American Idea

Like many technologies now being developed in Japan, MHD propulsion was first proposed by American scientists. Several researchers published papers on the subject in the early 1960s. A few years after the first papers appeared, Stewart Way, a Westinghouse Research Center consultant who collaborated with an engineer at the University of California at Santa Barbara, used an MHD thruster to propel a small model submarine. But Way used ordinary magnets rather than superconducting ones, and scientists concluded that the technology would have to await more-efficient methods of generating strong magnetic fields.

In 1968 a report on Way's research reached Yoshiro Saji, who had recently joined the faculty of Kobe Mercantile Marine University. Saji had been looking for some way to apply his specialty—cryogenics—to ships.

Saji had worked with liquid helium and knew of its role in superconductivity. With a boiling point of $-452.13°$ F ($-268.96°$ C), liquid helium is used to cool superconducting alloys down to their critical temperature —the temperature at which electrical resistance disappears.

Sailing Straight

Saji realized that the strong magnetic field required to boost the efficiency of MHD propulsion could be generated by a superconducting coil. He spent five years on theoretical work, and started actual experiments in 1973. In 1978 Saji and a group of colleagues succeeded in propelling a superconducting MHD model ship through a tank of seawater. There was only one problem: The miniature ship wouldn't go straight.

"No matter what we did, it always turned. It was really strange," Saji says.

The model was 4.5 feet (1.4 meters) long, with the superconducting coil and electrodes in a fin sticking straight down from the bottom of the boat. The coil was a simple oval immersed in a liquid-helium bath, and the electrodes were placed above and below the coil. In this arrangement, called an external-field system, the electromagnet does not surround a water-filled duct. Instead, the magnetic field is projected into the open water below the boat. That was the problem.

Earth's magnetism increased the strength of the magnetic field on one side of the fin and decreased it on the other, generating more propulsion on one side of the model. When Saji had the water tank reoriented along a line from magnetic north to south, the boat went straight.

To avoid the influence of Earth's magnetic field, the superconducting coil and the electrodes were laid horizontally on the flat hull of Saji's next model. In 1983 this 1.1-ton, 11.5-foot (3.5-meter)-long model achieved a speed of 30 inches (76 centimeters) per second.

Saji's work attracted the attention of Sasakawa, scion of a wealthy philanthropic family with long-standing ties to Japan's shipbuilding industry. Sasakawa asked for Saji's cooperation in taking the work under the aegis of the family-dominated Japan Foundation for Shipbuilding Advancement (now known as the Ship and Ocean Foundation). The foundation recruited scientists to take part in the design of *Yamato 1*.

Scientific committees assigned to tackle various aspects of the ship's design began their work in 1985. The committee designing the thrusters soon decided to adopt an internal-field system. The researchers were worried that the uncontrolled magnetic fields emanating from an external-field system might affect other boats and the marine environment.

The committee built a small internal-field thruster, loaded it in a model, and successfully sailed it through a tank of water. The thruster had a superconducting coil wrapped around a single water duct. For the actual ship, however, the thruster committee decided to arrange six ducts in a circle. The committee members claimed this would virtually eliminate magnetic-radiation leakage, because the stray flux emanating from one coil would be drawn into its neighbor.

Quelling Quench

While models demonstrated the MHD technology, scaling it up proved troublesome. The first sea trial, two years behind, was cut short when "quench" warning signals lit up in the ship's cockpit. Quench is the phenomenon in which a superconducting coil slips out of the superconducting state.

"Even if an unimaginably small portion of the coil slips into ordinary conducting mode, it generates an enormous amount of heat, which propagates throughout the system," explains Imaichi, who heads the thruster committee.

A sudden rise in temperature could destroy the coil, so *Yamato 1* is wired with sensors that constantly check for anomalies that might presage quench. "If the sensors are too sensitive, that interferes with ordinary operation," Imaichi says. That seems to have been the case on the maiden voyage of the *Yamato 1*. Apparently, heat leaking through a gap around some electrical lead cables set off the signals warning of imminent quench.

The signal glitch was a minor detraction from the day's success. In just seven years,

The Yamato 1's *unique hull design comes into full view when the ship is hoisted from the water. The ship has a maximum speed of 8 knots.*

the Japanese team had turned a laboratory curiosity into a full-scale moving boat.

Making it practical, however, is a bigger challenge. *Yamato*'s thrusters, refrigeration systems, and twin 2,000-kilowatt diesel generators virtually fill the 98-foot (30-meter) ship, leaving room for a crew of only 10 in a cramped cockpit. The equipment weighs 143 tons, or about 70 percent of the total vessel displacement of 204 tons. By comparison, the propulsion systems of current ocean-going freighters typically account for less than 10 percent of the displacement.

The maximum efficiency expected from *Yamato 1* is less than 4 percent. The efficiency of current commercial ships ranges from around 22 percent for hydrofoils to 60 percent for cargo ships. And a first-rate sculler can propel himself along faster than *Yamato 1,* which has a top speed of only 8 knots (9.2 miles per hour).

Enhancing Efficiency

For MHD propulsion to become practical, researchers will have to develop magnets that are much lighter and more powerful than current ones. *Yamato*'s coils now generate a maximum magnetic field of 4 teslas. Imaichi thinks the magnetic field can be raised to 30 teslas within the next two decades, and that this improvement will result in an overall efficiency of between 22 and 23 percent. This, he says, is comparable to the current efficiency of hydrofoils. "An MHD ship can be competitive with that kind of ship," Imaichi says. But to date, the maximum steady-state

Stealth-Submarine Research on Hold

Although the United States pioneered the development of magnet-powered ships in the early 1960s, the current American research effort in this area can be summed up in one word: negligible.

No U.S. Government agency has serious plans for building a ship or submarine with magnetohydrodynamic (MHD) propulsion. A few groups, however, have built MHD "thrusters" in the laboratory. In the largest project, researchers at Argonne National Laboratory in Illinois used a 21-foot (6.4-meter)-long superconducting magnet —30,000 times stronger than Earth's magnetic field—to propel water through an 18-inch-wide duct, simulating the action of an MHD jet in the open ocean.

Researchers at the Newport News Shipbuilding Company in Virginia and the Naval Underwater Systems Center in Rhode Island have conducted similar experiments, smaller in scale.

The conclusion from these studies? "There's no showstopper" that will make magnetic propulsion impractical, argues Michael Petrick, who led the Argonne team. The key to building an MHD-powered vessel, he says, is designing lighter and more-efficient magnets. Although the Argonne experimental thruster was several times more efficient than those in Japan's *Yamato 1,* it nevertheless wasted about 60 percent of the force generated by the magnet. Further, the stainless-steel magnet weighed 180 tons—too heavy for a ship.

Another limitation of the technology is that MHD ships will not work in areas with significant amounts of fresh water, which does not conduct electricity as well as salt water. Thus, it might be difficult for an MHD ship to dock at many major ports, says Mike Superczynski, an expert on superconducting magnets at the David Taylor Research Center in Annapolis, Maryland. Also, salt water would quickly corrode the electrodes in the thrusters, Superczynski argues.

Both the Argonne and Newport News Ship-building projects are now on hold, pending more government funding. "We've gone as far as we can go in the magnetic field achieved using superconductors has been about 20 teslas.

Incremental improvements will come from using superfluid helium, which can cool the superconducting coil down to −456.45° F (−271.36° C), allowing increased current and thus a stronger magnetic field. Superconducting niobium-tin-alloy filaments will also provide stronger fields, but the material is much harder to form into wires than the niobium-titanium alloy used in the present coils.

A leap forward could come with the refinement of new ceramic superconducting materials, which weigh as little as one-third as much as available metal superconductors. More important, with critical temperatures above −321.09° F (−196.16° C), the ceramics are "high-temperature" superconductors. This means they can be cooled with liquid nitrogen, which is less expensive and easier to use than liquid helium.

Taking these possible advances into account, Imaichi and other researchers have concluded that the ultimate efficiency of the technology will be somewhere around 50 percent, equivalent to the efficiency of current cargo ships. Hoping for the technological breakthroughs that will make MHD propulsion competitive with conventional ships, Sasakawa and his colleagues intend to build a second-generation MHD ship.

"It took years for Watt's steam engine to be put to practical use on a ship," Sasakawa explains. "If you don't have dreams, you can't make progress."

laboratory," says Newport News Shipbuilding's Rich Ranellone. "Our next step should be a demonstration at sea." Ranellone proposes installing a magnetic thruster on an existing submarine.

He shouldn't get his hopes up. The Defense Advanced Research Projects Agency (DARPA), which funds this type of work, doesn't have the money to keep MHD research afloat. "We've finished our program," says Captain Ted Rice, an undersea-warfare expert at the agency. "There are too many other things we need to attack."

Navy brass aren't excited about MHD propulsion either, agree experts close to the service. "The Navy is not even vaguely enthused about the idea," says Alan Berman, a staff member at the Center for Naval Analyses in Alexandria, Virginia. The biggest potential drawback: The huge magnetic fields emanating from an MHD-powered submarine would make it easy for enemies to detect, even if the submarine made no sound. An MHD-powered sub would also leave a trail of chlorine, created by the electrolysis of seawater. For now, it looks like the international race to build a magnet-powered ship has exactly one contender.

Robert Langreth

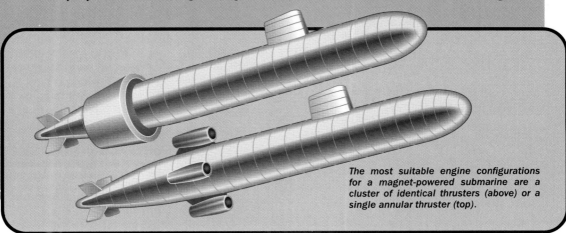

The most suitable engine configurations for a magnet-powered submarine are a cluster of identical thrusters (above) or a single annular thruster (top).

THE SUBATOMIC MENAGERIE   *by Christopher King*

Anoted economist once observed that "small is beautiful." If that is true, then the greatest beauty of all may reside in a realm that lies beyond our conventional vision: the world of subatomic particles. All that we are, all that we see and touch—our very universe, in fact—would not exist without these particles and the forces that bind them together. For hundreds of years, great thinkers and scientists have labored to understand this unseen world and to bring its occupants into view. Today, in sprawling laboratories that harness the power of gigantic machines, the effort continues. By learning more about the world of the subatomic, scientists hope to find definitive answers regarding the birth and evolution of our universe. They also seek to answer what is perhaps the simplest and most complex question of all: How does the universe work?

Beyond the Atom
Around the 4th or 5th century B.C., Greek philosophers developed the concept that all matter is composed of indivisible particles. Borrowing from a word meaning "something that cannot be cut," the Greeks referred to these particles as *atoms*. The idea that atoms constitute the fundamental building blocks of matter continued into the 1800s. In 1808 the British chemist John Dalton published *A New Theory of Chemical Philosophy,* in which he presented a list of chemical elements whose "ultimate particles," he proposed, were atoms.

As the 20th century commenced, however, researchers were beginning to suspect that atoms themselves were composed of even smaller, constituent particles. In the early 1900s, Ernest Rutherford, a New Zealand-born Englishman, proposed that most of the atom's mass is contained in a small, positively charged nucleus, which is orbited by negatively charged particles known as *electrons.* Investigation by later physicists provided a more complete picture of atomic structure: A nucleus—containing the positively charged *protons,* along with neutrally charged particles known as *neutrons*—orbited by negatively charged elec-

The computer simulation of two protons colliding (above) provides data on the energy, momentum, and mass of the collision products, which are part of the subatomic "zoo."

trons. The number of protons and neutrons determine an element's atomic weight. Hydrogen, for example, is the lightest element, with one electron and one proton, while uranium, the heaviest naturally occurring element, has 92 electrons and 92 protons. Atoms, in turn, group together to form molecules, which combine to constitute matter.

Physicists had cracked the atom, so to speak, but many more subatomic surprises lay in store. Through the efforts of experimentalists in the laboratory and the ruminations and calculations of theorists, new particles were brought to light. One was the *neutrino,* a neutral particle possessing little or no mass. Predicted in the early 1930s, another 20 years passed before neutrinos were proven to exist. Another early discovery was the *positron,* observed in 1932. This particle was seen to match the mass of the electron but also to carry an opposite, positive electrical charge. Thus was born the concept of "antimatter"—the idea that for all the particles composing matter, there exist corresponding "antiparticles." This further complicated the subatomic picture, and hinted at even more unseen physical entities.

Fundamental Forces

As theory and experiment continued to fill in the picture of how matter is formed, scientists arrived at the notion that four fundamental forces determine the behavior of particles, and thus, of all bulk matter. Together, these forces control all biological, chemical, and nuclear phenomena. They are the *strong force,* which binds together the protons and neutrons in the atomic nucleus; *electromagnetism,* which holds together the nucleus and electrons in an atom; the *weak force,* which controls the process of radioactive decay; and, finally, *gravity,* which draws any two objects with mass toward one another. While gravity is not very important at the atomic level (since the masses of individual atoms are so small), gravity does play a role in keeping the universe together.

Throughout the 20th century, the effort to understand the interactions between these forces and the subatomic world has produced a succession of new particles and

new insights. One such insight demonstrated that the protons, electrons, and neutrons within the atom turn out not to be the fundamental constituents of matter after all. There is an even deeper subatomic level.

This idea was developed in the early 1960s, by which time physicists had identified nearly 100 different particles. Many of these were collectively named *hadrons,* after the Greek word for "strong," since they are mostly the product of the strong force. In 1964 Murray Gell-Mann and George Zweig, two physicists at the California Institute of Technology (Caltech) in Pasadena, California, were independently attempting to sort out this hadron family. They proposed the existence of another group of particles that combine to form the protons inside the atomic nucleus. Gell-Mann gave the particles the fanciful name *quarks,* after a line from James Joyce's novel *Finnegans Wake* ("Three quarks for Muster Mark"). While quarks were far from nonsensical, Gell-Mann realized that his theory demanded that quarks possess very odd properties. One such property is an electrical charge that, unlike the whole charges possessed by protons and electrons, is fractional.

Quark theory eventually came to describe a whole family of these particles, whose properties and interactions were described in suitably fanciful terms, such as *color* and *flavor.* While the theory encountered some skepticism originally, it is entirely accepted today. "The theory of quarks, of what protons are made of and what holds atomic nuclei together, is very well established," says Frank Wilczek, a theoretical physicist at the Institute for Advanced Study in Princeton, New Jersey. "And in recent years there's been spectacular progress in testing it experimentally."

Hunting for Particles

Most of what scientists know about the subatomic world has been learned by accelerating particles to speeds approaching that of light, and then smashing them into targets or other particles to see what is thrown off in the resulting collisions. As the late Nobel laureate Richard Feynman noted, the process has been likened to taking two fine

Swiss watches, bashing them together, and trying to figure out how they work by examining the pieces that come flying away. Inelegant, perhaps, but very effective in high-energy physics. The first particle accelerator, built in the 1930s, was 4 inches (10 centimeters) across. Using electromagnetism, this first *cyclotron,* as it was known, accelerated protons that were then guided into a collision with a fixed target.

Today's accelerators are somewhat larger. The Fermi National Accelerator Laboratory—or, as it's more popularly known, Fermilab—occupies a broad plain in Batavia, Illinois. There particles are accelerated in an underground ring that is 3 miles (4.8 kilometers) in circumference. Roughly equal in scale is the facility at CERN (the French acronym for the European Center for Nuclear Research), the European accelerator complex on the French-Swiss border. At these sites, thousands of scientists from many nations participate in experiments in which carefully guided particle streams are driven into thousands of collisions per second, at energies nearly rivaling the Big Bang that created the universe. The collisions take place within huge particle detectors, which are larger than railroad freight cars and are crammed with sensing devices, where the telltale trails of streaking particles can be recorded and analyzed.

Within these giant accelerators, theories are put to the test. One cele-brated union of theory and experiment occurred at CERN in the early 1980s. Three theorists, Sheldon Glashow and Steven Weinberg of Harvard in Massachusetts and Abdus Salam of the Imperial College of Science and Technology in London, had proposed that two of the major nuclear forces, electromagnetism and the weak force, are actually facets of the same phenomenon. However, this so-called *electroweak* theory required the existence of three very heavy particles known as the *intermediate vector bosons*—two varieties of a W particle, and a particle known as Z. At CERN, Italian physicist Carlo Rubbia led an effort to produce these particles by colliding a beam of protons with a counterrotating beam of antiprotons.

In 1983, after millions of such collision "events" at CERN during the previous two years, Rubbia and his hundreds of physicist collaborators announced definitive evidence for the W and Z. In addition to producing particles, the electroweak theory also produced Nobel prizes. The 1979 physics prize was shared by Glashow, Weinberg, and Salam, while the 1984 prize honored Rubbia and his CERN colleague Simon van der Meer, whose technical designs had been essential in finding the particles. Today at CERN, Z particles, which were once only the stuff of theory, are now themselves being collided at ever higher energies to search for more-elusive quarry.

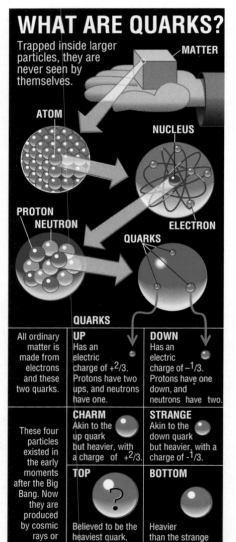

WHAT ARE QUARKS?

Trapped inside larger particles, they are never seen by themselves.

MATTER

ATOM

NUCLEUS

PROTON NEUTRON

ELECTRON

QUARKS

QUARKS

All ordinary matter is made from electrons and these two quarks.	**UP** Has an electric charge of $+2/3$. Protons have two ups, and neutrons have one.	**DOWN** Has an electric charge of $-1/3$. Protons have one down, and neutrons have two.
These four particles existed in the early moments after the Big Bang. Now they are produced by cosmic rays or high-energy accelerators.	**CHARM** Akin to the up quark but heavier, with a charge of $+2/3$.	**STRANGE** Akin to the down quark but heavier, with a charge of $-1/3$.
	TOP ? Believed to be the heaviest quark, with a charge of $+2/3$.	**BOTTOM** Heavier than the strange quark, with a charge of $-1/3$.

The first experimental proof for quarks came in the late 1960s and early 1970s at another celebrated lab, the Stanford Linear Accelerator in California. In contrast to the giant-ring accelerators, particles at this facility are shot along a 2-mile (3.2-kilometer) underground track. By smashing protons into electrons and evaluating the scattered energies, scientists determined that quarks are indeed present within the proton. Quarks, in fact, remain trapped inside the proton. As yet, no "free" quarks have been observed. Experiments in the late 1970s demonstrated the existence of yet another particle within the proton: the *gluon*. Gluons, carriers of the strong nuclear force, are the exchange particles that bind quarks together. They determine a quark's "color"— a property similar to electric charge.

At present, six types of quarks are now acknowledged, each with a characteristically whimsical name: "up," "down," "charm," "truth," "beauty," and "top." Only the "top" quark has not yet been observed experimentally, although physicists seem to be closing in on it.

Fitting It All Together

Quarks represent only one species within the subatomic "zoo." Particles are generally classed in two broad categories: *fermions,* (which, like the Illinois accelerator lab, are named after the Italian physicist Enrico Fermi), and *bosons,* named for the Indian physicist S.N. Bose. The fermions are further divided into quarks and *leptons.* Leptons include the electron, the muon, the tau particle, and their associated neutrinos. In all, hundreds of particles have been identified. There is also a whole menagerie of theoretical particles whose existence has yet to be demonstrated. These have names like *quark nuggets, photinos, sneutrinos, neutralinos,* and *polonyions.*

The hunt for particles is far from over. At the end of 1992, physicists at Fermilab were buzzing over results that seemed to suggest evidence for the top quark, although no formal claim has yet been made. At CERN, they're gearing up energies to search for the so-called *Higgs boson,* another long-sought particle—although it is not entirely certain whether CERN will have sufficient energy for the job.

The Higgs boson is definitely on the most-wanted list in Waxahachie, Texas, where the next generation of accelerator is taking shape in the form of the Superconducting Supercollider (SSC, currently under construction). This "Mother of All Accelerators," as it's been called, will fling particles around a 54-mile (87-kilometer) ring at energies as yet unattained in today's colliders. Assuming the $8.3 billion facility can survive congressional budget wrangling, its scientific promise seems limitless. Collisions at the SSC, scheduled to begin around the middle of the next decade, could assist in reaching what is perhaps the overriding goal of physics in the 20th century: the quest for a theory that will unify and explain all the fundamental nuclear forces.

Albert Einstein tried, and failed, to produce such a unified theory. Others have added their own efforts in the form of a "Grand Unification Theory" (GUT), a "theory of everything," "supersymmetry," or even "superstrings," in an attempt to explain the great variety of particles and the differences between the fundamental forces. Or, in the words of science writer Timothy Ferris, the challenge is "to draw the disparate elements of physics into one whole." And according to Nobel laureate Leon Lederman, discoverer of the muon neutrino, the Higgs boson may provide a key for the simplicity that has eluded scientists so far.

"We really do believe that a simple formula, a simple concept will explain everything in the physical universe," he says from his office at Fermilab, "including the origin of the universe—how the Big Bang took place and how the universe evolved from that calamitous event, what's happened since, and what will happen in the future. And when we find it, it will be so beautiful, so simple, that we'll be convulsed with laughter and say, 'Of course it's that—how could it be anything else!' And we'll print it on a T-shirt."

As physicists continue to explore the world of the subatomic, the actual physical results may be small. The scientific payoffs, however, will be anything but.

THE STRATEGY OF STRATEGIC MINERALS

by Peter Harben

Strategic minerals are national treasures. In tranquil times, one country's strategic mineral is simply another's foreign-exchange earner. During confrontational times, strategic minerals emerge as pawns in a political game in which producers use them as bargaining chips while consumers strive to reduce their worth. As nature distributed mineral resources around the world, little respect was paid to political boundaries or industrial

might—what is in the ground to be discovered is what can be developed. But when this lop-sided supply pattern involves a mineral that is crucial to a country's military strength and economy, the mineral becomes critical, and even "strategic."

What constitutes a strategic mineral may vary from country to country and from time to time. Strategic minerals have always been part of history. Copper from the Fertile Crescent of

The bronze helmet of an ancient Roman soldier contrasts markedly with the sleek lines of a F-15C aircraft. Nonetheless, these two items represent an unbroken chain of necessity—civilization's need for strategic minerals.

list of minerals considered strategic. Many critical factors contribute to a mineral's becoming strategic for a particular country. Extensive military use may render a mineral strategic even if it's relatively plentiful. Among other factors are the lack of domestic reserves, as well as the lack of known substitutes. If a small number of primary producers exist, or indeed, there is only a single supplier, the mineral in question will probably be considered strategic. Rapid technological advances may create a strong demand for a mineral. If, on the other hand, use of a mineral declines, production may drop off for economic reasons, and supplies of this mineral may become difficult to obtain. A mineral may be held "hostage" if the country that controls its production has an ideology hostile to the country that needs it.

Scarcity, for whatever reason, contributes greatly to a mineral's being designated strategic. If the mineral is found in remote locations, recovery becomes costly and difficult. This not only drives up the price, but keeps supplies low. Third World countries may lack the funds needed to explore their reserves or extract the resources if found. But these countries, wary of "colonialism," may hesitate to ask wealthier countries to provide the necessary capital. Foreign-trade controls are an additional element that may create scarcity where none actually exists. Severe regulatory restraints, increasing environmental restrictions, and important health and safety hazards may also artificially create scarcity.

U.S. Needs
The United States classifies a multitude of minerals as strategic, and, because many of these are scarce or nonexistent within its borders, the United States must significantly depend on imports. A few minerals that lack an adequate U.S. domestic supply include tin, beryl, cadmium, chromite, cobalt, columbium, diamonds, fluorspar, graphite, manganese, mercury, nickel, antimony, tantalum, and platinum-group metals. Of these, cadmium, chromite, manganese, nickel,

the Middle East was the strategic mineral 7,000 years ago. Then, during the Bronze Age some 3,000 years later, tin was added to copper. Pursuit of these minerals spread to parts of Europe, the Middle East, and East Asia. A hundred years ago in northern Europe, iron, coal, and limestone, critical to the Industrial Revolution, formed the foundation of the international industrial era of the 20th century. European nations strove to secure mineral supplies from their overseas colonies. World War II was, in part, caused when Germany, Italy, and Japan sought to secure mineral supplies unavailable within their existing borders.

Postwar technology relied on minerals, and increasing sophistication lengthened the and platinum-group metals qualify as strategic because of a limited number of producers or extended transportation routes.

No one believes the United States can be totally self-sufficient in minerals and metals. Nonetheless, government staff and private companies have made resource inventories of the country. Based on these inventories, scientists and bureaucrats must make provisions for exploiting somewhat marginal deposits such as chromite in the Stillwater Complex and manganese associated with the Great Lakes iron-ore fields. Having mineral-rich and friendly neighbors such as Canada and Mexico, however, has relieved U.S. concerns over nickel, potash, fluorspar, and strontium, among others.

The U.S. Government pursues a policy that retains a free-market economy but simultaneously reduces the strategic power of imported minerals. One method used by the United States to minimize the effect of strategic minerals has been the development of a policy called the National Defense Stockpile. This policy demands that sufficient stock of certain minerals be established and maintained to sustain a "conventional global war of at least three years' duration involving total mobilization of the economy." This "national attic" contains more than 90 commodities, ranging from aluminum oxide to zinc, valued at more than $9 billion.

Strategic Uses

Strategic minerals and metals help Western countries maintain a high standard of living and provide for national security. Obvious examples are the minerals used to manufacture steel and other alloys required for everything from flatware on the dinner table to armor plating on tanks. Manganese is ubiquitous, and needed in the production of virtually all types of steel, as well as products ranging from fertilizers to fragrances to ferroalloys. Dry-cell batteries consist of ground manganese dioxide, finely divided graphite, ammonium chloride, and zinc chloride, all packed into a zinc case.

Columbium, used in steels and superalloys, is required for pipelines, structural products, rocket subassemblies, and gas-turbine-engine components. Cobalt-bearing superalloys go into jet engines and high-strength magnets for electronics. Chromium is the basis for stainless steel, and nickel provides the additional strength and corrosion resistance vital for chemical and petroleum refining. Tungsten, vanadium, tantalum, and molybdenum enhance and strengthen steel while reducing corrosion. These minerals also increase steel's wear resistance and toughness, forming the basis for important alloys and compounds.

Titanium is used in everything from jet-turbine blades to lightweight bicycles to the white pigment in paper. Petroleum refining requires platinum-group metals, which are also essential components of catalytic converters and electronic circuit boards.

Bauxite, converted into a variety of alumina-rich refractories, abrasives, and chemicals, is also used to fabricate aluminum metal. This metal, in turn, becomes beverage cans, aircraft, food-wrap foil, and electrical components. Canning, electrical components, and transportation all need tin.

Besides being a girl's best friend, diamonds are a driller's best friend. Diamond drill bits cut through the hardest rocks during petroleum exploration and oil-well construction. Cadmium is used as an alloying agent and pigment; antimony, also a pigment component, is a flame retardant. Gallium is used in the manufacture of solid-state electronic components.

Geologic Reality

While the United States lacks many of the minerals and metals it consumes, South Africa and the republics of the former U.S.S.R. control large sources of these necessary minerals. Four materials, diverse in both their geology and their uses—diamond, chromite, bauxite, and manganese—illustrate how geology influences world politics, turning ordinary minerals into strategic ones.

Diamond distribution. Diamonds form deep below the Earth's surface under conditions of intense heat and pressure. They emerge at or near the surface within the rock called kimberlite, often forming a classic ice cream cone-shaped pipe. These pipes elude discovery because of their small areal exposure and their tendency to be buried. Kimberlite is

A diamond-tipped rotary blade easily cuts limestone into blocks (below) because of the diamond's superhard properties. Without diamonds, many manufacturing and drilling processes would be impossible. The global distribution of diamond-producing countries (map) does not weigh heavily in favor of the strategic requirements of Western nations.

extremely variable and complex, but can be roughly characterized as an ultrabasic rock (mostly peridotite) containing high-pressure minerals such as garnet and diopside, as well as chunks of country rock collected on the rock's long journey to the surface.

Sometimes, just sometimes, kimberlite contains diamonds. Some pipes have none at all, while others yield small, cloudy industrial diamonds. A select few contain the large, scintillating gemstones purchased by the very rich.

The rocks in which the kimberlite pipes are located, or emplaced, however, are those of stable Precambrian cratons generally older than 2 billion years. These stable, little-changing areas of the Earth's continental crust are found in southern Africa, Siberia, Scandinavia, Western Australia, the Colorado-Wyoming border area of the United States, Brazil, and Canada.

The rather specific geological environment in which diamonds occur means that large-scale diamond production is confined to 20 countries, more than half of them in Africa—Angola, Botswana, Central African Republic, Ghana, Guinea, Ivory Coast, Namibia, South Africa, Sierra Leone, Tanzania, and Zaïre. Just five countries—Australia, Botswana, South Africa, Zaïre, and the former U.S.S.R.—account for around 94 percent of the production volume.

At this time, North America produces no natural diamonds, although interest has been renewed in the Crater of Diamonds State Park in Arkansas. Because diamond production here is still regarded as unproven and economically marginal, environmental concerns may restrain development. A recent kimberlite find near Yellowknife in the Northwest Territories of Canada may prove to be North America's best diamond bet.

Chromite concentration. Commercial deposits of chromite, which is the sole chromium ore, are concentrated in belts of ultramafic rocks such as peridotite and dunite (rocks that have a low silica content, usually less than 44 percent). Most formed billions of years ago, when enormous bodies of magma crystallized and settled out in huge underground chambers. The resulting deposits, which contain virtually all (98 percent) of the world's chromite reserves, comprise stacked uniform layers stretching for dozens of miles.

The largest deposits occur in South Africa, which has well over 1 billion tons of chromite reserves; Zimbabwe; Finland; Montana in the United States; Canada; and Brazil. The Montana and Canadian deposits are not currently profitable to exploit. The other deposits, however, supply the bulk of the world's chromite production. Much smaller and younger deposits, shaped like elongated pods, are located in the former U.S.S.R., India, Albania, Turkey, Iran, and the Philippines. These important chromite reserves each contain 1 million to 10 million tons.

The concentration of chromite reserves in such large bodies restricts production to a belt that extends from southern and eastern Europe through the Middle and Far East to the southern Urals. Another belt in southern Africa encompasses South Africa, Zimbabwe, and Madagascar. Smaller deposits exist in the Philippines and

Brazil. Overall, the former U.S.S.R. and South Africa control almost 80 percent of the world's chromite reserves and 70 percent of its annual production.

When discussing strategic minerals, we should give special emphasis to the Bushveld Igneous Complex in Transvaal Province, South Africa. This vast mineral storehouse, by far the world's largest, contains little silica and much iron and nickel. The Bushveld granite contains deposits of tin, fluorite, rare earths, and tungsten, plus seams of iron ore, chromite, vanadium, nickel ore, magnesite, copper, and platinum-group metals. The Transvaal system, which forms 85 percent of the Bushveld complex's circumference, contains gold, lead, ironstone, and chrysotile asbestos. This complex may be one of the richest known in terms of its wide variety and large quantity of mineral wealth.

Bauxite bias. Present in virtually every rock type, aluminum is the most abundant metallic element in the Earth's crust. Bauxite, the commercial aluminum ore, however, is far less common. Fully 40 percent of world bauxite production comes from the Weipa

Chrome's durability and brilliant shine make it an aesthetically pleasing material for, among other things, buildings and automobiles. The world's major deposits of chromite, from which chrome is derived, are in remote and relatively inaccessible areas (map).

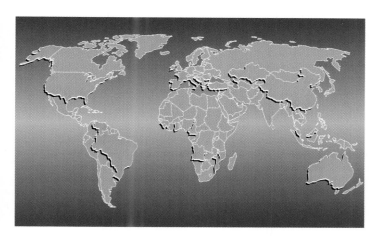

Bauxite is an aluminum ore found in many areas, including the United States (map). One of the most familiar metals in everyday life, aluminum is used in everything from aircraft and soda cans to foils and phone booths (right).

deposits in the Cape York Peninsula, northern Queensland, Australia. Nonetheless, bauxite distribution is still confined when it comes to specialized grades. For example, bauxite required for refractories is produced only in Guyana, China, and Brazil. Although the United States mines some bauxite in the southeastern states, its industries rely almost exclusively on imports from Guinea, Jamaica, Australia, and Brazil.

Manganese monopolies. Manganese, another common element in the Earth's crust, has been identified in more than 300 minerals, although high-grade commercial ore is confined to just a handful. Industry's desire for specific grades restricts still further the choice of commercial sources. For example, battery-grade manganese dioxide (MnO_2), which requires an MnO_2 content of nearly 90 percent with virtually no metallic contamination, is produced mainly in Gabon, Ghana, Greece, and Mexico.

The chief manganese-ore minerals—pyrolusite, psilomelane, braunite, and manganite—are all oxides (meaning their metal components combine with oxygen, much as rust is made of iron and oxygen). Oxide deposits are superior commercially because of their higher manganese content and their vastness. For example, geologists estimate that the deposits in the Ukraine and South Africa contain at least 3 billion tons each.

More than 30 countries produce the world manganese output of around 27 million tons. The largest suppliers are the for-
mer U.S.S.R., which produces more than 40 percent of the world's supply, and South Africa, which produces an additional 15 percent. Other countries supplying 1 million tons per year or more are Australia, Brazil, Gabon, India, and China. Reserves, however, are more restricted. South Africa holds more than 70 percent of the total, and the former U.S.S.R., 20 percent. Australia, Gabon, China, India, and Brazil divide the bulk of the remainder among them.

The entire manganese-reserve base for North America is in Mexico, particularly around Molango, Hidalgo State, midway between Tampico and Mexico City. Japan has little manganese, and, aside from the former U.S.S.R., virtually none exists in Europe.

Future Strategy

The National Defense Stockpile has proven to be an effective deterrent against would-be aggressors, and an economic safeguard for the United States. The stockpile has been used only 10 times since the end of the Korean War in the early 1950s. The most noteworthy use was during the Vietnam War, when President Lyndon Johnson dumped copper on the market in an effort to drive down the price. The stockpile remained untouched during the Persian Gulf conflict in 1991. With the apparent end of the Cold War, the questions now become: Has the stockpile outlived its usefulness? Have strategic minerals become a historical curiosity? The answers to these questions are as complex as the geologic settings in which these minerals are found.

To a certain extent, the stockpile itself was outmoded even during more-aggressive times. Various strategic minerals are produced in U.S. smelters with increased efficiency: cobalt and rhenium, copper and bismuth, gallium, germanium, and zinc.

The recycling movement sweeping the U.S. goes further than cleaning up the environment. For instance, recycled car batteries yield lead and antimony.

Other minerals that find their way back into the industrial manufacturing cycle include aluminum from cans, chromite from refractories, and platinum from automobile exhaust systems. Substituting one metal for another (or plastic, ceramics, or fiber optics for metal) reduces dependence on some strategic materials such as copper, while boosting the importance of others such as beryllium.

International diplomacy helps develop friendly suppliers through international-cooperation schemes such as the Caribbean Basin Initiative. U.S. mining companies continue to explore and exploit deposits overseas to ensure access to raw materials. For example, partners in the recently opened Escondida copper mine in Chile include the United States, Great Britain, and Japan. Creating synthetic minerals has greatly boosted domestic supplies. Synthetic diamonds are the most obvious example, as well as graphite, quartz crystal, silicon carbide, electrolytic manganese dioxide, and magnesia from seawater.

One aspect of the so-called peace dividend is that the United States may have time to modernize its definition of strategic minerals. Last year the Pentagon announced a plan to sell off approximately $4.8 billion of commodities in an effort to reduce the U.S. National Defense Stockpile by about 40 percent. Some commodities targeted—copper, silver, tungsten, lead, and zinc—are all produced in the United States. There appears to be less danger of supply interruptions for other targeted commodities such as asbestos and cadmium. The rationale for the decision to sell off part of the stockpile appears to be that some of the generated revenues would be used to purchase the high-tech metals and materials required in the production of more-advanced materials. Thus, it seems that minerals and metals will continue to be strategic; only the names will change.

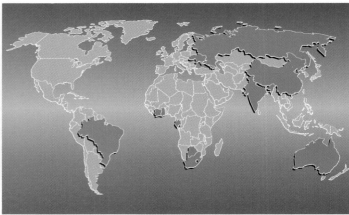

Dry-cell batteries, such as the type being loaded into the flashlight at left, are manufactured from manganese dioxide. Although manganese is a common element in many minerals, deposits rich in manganese are not so common (map). North America and Western Europe have virtually no important deposits of manganese.

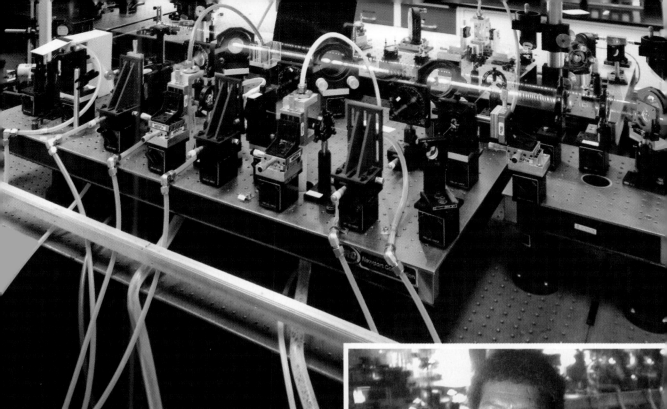

FEMTOCHEMISTRY:
Chemistry as it Happens

by David A. Pendlebury

S ome movies, at 90 to 120 minutes in length, last far too long, as disappointed film viewers around the world can attest. But imagine sitting through a film that lasts 32 million years. What's more, imagine that the only action in the entire movie occurs in just a single second.

Now you know how tiny a femtosecond is: A femtosecond is to a second what a second is to 32 million years. It is a nearly unbelievable, almost unimaginably short space of time. In fact, it's just a millionth of a billionth of a

Ahmed Zewail (inset), a leading authority on femtochemistry, has developed what can be described as the world's fastest camera. His apparatus (above), which combines an ultrafast laser with a spectroscope, can capture chemical reactions as they occur.

second. It also happens to be about the speed at which atoms and molecules come together or break apart to form new materials.

Until a few years ago, no one had ever witnessed a chemical reaction as it happened. Chemists could see what they put into a reaction, and they could see what came out at the other end, but precisely what happened in the flash of a few femtoseconds was simply beyond their powers of observation. That was a source of great frustration; after all, the moment when chemical bonds form and break—

the transition state—is the most fundamental event in all of chemistry. Swedish researcher Sture Forsen of Lund University likened the situation to an audience seeing only the very beginning and the very end of *Hamlet,* "The main characters are introduced, then the curtain falls for a change of scenery, and, as it rises again, we see on the stage a considerable number of bodies and a few survivors. Not an easy task for the inexperienced to unravel what actually took place."

Such was the state of transition-state chemistry until 1987. In that year, California Institute of Technology chemist Ahmed H. Zewail built what can only be described as the world's fastest camera, one with a shutter speed of just a few femtoseconds. Zewail had ingeniously joined together a long tradition of research in molecular-beam chemistry with the latest in ultrafast laser technology, developed by Charles V. Shank and his colleagues at AT&T Bell Labs. With his invention, Zewail became the first person in the world to finally capture the exact moment when molecules divide or unite. In that instant a new era of chemistry dawned.

At a champagne party held shortly after Zewail's discovery, a new word for this brand-new field of chemistry bubbled to the surface of the celebration: femtochemistry. Femtochemistry is already a substantial subfield in the realm of physical chemistry. Already, hundreds of researchers worldwide have produced thousands of papers about experiments using ultrafast lasers to probe the transition state. Some of these works have been the most-cited papers in physical chemistry published in the

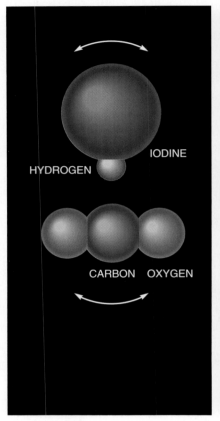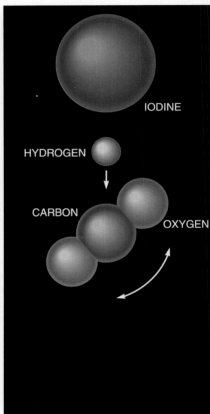

last several years. Not only is femtochemistry providing the most detailed information yet about reaction dynamics, but it also holds the promise of one day allowing researchers to control the course of chemical reaction in order to create completely novel materials.

Clocking the Chemical Bond
Zewail's "camera" is actually a combination of instruments, all of which are housed in a perfectly clean, dust-free room. The apparatus includes an ultrafast laser, which emits flashes of light in femtosecond pulses, and a spectroscope, which detects the presence of

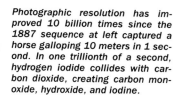

Photographic resolution has improved 10 billion times since the 1887 sequence at left captured a horse galloping 10 meters in 1 second. In one trillionth of a second, hydrogen iodide collides with carbon dioxide, creating carbon monoxide, hydroxide, and iodine.

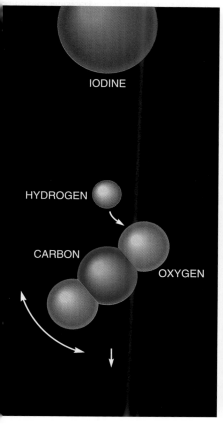

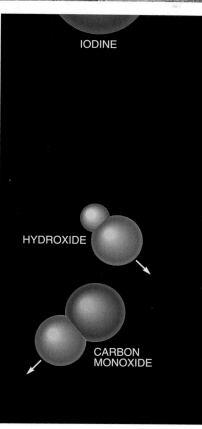

IODINE

HYDROGEN

CARBON

OXYGEN

IODINE

HYDROXIDE

CARBON MONOXIDE

pieces of the old molecule. To time their birth and determine their order of appearance, another beam of laser light, traveling just a few femtoseconds behind the first, hits the molecule's pieces as they are flying apart. The spectroscope detects which pieces are moving, based upon the light that they reflect. This second laser pulse (called the probe pulse) can be timed precisely at different intervals to reveal how long it takes for various chemical species to appear and in what order they do so.

The earliest work of Zewail and others examined these simple so-called unimolecular reactions. One of the first was the breakdown of the molecular cyanogen iodide (ICN) into iodine and cyanide. In another experiment, involving the dissociation of the molecule tetrafluorodiiodoethane ($C_2F_4I_2$) into tetrafluoroethene (C_2F_4) and iodine atoms (2I), it was discovered that one of the bonds in the molecule takes 100 times as long to break as the other bond.

Ultrafast-laser chemists soon began to investigate more complex reactions, such as bimolecular reactions, in which atoms and molecules combine to create new molecules, and the reaction dynamics of clusters, which involve a large number of interacting elements. More recently, femtochemists have studied the twisting (isomerization) of bonds, as well as hydrogen and electron transfer reactions.

different atoms and molecules by the spectrum of light that they give off when hit by the laser beam.

To take a picture of a simple chemical reaction, the laser beam is aimed at the target molecule, and the laser is activated. First, one pulse of light (called the pump pulse) strikes the molecule and energizes it. If that energy is strong enough, it causes the molecule to vibrate, and stretches the electron bonds that hold the molecule's atoms together so much that they break. Out of that destructive process, called dissociation, new chemical species are created from the

A Molecular Movie

The four "still photographs" shown on the previous two pages illustrate the results of a landmark experiment performed by Zewail and his colleagues. When the four illustrations are viewed in sequence, they constitute a movie that we'll call "Complicated Encounter."

In this film, hydrogen iodide and carbon dioxide play the leading roles. They have an exceedingly brief (1 trillionth of a second) encounter, but it becomes a bit more complicated than either imagined at the onset. Suddenly, the hydrogen iodide molecule finds its hydrogen atom slipping away into the arms of an oxygen atom that itself has broken things off with carbon dioxide. The hydrogen and oxygen atoms join together to become a couple called hydroxide. They then leave iodine to go its own way, and carbon monoxide to carry on its existence with only a single oxygen atom.

The femtosecond camera recorded the whole affair, in explicit detail.

Laser-customized Chemistry

"This ability to view molecular dynamics also suggests new ways of controlling reaction," Zewail has noted. "The prospect exists for fine-tuning the motion and reactivity of molecules. If successful in the coming decades, laser-customized chemistry may be developed." The technique of using lasers to control the course of a chemical reaction is known as bond-selective chemistry.

In early 1992 Zewail himself described a process using two simultaneous laser pulses to control the reaction of iodine molecules with xenon atoms to form xenon iodide. In this case the laser served not only to energize a reaction and to time its course, but it also actually steered the reaction to determine its outcome. Zewail's team found that the formation of xenon iodide could be controlled by adjusting the time delay between the two laser pulses. In effect the reaction could actually be switched on or switched off. Theoretical work on the control of chemical reactions has been considered by many research groups, and numerous schemes have been proposed for femtochemistry control.

From Spectroscope to Imaging

New femtosecond-resolution experimental tools and techniques are continually being developed. For example, Zewail has proposed the creation of a new type of femtosecond camera—one that would replace the optical-probe pulse with an electron pulse. This method, called ultrafast diffraction, would allow researchers to map with amazing precision the movements of individual atoms in molecules during the course of quite complex reactions.

Last year, the Caltech group reported the first successful imaging of molecules using this technique. By providing new knowledge of structural change over time, ultrafast diffraction may open the door to applications through precise control of the course of chemical reactions.

Near the End of the Race Against Time

Chemists have long held a keen interest in stories of molecular union and division. Their ability to recount these stories and to describe how different chemical actors play their roles is seen as a high mark of intellectual achievement in chemical research. Indeed, the Swedish Academy of Sciences has twice awarded its Nobel chemistry prize to researchers in the field of reaction dynamics —in 1967 to Manfred Eigen of Germany and Ronald G. W. Norrish and Sir George Porter of Great Britain; and in 1986 to Americans Dudley R. Herschbach and Yuan T. Lee and Canadian John C. Polanyi. Today's femtochemists follow in the footsteps of these trailblazing scientists.

In March of 1993, at a worldwide conference held in Berlin focusing on femtosecond chemistry, the 1967 Nobel laureate Sir George Porter, who with Norrish more than 40 years ago pioneered flash-photolysis techniques and achieved resolution of a chemical reaction at the then-record speed of a millisecond, or a thousandth of a second, observed: "The study of chemical events that occur in the femtosecond time scale is the ultimate achievement in half a century of development and, although many future events will be run over the same course, chemists are near the end of the race against time."

SCIENTIFIC FRAUD *by Linda Marsa*

On an early spring afternoon in March 1987, two scientists hunched over a small desk in a hotel room in Frankfurt, Germany—the pair pressured by their governments to end more than two years of bitter battles over who first isolated the AIDS virus. Robert Gallo, one of the most powerful and protected superstars at the National Institutes of Health (NIH), and Luc Montagnier, a virologist with Paris's Pasteur Institute, hammered out a "definitive scientific history" of their stunning scientific achievements.

The account that emerged from the hotel-room meeting spawned an agreement between the French and American governments. Montagnier and Gallo would share the credit for the discovery, and their governments would jointly patent the rights to the test for detecting the AIDS virus. Officially, the occasion marked the end of a controversy that erupted in 1985, when evidence surfaced that Gallo's

AIDS-causing HIV virus was virtually the genetic twin of a virus sent to him in 1983 by Montagnier. The matter rested, however until 1989, when the *Chicago Tribune* published a 50,000-word exposé on Robert Gallo that finally prodded federal investigators to probe more deeply into the events in his laboratory in 1983.

Now one of the more-ignominious chapters in the annals of science appears to be grinding to a conclusion. A blue-ribbon scientific panel, set up to monitor an internal NIH investigation of the affair, accused Gallo of "intellectual recklessness of a high degree." The panel members, nominated by the National Academy of Sciences (NAS), criticized Gallo for his failure to acknowledge having grown and studied an AIDS virus sent to him by the French. Gallo is also the target of a federal inquiry investigating charges of perjury and patent fraud related to his patent application for the AIDS test.

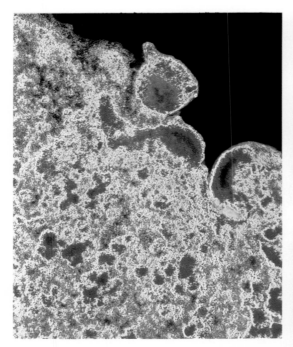

American clinical researcher Robert Gallo (above right) and French virologist Luc Montagnier (right) are engaged in a high-profile battle over who was the first to isolate the human immunodeficiency virus (above, shown magnified 37,800 times), which causes AIDS.

Gallo is one of the giants of science and the only scientist ever to win two Albert Lasker awards, perhaps our nation's most prestigious biomedical honor. The battle for scientific supremacy, however, destroyed more than egos. While Gallo and Montagnier were wasting precious time taking potshots at each other across the Atlantic—instead of using their considerable talents to find a cure—this plague was claiming thousands of lives on both sides of the ocean.

And on this score, the French are hardly saints. In 1985 health officials in France stalled licensing the American test for the AIDS virus in order to give French scientists a chance to devise their own. This action had devastating results. In the intervening five months, virtually half the hemophiliacs in France became infected with HIV through the untested blood supply; many have since died. The unconscionable act eventually erupted into a national scandal that threatened to topple the Mitterrand government.

Breach of Protocol?

In the United States, the Robert Gallo debacle is just the latest in a rash of recent scandals that have rocked the scientific community and tarnished its once-pristine image. Incidents of simply sloppy science, misconduct, plagiarism, manipulation or faking of data, and outright criminal behavior have made front-page news with alarming regularity. Even Nobel laureate David Baltimore's career suffered when he stubbornly refused to admit the possibility that a colleague had committed fraud in a scientific paper Baltimore had coauthored. But the Gallo affair, perhaps more than others, underscores the fact that when scientists go astray, it represents no mere breach of protocol. People's lives are put at stake.

Indeed, these disturbing revelations of wrongdoing by reputable researchers have eroded public confidence, prompted congressional oversight committees to launch costly investigations, and forced universities and federal agencies to develop more-rigorous policing mechanisms to flush out the charlatans. Some worry about the fate of the next generation of scientists. And others question what this means for the future of science, a collaborative enterprise that must be conducted in an atmosphere of trust in order to thrive. "Honor is integral to the scientific method," says Bernadine Healy, M.D., director of the National Institutes of Health. "Without it, science would crumble."

Whether misconduct is more prevalent, however, cannot be documented. The agencies that dispense many of the federal research grants—the NIH, which has a budget of $9 billion and supports about 100,000 scientists, and the National Science Foundation (NSF), which runs on $2.6 billion a year—do not keep statistics on fraud. But out of about 18,000 grants the NSF made in 1991, only 52 cases of misconduct were reported to the agency, according to Donald Buzzelli, a staff associate for oversight in the NSF's Office of Inspector General.

But scientists themselves report the actual incidence of misconduct is much higher. In a November 1991 survey of 1,500 scientists conducted by the American Association for the Advancement of Science (AAAS), more than a quarter of respondents said they had witnessed faking, falsifying, or outright theft of research in the past decade. "There's no hard evidence, but my gut feeling is, the problem has gotten much worse in the past five years," says Walter Stewart, an NIH staff scientist who has participated in several fraud investigations.

Of course, charges of cheating, corner cutting, and deception are not new. Many great scientists —Galileo, Mendel, Newton, and Dalton, among them—fudged or concocted data to make their theories more compelling or to demolish the arguments of their rivals. But on the whole, science used to resemble the priesthood; individuals were called to the profession, which operated on the honor system. Scientists were, says one congressional aide, "the white-coated guardians of truth searching after the grail of knowledge."

Researchers chafed at outside intrusions and insisted their self-correcting system of internal checks was enough to catch any miscreants. Peer review panels scrutinize grant applications before any money is awarded. Editors of scientific journals send articles submitted for publication to "referees," other experts in the field, to determine the significance and validity of the findings. And once articles are published, researchers attempt to replicate the results to double-check their accuracy.

Fraud or Sloppy Science?

In isolated instances, such as the cold-fusion controversy, the system still appears to work. In March of 1989, the University of Utah announced that two chemists, Martin Fleischmann and Stanley Pons, had harnessed the source of the Sun's energy and had produced cold fusion in a test tube of heavy water. Excited chemists and physicists formed teams to confirm these astonishing findings. As the weeks passed, though, the scientists could find no evidence of fusion using this procedure, and concluded the Utah experiment was flawed.

"My guess is, when they made their claim, they genuinely believed it. Therefore, they were guilty of sloppy science— not fraud," says Robert Park, a professor of physics at the University of Maryland and director of the Washington, D.C., office of the American Physical Society. But the chief

Galileo Gali-liar?

Galileo Galilei (1564–1642) is considered the founder of the modern scientific method. But he wrote about experiments that were so difficult to reproduce that many doubt he actually conducted them.

reason scientists tried to replicate cold fusion is because these experiments could be done on a tabletop with their lunch money.

In today's financially constrained climate, though, scientific researchers rarely repeat each other's work. "Nobody gets funding to do replications, so science is not the self-cleansing apparatus it once was," says Jules Hallum, director of the NIH's Office of Scientific Integrity (OSI). Flawed or even flagrantly spurious science has slipped through the cracks in these supposedly fail-safe mechanisms.

Star-crossed Science

Evidence indicates Johannes Kepler (1571–1630), the father of modern astronomy, doctored his calculations to bolster his theory that the planets move in elliptical orbits, not in circles, around the Sun.

decades cost plenty. The electron microscopes, ultracentrifuges, and the gene sequencers that have emerged in the past decade as standard laboratory equipment cost well into five figures. Sophisticated imaging devices, like PET scanners, cost upwards of $1 million. Laboratories like Baltimore's, where as many as 60 people—graduate students, postdoctoral fellows, and technicians —struggle for space, equipment, attention, and glory, require hundreds of thousands of dollars a year to run.

Deception, by its nature, must be clandestine. So the job of keeping science clean has fallen to whistle-blowers. Yet, far from welcoming or supporting whistle-blowers, some research institutions treat them shabbily. Sometimes they even drive them out of the profession—raising serious doubts as to whether science can be trusted to govern itself. Something has gone awry. But what?

Science Is Big Business

"This is all about money," argues Robert Bell, an economics professor at Brooklyn College and author of *Impure Science,* a troubling—and exhaustively researched— indictment of shoddy scientific practices. "Like the Pentagon's defense contractors, the science community has evolved into another patronage system which enriches those at the top. Universities have a vested interest in not finding anyone guilty of fraud. Because if they do, they may have to return the delinquent researcher's grants. When someone blows the whistle, universities set up investigatory panels, which are inevitably kangaroo courts that cover up abuses."

Indeed, science is big business, and all the awesome advances of the past two

Unfortunately, this dramatic escalation in costs for equipment and research comes at a time when funding sources are dwindling. Nobel laureate physicist Leon Lederman, the former president of the AAAS, conducted a recent survey of researchers at 50 universities. He calculated that 1990 federal funding for scientific research at these institutions was only 20 percent higher than in 1968, while the number of scientists has doubled. Small wonder that the jockeying for grants often looks more like Roller Derby than rivalry among highly educated professionals.

Pressure for the Home Run

Almost by default, top scientists have been cast in the role of empire builders. They jet to scientific meetings around the world to network with equally eminent colleagues or woo big university donors, while their underlings grind out massive volumes of research and grant proposals to keep the laboratory engines lubricated with money. Rarely do science's superstars conduct the hands-on research that made them famous or mentor the new crop of young scientists or even closely monitor the results of every experiment conducted in their laboratories.

Added to this is the fierce competition to make a key discovery. And progress up the ranks depends more on the number of papers a scientist has published than on competence. "The game of acquiring a long list of publications is a relatively new development as evidenced by the fact that just two decades ago, the current problems with paper inflation were unthinkable," note William Broad and Nicholas Wade in *Betrayers of the Truth*. "In 1958, James Watson"— who won the 1962 Nobel Prize for deciphering the structure of DNA in 1953— "had on his curriculum vitae eighteen papers. . . . Today, the bibliography of a candidate facing a similar climb often lists 50 or even 100 papers."

Yet nature refuses to behave the way researchers would like. Often experiments don't turn out the expected results, so scientists are forced to try a fresh approach—and pray they get more money to keep going.

But the pressures to hit a home run each time tempt scientists to cut corners, to round off numbers so that results appear more impressive, to overlook anomalous findings that would put the data in a less favorable light—or to just cheat. While "most scientists are scrupulously honest and deplore scurrilous behavior," says the NIH's Healy, they are nevertheless reluctant to inform on others. "They don't rally to alter corrupt situations because they see whistleblowers get burned," says Bell.

Results Too Perfect

Like Robert Sprague. Just before Christmas in 1983, the psychology professor and director of the Institute for Research on Human Development at the University of Illinois made the most agonizing decision of his life. He sent a lengthy letter to the National Institute of Mental Health (NIMH), accusing Stephen Breuning, a former protégé and close friend, of falsifying data.

Breuning, while a psychologist at the University of Pittsburgh, had carried out drug studies on severely retarded children who often mutilated themselves by banging their heads against the wall. These children were calmed with powerful tranquilizers known as neuroleptics. But neuroleptics had an unfortunate side effect: They seemed to cause involuntary muscle spasms such as flapping arms or wagging tongues.

Breuning questioned whether these drugs were beneficial. Experiments he conducted between 1980 and 1983 indicated they did more harm than good, and that controversial stimulants, including Ritalin, far more effectively controlled the wild, self-destructive behavior with fewer adverse side effects. Colleagues praised his work, which helped shape public-health policy for several years to come.

But Sprague felt there was something fishy about Breuning's work. There were glaring inconsistencies, and his results were just a little too perfect. In the real world, nobody

A Matter of Some Gravity

Isaac Newton (1642–1727) crunched numbers to make the predictive power of his universal gravitational theory carry more weight. Scientists have since noted he "adjusted" his calculations on the velocity of sound and on the processions of the equinoxes so they would support his theory.

produces 100 percent correlations. Poring over Breuning's data, Sprague realized that his friend had cheated. Sprague was heartsick—and outraged.

After he sent his detailed dispatch outlining three separate incidents of fraud to the NIMH, Sprague expected swift action. In fact, the NIMH ordered the University of Pittsburgh to convene a panel to look into one of Sprague's allegations. Confronted with the charges, Breuning consequently confessed to filing research reports containing false statements.

But Sprague didn't expect that *his* life would turn into a Franz Kafka novel: that the NIMH would make *him* the target of a probe or sharply cut his grants. NIMH officials stoutly deny these incidents had any relation to the Breuning case, even though the institute had fully funded Sprague for the preceding 17 years' worth of research. Nor did the scientist imagine this gut-wrenching affair would drag on for more than three years before a glacially slow NIMH panel rendered its final verdict: "Virtually all of Breuning's work was fabricated and that Sprague's work and accusations were beyond reproach." The University of Pittsburgh was forced to return $163,000 in federal grants it had received, and Breuning was subsequently convicted on criminal charges for falsifying medical research.

Scientific Watergate

What happened to Margot O'Toole is even more egregious. In 1986 she worked as a postdoctoral fellow in the MIT laboratory of Thereza Imanishi-Kari. That year, Imanishi-Kari coauthored, with the director of MIT's Whitehead Institute, David Baltimore, a scientific paper published in the journal *Cell*.

Mendelian Misconduct

Abbé Gregor Mendel's experimental results (1865) were so perfect that later researchers were convinced he falsified his data, which formed the basis of modern genetics.

The article purported to show that antibodies expressed by one mouse could be made to mimic those of another mouse. O'Toole accidentally came across Imanishi-Kari's laboratory notes and noticed serious discrepancies. "It was obvious from these records that the experiment had not yielded the published results," she says. O'Toole took her concerns to Imanishi-Kari, and later reported her suspicions to Baltimore and MIT officials. Baltimore curtly dismissed her claims, calling O'Toole a "discontented" postdoctoral fellow. Suddenly O'Toole became a pariah in the scientific community: She lost her MIT laboratory position and could not find another one. Finally she was reduced to taking a job answering phones for her brother's moving company.

O'Toole's ordeal spotlights how science's allegedly self-correcting mechanisms broke down at every checkpoint. It never occurred to the referees assigned to review the paper that there could be anything even slightly shady in an article coauthored by David Baltimore, one of the world's most distinguished scientists. Fellow scientists likely wasted countless hours and money trying to build on the bogus findings. Investigators at Tufts University, MIT, and, initially, at the NIH performed little more than a perfunctory probe of O'Toole's charges. As *The New York Times* wrote in a scathing editorial, "the initial investigations of Dr. O'Toole's complaints smacked of an old-boy network drawing up the wagons to protect scientific reputations."

The truth came to light only when the case fell under the noses of NIH scientists Walter Stewart and Ned Feder and the House Subcommittee on Oversight and Investigations, headed by Representative John

Dingell. But Baltimore can blame only himself for the scale of the scandal. His defiant insistence that his critics didn't know what they were talking about escalated this relatively minor transgression into what is dubbed a "scientific Watergate." He even orchestrated a letter-writing campaign among his colleagues to stop Dingell's inquiry. And Baltimore retracted the article and halfheartedly apologized to O'Toole only after the Secret Service forensically analyzed Imanishi-Kari's laboratory notes, proving her data entries were falsified at a later date.

After five years of vilification before she was finally vindicated, would O'Toole do it again? "Absolutely. Because at the root of this were the most fundamental principles of the entire profession," says O'Toole, who now works for a biotech company in Cambridge, Massachusetts.

What's Being Done?

These jolting scandals may ultimately prove salutary. A number of hopeful signs demonstrate that the obviously shaken scientific community—acting out of a sense of enlightened self-interest—intends to clean house and stop the university cover-ups, the federal foot-dragging, and the witch-hunts of whistle-blowers, all of which allow abuses to flourish.

The OSI's staff may grow from 19 investigators to 28, including three lawyers, and, to ensure objectivity, the office may be moved away from the NIH. The NSF has issued stricter guidelines regarding misconduct, and is keeping a tighter rein on universities. And a 22-member panel of the National Academy of Sciences (NAS) in Washington, D.C., has worked for more than two years to hammer out the principles of good scientific conduct.

The NIH also now requires ethics programs in graduate students' curricula. "We operated under the false belief that young scientists would learn ethics in the laboratory by osmosis or by being exposed to good role models," says Stephanie J. Bird, a special assistant to MIT's associate provost, recruited to develop ethics programs there. "But now there's a recognition that ethics need to be explicitly articulated." Other schools, including Harvard and Dartmouth, now have similar ethics programs.

In 1989 the American Association for the Advancement of Science gave its Scientific Freedom and Responsibility Award to Robert L. Sprague. In accepting one of science's highest honors, Sprague observed that it wasn't surprising some scientists cheat. After all, they're only human. "What is surprising," charged the soft-spoken, bespectacled psychologist, who hardly resembles a rabble-rousing renegade, "is that the system of science actually works against a speedy, appropriate adjudication of suspected misconduct. This is intolerable in a civilized society."

Skullduggery

The Piltdown Man is generally considered the greatest scientific hoax of all time. In 1908 a part of a skull hailed as proof of the missing link between apes and humans was unearthed in an English gravel pit on Piltdown Common, in Sussex. In the 1950s, however, researchers using modern dating techniques revealed the skull was actually an ape jaw with part of a human skull attached that had been stained to appear old.

Chemistry

DIAMOND PRODUCTION

Synthetic diamonds were the subject of intense scientific research in 1992. Rustum Roy, a material scientist at Pennsylvania State University, announced a new solid-state process to produce synthetic diamonds by exposing preformed carbon composites to moderately high temperatures of 1,100° to 1,830° F (600° to 1,000° C) in the presence of a microwave plasma of atomic hydrogen. Compared to conventional techniques, this new procedure represents a simpler, cheaper route to making shaped diamonds or diamond-composite objects.

For years, researchers have failed to make a substitute to match diamond's hardness, a characteristic derived from the mineral's unique network of strong carbon-carbon covalent bonds. Yip-Wah Chung of Northwestern University, Evanston, Illinois, announced development of a material he considers harder than diamond. The new material is composed of a thin film of a carbon nitride.

Synthetic diamonds and diamond thin films are used in industry to make cutting tools for such varied jobs as cutting metals and slicing architectural stone. Chung now plans to coat steel-cutting tools with this superhard material, which was found wear-resistant in friction tests.

Scientists at the AT&T Bell Laboratories, Murray Hill, New Jersey, and Crystallume, Inc., Menlo Park, California, reported that synthetic-diamond films can match the heat-carrying capacity of real gem-quality diamond crystals. The discovery could lead to applications of synthetic-diamond films as heat spreaders in preventing thermal damage to microelectronic circuits.

MOLECULE OF THE YEAR

Nitric oxide (NO), a small, gaseous molecule, has gained notoriety as an ingredient of smog and acid rain. Its reputation changed for the better in 1992, when, after years of accumulating evidence, scientists established that NO functions as a bioregulator and a neurotransmitter, playing a key role in different biological processes of the body. Scientists found that small amounts of NO can transmit nerve impulses, dilate blood vesicles, and kill foreign organisms and tumor cells. For these contributions the journal *Science* declared NO the 1992 "Molecule of the Year."

The exact mechanism of NO's activities still remains elusive, however. Jonathan S. Stamler, Department of Medicine, Harvard Medical School, proposed that the uncharged, free-radical form of NO may not be the only species responsible for its biological functions. He suggested that other intermediate compounds containing the NO group may also play a role in the mechanism.

In other developments relating to NO, Michael A. Marletta of the University of Michigan, who in 1985 discovered the key role NO plays in immune-system response, determined the structure of the enzyme, known as NO synthase (NOS), which triggers or catalyzes production of NO in mammalian cells. Scientists also investigated the possibility that NO may aid learning and memory processes. Scientists hope someday to develop drugs that turn NO on and off or control its release in cells.

Because NO is a free radical with a half-life of only a few seconds, it has been difficult for scientists to measure it in single cells. Tadeusz Malinski and Ziad Taha of Oakland University, Rochester, Michigan, developed a unique microsensor that makes it possible to measure NO concentrations both at the outside surface of single cells and in their interior. Such single-cell measurements of NO will greatly improve our understanding of the role of NO in various biological processes.

NOMENCLATURE PROPOSALS

Names for the three heaviest known elements (atomic numbers 107, 108, and 109), discovered between 1981 and 1984 in Darmstadt, Germany, were proposed by German scientists. Element 107 was named nielsbohrium (Ns), after the Danish physicist Niels Bohr; element 108 was named hassium (Hs), after the Latin name (Hassia) for the German state of Hesse (where Darmstadt is located); and element 109 was named meitnerium (Mt), after Austrian physicist Lise Meitner. The names are yet to be accepted by the nomenclature commission of the International Union of Pure and Applied Chemistry (IUPAC).

German scientists have proposed naming elements 107 and 109, two of the heaviest known elements, after renowned physicists Niels Bohr (left) and Lise Meitner (right), respectively.

FULLERENE ADVANCES

Another hollow, cagelike structure similar to fullerenes was found in 1992. A. Welford Castleman, Jr., and his colleagues at Penn State University announced the existence of a second class of buckyball-like clusters called *metello-carbohedrenes*, or *met-cars*. Despite their structural resemblance to buckyballs, met-cars have different properties.

The first met-car discovered, Ti_8C_{12}, containing eight titanium and 12 carbon atoms, was prepared by vaporizing titanium metal in a laser vaporization plasma reactor. The titanium plasma undergoes reaction with hydrocarbons such as methane, ethylene, and benzene. Ti_8C_{12} has a pentagonal dodecahedron structure consisting of 12 pentagons, each with two titanium atoms and three carbon atoms at its vertices. Welford and his team have since discovered other met-cars with transitional-metal atoms such as vanadium, hafnium, and zirconium.

Their unique optical and electrical properties are expected to make met-cars useful in data-storage devices and high-speed computers. Because transitional metals such as titanium are good catalysts, met-cars could also be used as catalysts to promote chemical reactions.

Among a number of developments in fullerene-related science, the presence of the fullerenes C_{60} and C_{70} in a 600 million-year-old rock established for the first time that these molecules exist in nature. Also, a simple recipe to make large quantities of previously scarce fullerene tubules was announced by researchers at the NEC Corporation's Fundamental Research Laboratories in Tsukuba, Japan. Fullerene tubules (carbon fibers) are better than graphite fibers in strengthening materials, and, because they are electrically conductive, may have potential applications as molecular-scale "wires" and other tiny electronic components.

BEEFY-MEATY-PEPTIDE

In other areas of chemistry, Arthur M. Spanier and his colleagues at the United States Department of Agriculture (USDA) Southern Regional Research Center in New Orleans, Louisiana, found the naturally occurring peptide that is responsible for the meaty taste of beef.

The peptide—called BMP for beefy-meaty-peptide—consists of eight linked amino acids produced from a larger parent protein during the aging of the beef after slaughter. Removal of one or two amino acid links from BMP destroys its special taste.

Discovery of BMP could make it possible for people to enhance meat flavor in many TV dinners, soups, and other foods. The USDA group plans to look for similar compounds in pork and poultry.

Vinod Jain

Mathematics

HOW DO WE KNOW THAT ANYTHING IS TRUE?

PROVE IT!! These words are spoken by children on a playground, lawyers in a courtroom, and any other people in conflict. But these words have a very particular meaning for mathematicians, who use proofs to verify new results. The need for proofs is stronger in mathematics than the evidence required in other disciplines. For instance, lab sciences depend on the accuracy of the machinery being used, and social sciences depend on the way that a survey is designed. In these disciplines, no one can say for certain whether a theory is right or wrong. In math, the rules are well established and not subject to any experimental error. A proof is either true or it is not.

The modern approach to proofs finds its roots in books written around 300 B.C. by the Greek mathematician Euclid. In a series of books titled *Elements*, Euclid set forth the foundations of geometry in a very systematic way. In so doing, he provided the framework all mathematicians have since followed: state the fundamentals (called axioms) in as simple a way as possible; use the rules of logic to deduce theorems and applications of these fundamentals; and make sure that every step follows from the previous step. Modern mathematicians sometimes are informal in the way that they follow these rules, but when they are challenged, they always go back to Euclid's approach.

DETERMINISM

A very common belief about the nature of proofs, held up until this century, is called *determinism*. Roughly stated, it says that any statement that you can make should be either provable or disprovable if our axiomatic system is sophisticated enough. Thus, for example, a statement is proposed that any even integer can be written as the sum of 2 primes. This is known as the Goldbach conjecture, and it is a question that is unknown. We can check the question for a lot of values $(4 = 2 + 2, 6 = 3 + 3, 8 = 3 + 5, 10 = 5 + 5, 12 = 5 + 7, \ldots)$, but we can never check all of the infinite number of cases in this way. We can prove a lot of other results about integers, such as the fact that all integers can be uniquely written as a product of prime factors. Thus, it seems reasonable to assume that we could either prove the Goldbach conjecture or find a counterexample, the idea behind determinism. A clever mathematician, given enough time, should be able to either prove or disprove the claim.

THEOLOGY OF MATHEMATICS

In the 1930s an Austrian mathematician named Gödel made a stunning breakthrough that shattered the determinism myth. He showed that no matter how sophisticated our axiomatic system became, there would be statements that could be formulated that could not be proved or disproved within the system. This result not only disproved determinism, but it also made it clear that it is impossible to tell which statements in the system would not have a proof or counterexample. Thus, the Goldbach conjecture may or may not be provable within our current assumptions (or any other consistent axioms that we might want to add), but unless we come up with a proof or a counterexample, we may never know. This result has sometimes been called the "Theology of Mathematics" because it shows that our universe is inherently unknowable.

One of the mathematical consequences of Gödel's result is that we know that some of the great questions in math may never be known or disproved. Among these, we could include the Goldbach conjecture or the distribution of the digits in the infinite expansion of pi. The step-by-step method that had been so successfully exploited since Euclid's time was not the complete answer to deciding whether a statement is true. This did not mean that Euclid's method is not valid, just that it is not the whole story.

In the past few years, mathematicians have considered the question of how we can know if very long proofs are actually true. Some sophisticated proofs have required thousands, and even tens of thousands, of pages to write. In the old Euclidean way of doing things, every sentence and punctuation mark of those 10,000 or so pages would have to be checked in order to verify that it is true. If it is wrong in even one place, that error could be enough to make the whole result collapse. Although painstaking, the results that require so much work are important for mathematicians to know.

SHORTCUT FOR MATH PROOFS
Early in 1993 a major theoretical breakthrough was accomplished by Rajeev Motwani, Ph.D., assistant professor at Stanford University, and colleagues from the University of California at Berkeley and AT&T Laboratories. These researchers were able to devise a way to encode a long proof so that any mistake, even in the smallest detail, would appear throughout the encoding. Thus, theoretically, even the longest proof known would be able to be checked almost as easily as the shortest proof. In reality, the mistake doesn't appear everywhere, but it appears so frequently in the computations that a fairly quick spot check will reduce the probability of missing the mistake to essentially zero. This method is far superior to being forced to check every single argument; indeed, it is viewed as a tremendous step forward in verification of long proofs.

This method has far-reaching implications in computer science. Certain problems, such as designing the best telephone network, are so difficult that they are thought to be impossible to solve. Computer scientists had hoped to come up with algorithms that would give at least reasonable approximations to the solutions. As Dr. Laszlo Babai told *The New York Times*, "We are not interested in the best possible solution. If we can get within 1 percent of an exact solution, that's good enough." But the work by Motwani shows that it is just as difficult to get a reasonable solution as it is to get an exact solution, so the computer scientists will have to think of alternative ways to tackle these difficult problems using techniques other than the straightforward search for an algorithm.

These new results have not overshadowed Euclid, or shown his techniques to be useless. His achievement still stands as one of the great advances of all time. We now have a better understanding of the limitations of his method, and we are better able to deal with the verification of long and difficult proofs. We also have a much better idea of how complicated our universe really is, and that we will not be able to understand everything no matter how hard we try.

James A. Davis

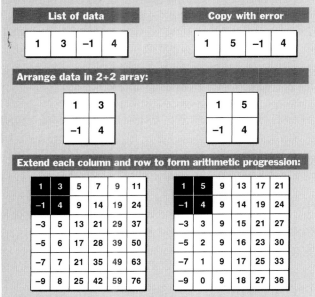

Speedy Checking of Complex Proofs

To check whether two short lists of data are identical without comparing all the data, the numbers are arranged in a two-by-two array and each row and column extended to form simple arithmetic progressions. An error is revealed if any pair of boxes in the same position do not match. Checking any pair will catch an error two out of three times; checking two entries will catch errors 9 times out of 10.

List of data				Copy with error			
1	3	−1	4	1	5	−1	4

Arrange data in 2+2 array:

1	3
−1	4

1	5
−1	4

Extend each column and row to form arithmetic progression:

1	3	5	7	9	11
−1	4	9	14	19	24
−3	5	13	21	29	37
−5	6	17	28	39	50
−7	7	21	35	49	63
−9	8	25	42	59	76

1	5	9	13	17	21
−1	4	9	14	19	24
−3	3	9	15	21	27
−5	2	9	16	23	30
−7	1	9	17	25	33
−9	0	9	18	27	36

To compare two lists with a million entries each, they can be arranged into 20-by-20 arrays. Random comparison will now catch errors 999 times out of 1,000 if 10 comparisons are made and 999,999 times out of a million if 20 comparisons are made. Checking long mathematical proofs can be streamlined in part by using a similar process of selective comparison.

Source: Dr. Laszlo Babai

Physics and Chemistry

In 1992 the Nobel Prizes in Physics and Chemistry honored two scientists whose insights into the behavior of very tiny particles set the stage for very substantial scientific advances.

The prize in physics was awarded to Georges Charpak, Ph.D., for inventing a particle detector that has been indispensable in more than 20 years of key discoveries in high-energy physics.

The Nobel Prize in Chemistry honored theoretical chemist Rudolph A. Marcus, Ph.D., for his contributions to the theory of electron-transfer reactions.

THE PRIZE IN PHYSICS

In seeking clues about the origins of the universe and the relationships between such fundamental forces of nature as gravity and electromagnetism, high-energy physicists employ immense and complex machinery to search for what is unimaginably tiny. Particle accelerators drive carefully controlled beams of subnuclear particles, such as protons and electrons, into high-speed collisions. Each collision in effect re-creates on a miniature scale the "Big Bang" that created our universe. In the resulting fireballs, the hunted particles—some of which have not existed in nature since the Big Bang—fly off from the collision and fleetingly declare their presence.

The problem for physicists is recognizing the signature traces of a given particle out of what may be several hundred particles spawned in each collision experiment. Earlier in this century, particle physicists relied primarily on photographic methods, studying photos taken in "bubble chambers" of superheated liquid hydrogen, seeking the telltale trail left behind by a streaking particle. Such methods were slow, however, and had poor resolution.

In 1968 Georges Charpak, working at the sprawling accelerator laboratory known as CERN (the French acronym for the European Center for Nuclear Research) on the French-Swiss border, devised a solution to the problem. He invented a new kind of particle detector, the "proportional wire chamber." The gas-filled chamber contains an array of charged wires spaced 0.039 to 0.078 inch (1 to 2 millimeters) apart, each wire linked to amplifiers and recorders. Charged particles passing through the chamber ionize atoms of gas, freeing electrons. The electrons then move toward the nearest charged wire. By tracking the electrical signals on each wire as well as on adjacent wires in the crisscrossing network, Dr. Charpak's invention made it possible to obtain a three-dimensional record of a particle's path. Another advantage was that signals within the detector could be sent directly to a computer for analysis. These developments, and Dr. Charpak's subsequent refinements, improved the detection rate to a very high

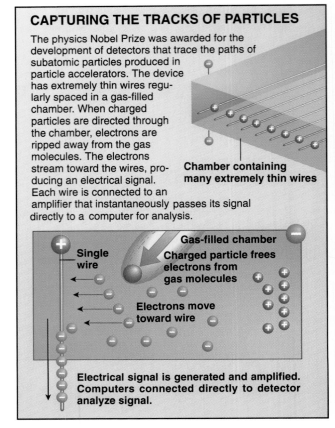

CAPTURING THE TRACKS OF PARTICLES

The physics Nobel Prize was awarded for the development of detectors that trace the paths of subatomic particles produced in particle accelerators. The device has extremely thin wires regularly spaced in a gas-filled chamber. When charged particles are directed through the chamber, electrons are ripped away from the gas molecules. The electrons stream toward the wires, producing an electrical signal. Each wire is connected to an amplifier that instantaneously passes its signal directly to a computer for analysis.

Chamber containing many extremely thin wires

Gas-filled chamber

Single wire

Charged particle frees electrons from gas molecules

Electrons move toward wire

Electrical signal is generated and amplified. Computers connected directly to detector analyze signal.

Dr. Georges Charpak (near right) won the Physics Nobel Prize for his invention of a powerful particle detector. The Chemistry Nobel Prize was awarded to Dr. Rudolph A. Marcus (far right) for his contributions to the theory of electron-transfer reactions.

degree of resolution and speed—as high as a million particles per second per wire.

Two of the key discoveries in particle physics—the finding of the "charm quark" in 1974 and the discovery of the particles designated "W" and "Z" in 1983 —were accomplished thanks to Charpak's 1968 invention. These discoveries were rewarded with separate Nobel Prizes in Physics, in 1976 and 1984, respectively.

THE PRIZE IN CHEMISTRY

The Nobel committee honored Rudolph A. Marcus of the California Institute of Technology (Caltech) in Pasadena for his theories concerning what is perhaps the simplest of chemical reactions: the transfer of an electron between two molecules. Although simple, the reaction is central to a number of key processes in nature, such as plant photosynthesis, chemiluminescence (the emission of light from a chemical reaction), and the energy-generating processes in every living cell. Electron-transfer reactions also underlie such technologically significant processes as the generation of current in batteries, the conductivity of electrically conducting polymers, and corrosion.

In a series of papers published between 1956 and 1965, Dr. Marcus developed the theoretical and mathematical concepts that now collectively bear his name: the Marcus Theory. This model assumes that when two molecules interact in a solution, their structure, as well as the structure of the nearby molecules in the solution, is changed. This change

temporarily increases the energy of the molecular system, enabling electrons to jump between molecules. Dr. Marcus also predicted that increasing the molecular "driving force" would, in some instances, actually slow down a chemical reaction. This idea went against what most chemists expected, and many remained skeptical until the theory was verified by experiment in the mid-1980s.

Dr. Marcus's theories now provide chemists with crucial information for predicting and characterizing electron-transfer reactions. This knowledge is essential for applications such as designing batteries and biosensors, and for exploring such natural phenomena as photosynthesis and cellular metabolism. Having studied the chemistry of electron transfers for more than 40 years, Dr. Marcus shows no signs of slowing down. He is now investigating reactions that involve the transfer of electrons over longer distances.

Dr. Georges Charpak was born in Poland in 1924. He obtained his Ph.D. in 1955 at the College de France, Paris. He has worked at CERN since 1959, and is also a professor at the École Supérieure de Physique et Chimie in Paris.

Dr. Rudolph A. Marcus was born in Montreal, Canada, in 1923. He earned his bachelor and doctoral degrees from McGill University in Montreal. Dr. Marcus was a professor of chemistry at the Polytechnic Institute of New York in Brooklyn, and later at the University of Illinois, before joining the faculty of the California Institute of Technology.

Christopher King

REVIEW

Physics

PUTTING THE SQUEEZE ON SPECTROSCOPY

Scientists striving for a quieter, gentler laser beam have developed a light source with noise (random fluctuations of electromagnetic radiation) below the limit prescribed by quantum mechanics. Known as squeezed light, this special form of electromagnetic radiation allows physicists to study atomic structure with unprecedented precision.

Eugene S. Polzik, John Carri, and Jeffrey Kimbell of the California Institute of Technology (Caltech) in Pasadena created a squeezed-light source to detect an electron's particular energy-level transition in a cesium atom. Their squeezed light contributed less than half the noise normally associated with such spectroscopic measurements.

Unlike previous forms of squeezed light, the Caltech source can be operated over a wide range of frequencies. For this experiment, researchers used light at wavelengths around 852 nanometers.

To generate squeezed light, physicists reduce noise in one component of the wave, either the amplitude or phase, in exchange for increased noise in the other component. In a sense, squeezing light is like squeezing a tubular balloon in the middle to make the ends bulge. While one part shrinks, another part increases. The payoff of greater precision comes from measuring and detecting only with the squeezed component.

By redistributing the noise, scientists are able to bypass the so-called quantum limit, a consequence of Heisenberg's uncertainty principle. Simply put, the uncertainty principle does not allow the energy of any system to be specified completely. Even a vacuum contains noise in the form of small, random fluctuations in the electromagnetic field. These ever-present background fluctuations create a minimum amount of noise—the quantum limit—in all light, including that from the finest lasers.

Caltech physicists Eugene Polzik (left) and Jeffrey Kimbell have created a squeezed-light source that permits atomic measurements of unprecedented precision.

Most applications of squeezed light center around spectroscopy or motion detection. Squeezed-light interferometers split a laser beam into two parts, which are reflected off different points and then recombined to form interference patterns of alternating light and dark bands. Changes in the interference pattern can indicate motion and be used to measure acceleration.

THE GREAT ESCAPE

In the subatomic world, tunneling refers to the passing of electrons through barriers of material classically forbidden to them. When it tunnels, an electron behaves as a wave rather than a particle, which in classical physics cannot penetrate the barrier.

By confining electrons to quantum dots (cells only a few nanometers across), researchers can test the predictions of quantum mechanics and control the energy levels of the trapped electrons. Like electrons bound to an atomic nucleus, the captive electrons in a quantum dot occupy discrete energy levels. Quantum dots show promise as sources of precise measurements of electrical current and extremely sensitive transistors capable of regulating large currents.

As reported in the May 18, 1992, *Physical Review Letters*, Raymond C. Ashoori and his colleagues at AT&T Bell Laboratories in Murray Hill, New Jersey, embedded a quantum dot inside a semiconductor structure—gallium arsenide sandwiched between two layers of aluminum gallium arsenide. By varying the voltage applied to the device, electrons tunneled into the dot only when an electron energy equaled one of the discrete levels of the quantum dot. The Bell Laboratories team measured the energy spectrum by detecting the current that flows as a single electron enters or exits.

HITTING THE SNUs BUTTON

Results from the solar-neutrino experiment called GALLEX failed to resolve the long-standing mystery of missing neutrinos, those supposedly massless particles created by the fusion of hydrogen and other nuclei in the Sun. As with earlier experiments, researchers found fewer neutrinos than predicted by theory, although they were able to detect a previously unmeasured low-energy neutrino.

A European-U.S.-Israeli collaboration at the Gran Sasso Laboratory in Italy, the GALLEX experiment captures neutrinos in a massive underground tank of gallium chloride. In June 1992, after a year of collecting data, researchers submitted two papers to *Physics Letters B*, announcing an average capture rate of 83 ± 21 solar-neutrino units (SNUs). One SNU (the term SNU is a shorthand way for expressing the ratio of solar neutrinos detected) equals 10^{-36} neutrinos captured per atom per second. According to the standard theory of electroweak interactions that governs neutrino behavior, capture ratios should range from 124 to 132 SNUs.

This is not the first time missing neutrinos have confounded physicists. In 1967, when the original solar-neutrino experiment was conducted in a chlorine detector in a South Dakota gold mine, only one-third the neutrinos predicted by theory was captured. In 1988 a water-based detector in Japan called Kamiokande II confirmed only half of the predicted neutrinos. Two years later came SAGE, the Soviet-American Gallium Experiment, utilizing a gallium detector similar to GALLEX. Initially it saw no neutrinos, but subsequent runs yielded capture rates approaching the GALLEX range.

The neutrino shortage is forcing physicists to either reassess their understanding of how the Sun emits light, or revise the standard theory. According to the standard theory, there are three classes of neutrinos: electron type, muon type, and tao type. All are massless, travel at the speed of light, and do not change from one type to another.

Physicists and astronomers believe that nearly all of the Sun's energy comes from the fusion of protons into helium, a reaction that produces the low-energy electron neutrinos detected by GALLEX. A thorough understanding of the Sun's fusion processes will help scientists develop fusion-power plants on Earth, which would be safer and cleaner than existing fission nuclear reactors.

Physicists have come to realize that the early neutrino experiments failed because chlorine- and water-based detectors are sensitive only to high-energy neutrinos, not low-energy electron neutrinos. The low neutrino capture rates seen with GALLEX and SAGE, which both use detectors that are sensitive to electron neutrinos, proved even more baffling.

One explanation for the lack of neutrinos is the MSW theory, named for physicists Stanislaw Mikeyev and Alexei Smirnov of the Soviet Academy of Sciences in Moscow, and Lincoln Wolfenstein of Carnegie-Mellon University in Pittsburgh, Pennsylvania. They propose that electron neutrinos change into tao or muon neutrinos before they reach the Earth. Such transformations cannot occur unless neutrinos have mass.

The next, and perhaps final, phase in the ongoing neutrino roundup is testing the MSW theory by constructing a detector that captures all three kinds of neutrinos. Two experiments—the Sudbury Neutrino Observatory (SNO) in Ontario, and Super-Kamiokande, a larger, improved version of Kamiokande II, should solve the problem of missing solar neutrinos in a few years.

Therese A. Lloyd

PLANTS and ANIMALS

CONTENTS

THE PLIGHT OF THE MANATEES *by Gode Davis*

Cued by the flash of Ed Gurstein's hand signaling in the water, Stormy, a 1,200-pound (545-kilogram), seven-year-old manatee, swims with an elephantine elegance to place his head in the wire hoop suspended in his tank. Suddenly a tone is sounded in the water. Then a light comes on at the tank's other end, signaling Stormy to swim out of the hoop and bump his lips against one of two paddles suspended on either side of the light. Gurstein waits to learn which paddle Stormy will select—the manatee swims to the left paddle if he did hear the tone, and to the right paddle if he did not.

Beginning in early 1992, Florida Atlantic University marine researcher Gurstein and Geoffrey Patton, a senior biologist at Mote Marine Laboratory in nearby Sarasota, have been conducting hearing-related experiments (at the Lowry Park Zoo in Tampa) with Stormy and another manatee, Dundee, as subjects.

"First we put the manatees in a quiet environment to determine their absolute hearing, then we explore their masked thresholds by

Some 1,800 manatees live in the waterways of Florida, a habitat they share with an ever-increasing number of speedboats. Some manatees bear scars from mishaps with boats or their propellers (above); others have died from such encounters.

including natural noise, and then we determine each animal's ability to localize where a sound is coming from," Gurstein explains. By testing manatee hearing, Gurstein and Patton are trying to find out why nearly every Florida manatee seems to have collided with a boat.

"One manatee in Florida has been hit at least 12 times, judging from his scars. Why don't they learn to avoid boats?" Patton asks.

"It's possible they can't hear the boats approaching," says Gurstein. Although Florida's manatees are endangered on several fronts, boat-related mortality is probably the greatest single threat to their population. In 1991 at least 174 manatees died in Florida (the death toll has been steadily climbing in recent years), and according to the state's Department of Natural Resources, which determines the cause of death in each case, 53 of those died of injuries caused by boats.

The hearing research is part of an effort to protect manatees, submarine-shaped aquatic mammals that grow up to 14 feet (4.24 meters) long and weigh up to 3,500 pounds (1,600 kilograms). Under a recovery plan for manatees that began in 1983, federal and state officials are conducting basic research to determine the manatee's life history, behavior, and habitat requirements, and are taking sometimes controversial steps to protect habitats, reduce mortality, and educate the public.

By means of aerial surveys, scientists have determined that at least 1,800 indigenous manatees roam the extensive network of waterways and coastal rivers of Florida and southern Georgia. Still, given the difficulty of spotting the animals under murky water, no one is certain how many there are, or whether the population is increasing or decreasing. With so many deaths, their numbers are probably not increasing. Hunting and poaching of manatees has been effectively curtailed by relatively recent legal proscriptions in Florida and neighboring states. And while the animal is seldom threatened by sharks and alligators, human-caused hazards are certainly making the manatee's existence quite precarious.

Nature's Sirens
Florida's manatees, probably one of two subspecies of the West Indian manatee, belong to the order of Sirenia, the only aquatic

mammals that subsist on vegetation. The name Sirenia originates from the mythical Sirens in Homer's epic poem *The Odyssey*—whose loud, mesmerizing songs were said to lure Greek sailors to their demise. Other living Sirenians include the West African manatee, the Amazonian manatee (whose range appears to overlap with the "other" West Indian manatee, which is found in Caribbean waters and along the coast from southern Mexico to northern Brazil), and the dugong —a related species said to resemble a slimmed-down manatee. Dugongs inhabit Indo-Pacific coastlines, including Australia's ocean shoals, where they are harvested legally by aboriginals, despite dwindling numbers. A fifth Sirenian, Steller's sea cow, was

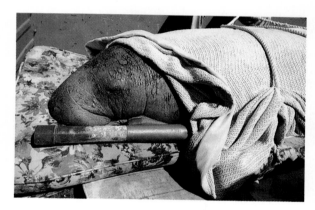

Many Americans are outraged by the senseless deaths of nearly 200 manatees each year. If rescued in time, injured manatees can usually be returned to their natural habitat after treatment.

hunted to extinction by whalers and sealers in the Bering Sea during the 19th century.

Like other Sirenians, Florida's manatees are distant descendants of the elephant, said by scientists to have forsaken land for water millions of years ago. Although possessing a whiskered muzzle and lacking external ears, the manatee's deep-set, tiny eyes, its Mona Lisa anatomical smile, a unique pattern of hair growth (hair over wrinkled, leathery skin), and its gray, bloated, sausagelike body resemble how we might think aquatic-adapted elephants would look. Indeed, a manatee's tapering upper lip en-

Communities along certain Florida waterways have begun to post signs on the manatee's behalf (left). Research has shown that the manatee is more intelligent than its sluggish behavior might suggest.

Habits and Habitats

Manatees can be found in fresh, brackish, or salt water. Traveling freely from one body of water to another, they may require fresh water for drinking. Adult manatees are herbivores, ingesting daily as much as 1 pound (0.5 kilogram) of aquatic grasses for every 10 pounds (4.5 kilograms) of body weight. Fond of habitats with abundant surface vegetation, such as water hyacinths, as well as a variety of bottom-growing vegetation, these shallow-dwellers behave like "sea cows," spending a quarter of their time grazing.

Because they have a narrow range of water-temperature tolerance (with a minimum being about 68° F, or 20° C), indigenous manatees tend to remain in Florida's warmest waters during the winter months, preferring natural springs such as Blue Springs, Crystal River, and Homosassa Springs, and perhaps ironically, also congregating in the heated discharge waters of electrical-power plants. Once the gulf and ocean waters seasonally warm, those West Indian manatees native to the United States travel throughout their geographic range, which includes both the east and west coasts of Florida and parts of Georgia waters. Occasional manatees, a few fitted with radio-tracking devices, have been sighted as far north as Virginia and the Carolinas, and as far west as Texas. Using such telemetry tracking, about 20 wild manatees have been followed continuously for weeks and months. The manatees under surveillance have been fitted with harnesses above their tails; a stiff tether extending from the harness holds an electronic transmitter above the waterline. The electronic signals are picked up by satellite, enabling scientists to find out where these animals congregate and how long they stay there. That information, along with the findings from manatee carcasses and other data, is used to make decisions on which habitat areas are essential to the manatees.

Information derived from radio-tracking devices, as well as captive-breeding programs in simulated ocean areas such as at the Miami Seaquarium, have also enabled scientists to better document a manatee's life cycle. "Manatees usually become mothers before reaching their ninth birthdays,"

gaged in vegetation feeding or cleaning irritating matter from its mouth brings to mind an elephant's trunk. Like elephants, the manatee is also a tactile-sense-oriented creature—often using its sensitive facial hairs and trunklike upper lip to investigate objects in the immediate surroundings.

Fatigued after surviving tedious Atlantic voyages, seafaring explorers (including Christopher Columbus) once mistook Florida's manatees, not for elephants, but for mermaids. Wrote Columbus: "They were not as beautiful as they are painted, although to some extent they have a human appearance in the face." Herds of sea cows once roamed in prodigious numbers, but when aboriginal natives and Europeans alike realized that the manatee was a slow-moving, docile, and ever so vulnerable gentle giant, the creatures were slaughtered for their hides, bones, fat, and meat for centuries. In fact, though illegal to be harvested, tasty, lamblike manatee meat is a delicacy in some countries.

says Jessica Kadel, a Mote staff biologist specializing in manatee behavior. Female "cows" give birth to a single calf after a 13- to 15-month gestation period. During the first explosive contraction, the baby makes its initial appearance as an elastic grayish bubble about 12 inches (30 centimeters) in diameter; it grows bigger and more oblong during each successive contraction, until the amniotic sac bursts and, moments later in a frenzy of movement, the baby manatee is born, tail fluke first. Darker-colored than adults, newborn calves measure between 4 and 5 feet (1.2 and 1.5 meters) from muzzle tip to fluke end, weighing between 60 and 70 pounds (27 and 32 kilograms).

"Babies stay with their mother, often without the company of other manatees, for up to two years," Kadel says. During this time, they not only learn to feed on the plants that comprise their vegetarian meals, but, according to Kadel, "by tagging along with mom, the calves memorize the migratory routes crucial to survival." After calves leave their mothers, they spend much of their adult lives either alone or in transient groups—swimming, feeding, or basking at the water's surface or underwater. When bottom-feeding, manatees must surface for air every 10 to 15 minutes.

Manatees can live up to 50 years. Although human development and pollution often degrade and destroy wildlife habitats and indirectly affect manatees along their migratory routes, the animal, owing to its protected endangered-species status (under the federal Endangered Species Act of 1973), would seem to have few worries. Relatively free from natural predators (except for large sharks in the open ocean, which tend to avoid the seaborne elephant's powerful tail), the docile creatures are occasionally free to play with other manatees—sometimes "kissing," mouthing, bumping, and

Wildlife officials have embarked on several programs to at least maintain Florida's manatee population. One strategy entails moving individual manatees (below) from busy waterways to less congested ones.

chasing each other. Another form of manatee play can be less innocuous. Manatees seem prone to mouthing ropes, hooks, and other fishing gear from anchored boats, entangling their flippers in crab-pot lines and fishing nets, or getting crushed in canal locks or floodgates. While ordinarily cruising at 2 to 4 miles (3.2 to 6.4 kilometers) per hour, and capable of sprinting from danger with rapid strokes of its rudderlike tail, "a manatee's tactile inquisitiveness can get it into potentially lethal situations," admits Gurstein, but he remains unsure if this insidious behavior or a lack of hearing ability is most responsible for the greatest manatee mortality factor: boat and barge collisions.

Sacred Cows?

The 1991 manatee death toll attributed to boat and barge collisions was the highest yet, but up only slightly from the previous year's figures, when 47 died, with many more maimed or scarred. "Most manatees have collided with boats repeatedly," Gurstein says. The primary culprits are "increasing numbers of boaters in crowded Florida waters who don't abide speed restrictions in their large boats," according to Kadel. Under the recovery plan, as many injured

Many people have grown fond of the manatee and its gentle ways. Baby manatees are particularly endearing. At marine laboratories, orphans are doted upon by their human caretakers (below).

manatees as possible are transported to sanctuaries like the Miami Seaquarium, where marine-mammal veterinarians try to nurse them back to health before fitting them with tracking transmitters and reintroducing them into the wild. But by far the most controversial aspect of the recovery plan is the enforcement of "speed restrictions"—the attempt to slow down boats near manatee habitats. Reduced boat-speed zones have been or are being put into effect in many Florida counties with large manatee populations, including Volusia County, where Rick Rawlins owns the Highland Park Fish Camp.

"Five to 7 miles [8 to 11 kilometers] per hour? The rules would add as much as five to six hours to a day of fishing. My customers are leaving me," argues Rawlins, part of a vocal boating and waterskiing contingent who believe that Florida's sea cows have become "sacred cows"—protected under the mistaken premise of being endowed with anthropomorphic traits. Rawlins has formed a group called Citizens for Responsible Boating to fight the regulations; in several Florida counties, the speed restrictions have actually been reversed due to the group's efforts.

Perhaps the greatest irony for Florida's only resident Sirenian is that its continued survival may depend, not upon an approaching boat's speed, but upon whether the creatures can hear the boats approaching. While attributing humanlike traits to manatees is questionable, a certain innate intelligence and demonstrated long-term memory skills in the species appear to be well documented.

"We're not sure how this intelligence relates to hearing ability, except that, unlike certain intelligent marine mammals like porpoises, manatees don't appear to use sound for echolocation in turbid waters—even though they emit intriguing squeaks or squeals during play or when alarmed," according to Gurstein. While results of the hearing tests continue to remain inconclusive, if the manatees' ability to hear can be determined, "the solution could boil down to some simple plastic device on the hull that would vibrate at the right frequency and cue the animal," Patton says.

THE MAGIC OF TOPIARY

by Richard Wolkomir

Edward Scissorhands cheated.

If you saw the movie, you know that Edward carved a bush into a tyrannosaur in what appeared to be a flurry of flying twigs. But real topiary must grow. It can take years, even decades. Since its beginnings in the ancient world, topiary has demanded Zen-like patience. Still, Edward was onto something. His leafy sea serpent and elk pretended to be traditional "woody" topiary, rooted in the earth. Actually, they were a new breed: "stuffed" topiary, which, instead of growing out of the ground, is assembled. The new topiary mostly grows in sphagnum moss, stuffed into a frame of steel rods or wires.

I learned all this at a national topiary conference at the San Diego Zoo, co-sponsored by the American Ivy Society. North America turns out to be big-time topiary habitat. "People are passionate about it!" exclaims Pat Hammer,

former topiary specialist at Longwood Gardens in Kennett Square, Pennsylvania, who has recently moved to California to start her own topiary business. The city of Columbus, Ohio, has just converted a downtown park into a life-size topiary re-creation of Georges Seurat's painting *A Sunday Afternoon on the Island of La Grande Jatte*. The scene, a landscape of a painting of a landscape, is composed of 47 topiary figures, eight boats, three dogs, and a monkey, all of yew. And topiaries on a smaller scale are scattered about the country. Horticulturists at the conference tell me that topiary is any plant that can be grown in an unnatural shape, including most lawn shrubs. I jog through a San Diego suburb, Del Mar, and sure enough, every front yard displays cypresses and junipers snippered into balls, squares, pyramids, or pompoms—the "poodle" cut.

Geometric topiaries have been around since Colonial Williamsburg. But lately the Disney theme parks, which teem with statuelike topiary, have inspired homeowners here

Topiary, the art of training plants to grow in unnatural, ornamental shapes, achieves a certain degree of realism in a fox hunt (above) where the various figures have been shaped from yew shrubs.

The "Frog Footman" at left combines ivy (hair), moneywort (jacket), and ajuga (pants), all of which grow around a metal frame filled with sphagnum moss. The "Bear Photographer" (above) has a distinct floral touch.

and there to trim their front-yard hedges into privet Plutos or Donald Ducks. It is poor man's statuary. That sort of precision pruning, however, requires a great deal of patience, not a typical American trait. Thus, we now have the new "stuffed" topiary: blast-your-socks-off vegetable statues in a mere matter of hours.

Mail-order catalogs offer tabletop figures that range from pregrown potted ivy wreaths to ficus cupids with bows and arrows. Stores like Kmart sell little pottery bunnies and pussycats and the like, which sprout into microtopiaries as you water them. Topiary businesses have sprung up across the country, from Topiaries Unlimited, in Pownal, Vermont, to O'Farrior Topiary, in Glendora, California, where customers can purchase, among other figures, a mossy dancing duo—Fred O'Bear and Ginger O'Hare—preassembled and ready to go. Many topiary companies also make jumbo stuffed topiaries for malls and high-rise lobbies, and some sell stuff-it-yourself frame kits. Mia Hardcastle, whose company in Tampa, Florida, Topiary Inc., created figures for *Edward Scissorhands,* says her team of five welders ships out about 500 topiary frames a week, at prices from $5 to $5,000.

The San Diego conference inaugurates the zoo's new displays of a big-as-life botani-

cal rhinoceros, a sun bear, a tortoise, a hummingbird, and a lion, as well as assorted gorillas, flamingos, and koalas—all made of stuffed frames. The show begins at the entrance to the zoo, where two life-size ivy elephants face off, trunks upcurled, presumably whooshing out chlorophyll.

"Hey, check out those elephants!" a dad in shorts exclaims, motioning his twin toddlers toward the vegetable pachyderm, his camcorder already whirring.

"He's got weeds growing out of him," one blonde matron tells another, pointing to a small purple flower growing amid the ivy on one elephant's thigh like a herbaceous hickey. Zoo gardener Linda Dobbins says people usually complain about the elephants' gleaming polyresin tusks: "They call up angrily, because they think the tusks are real." But what really make the three-ton elephants look real enough to pick up peanuts are their metal skeletons.

To create the skeletons, the zoo's staff sent the dimensions of an African bull elephant to the Longwood Manufacturing Company, an industrial machine shop in Pennsylvania. Months later, via truck, they received two 1-ton welded elephant skeletons made of 3-inch (7.5-centimeter) steel pipe covered with an epidermis of steel mesh. Zoo gardeners supervised by horticulturist Chuck Coburn snaked plastic tubing through the frames to create an automatic watering system. Then they stuffed each skeleton's hollow core, where real elephants have lungs and pancreases, with garbage bags full of lightweight foam packing. Around this, to fill out each elephant, they tamped 83 bales of sphagnum moss. In the moss, they implanted 7,500 individual pots and plugs of ivy. Plastic tusks and pine eyes, with metal lashes, completed the illusion.

Ersatz elephants with ivy sprouting from steel frames and pop-up watering devices may be considered topiary for the Twilight Zone. Classic topiary, by contrast, began with just a little bush. Year after year a gardener gently bent its branches, snipping here, letting it grow there. One day it was a peacock.

An Ancient Art

Pliny the Elder credited the art's invention to a friend of the emperor Augustus, circa A.D. 1. But in his book *Topiary,* British writer A. M. Clevely says the imperial city's gardeners were likely Greek, Syrian, Egyptian, or Hebrew slaves who probably introduced topiary from other lands. In any case, it was a hit. The Romans coined the word *topia* for such an ornamental garden, and *topiarius* for the gardener. According to Pliny, gardeners of his time clipped cypress "to make representations of hunting scenes, fleets of ships and imitations of real objects." Thirty years or so later, Pliny the Younger's Tuscan villa was filled with box trees cut into forms, including letters, he wrote, that spelled out the "name of the gardener or his master."

During the Dark Ages, when survival took precedence over leafy obelisks and pruned puppy dogs, topiary, like book learning, managed to live on in monasteries. By 1452 topiary had become chic again. Italian architect and scholar

Using the actual physical dimensions of a leaping deer, a welder (above) creates a topiary frame. Before any plants are attached, the frame must be fitted with an automatic watering system.

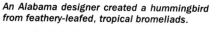

An Alabama designer created a hummingbird from feathery-leafed, tropical bromeliads.

Leon Battista Alberti (an early explorer of the rules of perspective) designed a garden for a Florentine villa with topiary "spheres, porticoes, temples, vases, apes, donkeys, oxen, a bear, giants, men and women, warriors, a witch, philosophers, Popes, and cardinals." Inspired by the Italians, French horticulturists created parterre gardens, where low hedges swirl and knot in intricate, embroiderylike patterns. As Clevely notes, "By the turn of the 17th century the design of gardens and garments could almost be considered a single craft." This form of topiary—imposing a mathematical logic upon nature—reached its apotheosis at Versailles in the late 17th century. It spread through Europe, and as far off as the gardens of Chinese emperors.

England, on the other hand, in topiary as in everything else, had a weakness for the eccentric. Lions, dragons, and other botanical beasts stalked Henry VII's Richmond Palace gardens. A 1618 British horticultural manual advised: "Your gardener can frame your lesser wood to the shape of men armed in the field, ready to give battle; of swift-running grey hounds, or of well-scented and true-running hounds to chase the deer or hunt the hare."

The reaction toward "natural" landscaping that set in during the 1700s was so extreme that at least one gardener planted dead trees and created artificial molehills. The urge to prune soon returned. England today is dotted with privet locomotives, yew lyrebirds, and tunnels of hornbeam.

Introduction to America

Topiary migrated early to America and evolved rapidly. In 1872 Massachusetts textile mogul Thomas Brayton, a no-nonsense Yankee, bought a big white clapboard house overlooking Narragansett Bay and promptly built a train platform at the yard's edge so he could commute to his factory in Fall River. The grounds consisted of 7 acres (2.8 hectares) of open meadow and orchard. In 1905 he picked out an apt laborer at his cotton mill, Joseph Carreiro, an immigrant from the Azores. "I've got a summer house —I want a little flower garden and a little vegetable garden," Brayton said, and added, "Do what you want."

Fertile words. Carreiro, whose grandfather and father had worked in an Azores botanical garden, took Brayton at his word.

Infant shrubs appeared in the meadow. Over years, they grew into a garden of 59 geometric shapes and figures and 21 beasts. Some were inspired by pictures Carreiro saw in his children's schoolbooks: a giraffe, a lion, peacocks, a unicorn. A picture on a package of dates engendered a topiary camel. There was an elephant, and later a donkey—because a wealthy Providence visitor, an ardent Democrat, demanded representation for her party. The gate was guarded by a life-size policeman with a leafy chest.

When Thomas Brayton's wife died in 1917, he closed the house and stayed away.

Carreiro, living with his family in a cottage on the estate, kept making topiary. The novelist Mary McCarthy, a frequent visitor there, said the privet beasts were so huge that, walking into the garden, the impression "was of a sheared family of Mesozoic creatures."

McCarthy was a friend of Thomas Brayton's daughter, Alice, who used her father's lush, strange topiary garden as a sort of social lure. After Brayton died in 1939, Alice became the estate's owner, dubbing it Green Animals. The 61-year-old spinster had resolved to do what she wanted, which was to continue cutting a figure in Newport society. She offered a unique attraction: topiary garden parties.

"She'd have a bar set up at either end of the garden, so people could get a martini, wander through the topiary, get another martini, and wander back again," says Crisse MacFadyen, grounds manager at Green Animals. The place became famous. In the 1940s Jacqueline Bouvier "came out" among the Brayton topiary.

On a scorching summer afternoon, I sit in the garden talking about topiary with George Mendonca, Joseph Carreiro's son-in-law, who is 83. He came to Green Animals to help clean up after the 1938 hurricane, married one of Carreiro's daughters, and in 1945 succeeded his father-in-law, becoming a master topiarist in his own right. Mendonca is now "semi-unretired" so he can continue to help restore original detail to the topiary, which had declined—especially in 1954, when another hurricane blew through, decapitating the giraffe, lopping off a section of its 22-foot (6.7-meter) neck.

"My darling giraffe—I'll never live to see it grow back," Alice Brayton lamented at the time, leaning out her upstairs window.

"There's something I can do, but I don't want to, because it'll have a short neck," Mendonca told her.

She grinned wickedly.

"Don't you know Rhode Island giraffes have short necks?" she snapped.

Even a short-necked topiary giraffe is a long time forming. Mendonca shows me a bear that he is in the process of growing next to his cottage, still a shapeless yew shrub. "I'm on that three years already, and I still

At the San Diego Zoo's Flamingo Lagoon (above), birds on steel legs share the shore with the real thing. The topiary giraffe at right has made a gradual comeback after losing its neck in a hurricane in 1954.

don't have much," he says. "You can't carve it out, you grow it out—and then keep cutting away whatever grows that isn't bear," he says.

Green Animals is now one of the country's three leading topiary gardens open to the public. The others are Longwood Gardens and the stunning Ladew Topiary Gardens near Towson, Maryland, which had a different genesis. It grew out of a man's lifelong devotion to goofing off. The man in question was born with a silver spoon in his mouth in Manhattan in 1887, when Fifth Avenue families like his had liveried footmen. He liked roller-skating in Central Park, but school was a bore. Young Harvey Ladew believed his absences from school, touring Europe, helped his education: "I can order breakfast in about five languages."

Harvey grew up to resemble the actor William Powell, but with the soul of Peter Pan. He resolved to spend his first 50 years playing, his final years working. Except for the after-50 part, he stuck to it. His passion was fox hunting. In the late 1920s, he bought an old 250-acre (100-hectare) farm

in the hunt country of Maryland. In England —where he spent months each year riding to the hounds—he had an epiphany.

"Hacking to an early meet, I rode along beside a very tall yew hedge bounding a large estate," he wrote. "Along the top of the hedge was a realistic Topiary fox hunt: a fox closely pursued by a pack of hounds." By the time he died in 1976, Ladew had turned his farm into a topiary extravaganza.

I visit the Ladew estate on a misty summer morning. On the front lawn, a horseback rider is leaping a hedge, following five hounds streaking across the grass after a fleeing fox. It takes a moment to realize none of these figures is actually going anywhere.

In all, Ladew created 15 gardens, like separate rooms with high hedge walls, around a central circular lawn. Some of these topiary walls contain windows and seem hung with swag curtains. Others are shaped into pyramids or arches or oddities, like three balls balancing atop one another. One wavy hedge has 12 swans floating serenely along the top.

A Heart Pierced by an Arrow

A doorway in another hedge leads into a room with hemlock walls. Inside, shading a table, is an umbrella of white roses and purple wisterias. In a nearby pond, a topiary junk floats by with real red canvas sails, watched from the shore by a yew Buddha beyond a crab-apple arch. Just a short way off is a "sculpture" garden, populated by green sea horses and lyrebirds, a unicorn, a butterfly, and a heart pierced by an arrow. Beside Winston Churchill's top hat stands a topiary of a giant hand making a V for Victory out of yew.

After seeing so much traditional topiary, I am unsure what to make of the new, unattached-to-the-ground, "stuffed" kind, with its steel frames packed with moss. But the elephants at the San Diego Zoo look so real they seemed ready to trumpet. And Pat Hammer, who recently published *The New Topiary*, tells me that stuffed topiary is traditional, too: it goes back to the Victorians.

At the Alice's Wonderland display at Longwood Gardens in Kennett Square, Pennsylvania, children join a topiary of the Mad Hatter and the March Hare for a tea party.

Florists in the 1800s made "set pieces" —shaped wire frames stuffed with sphagnum moss and planted with flowers and ivy. "The Smithsonian has hundreds of those frames in its collection," Hammer points out. A funeral display might feature a "broken heart" or an "empty chair." "In those days, railroads had hotels along their rights-of-way, and in their lobbies were big floral replicas of locomotives," Hammer says. "Between flower seasons the newly rooted ivy remained in place."

Hammer's company, Samia Rose Topiary in Encinitas, California, is involved with projects from Seattle to Atlanta. As stuffed topiary's Joanie Appleseed, Hammer has helped get the new form started at sites like Sesame Place, a theme park outside Philadelphia. Chris McCarron, the park's grounds manager, shows me Oscar the Grouch, in Telecurl ivy, sitting in a jalopy from the Sesame Street movie *Follow That Bird.* Oscar rests an arm on the jalopy's door.

"It's hard to keep that arm moist," McCarron says, probing the moss.

Nearby is Cookie Monster, in dark Ivalace. "It has curly edges, so it translates to fur," McCarron tells me, adding that he uses plastic eyes. Greeting kids as they enter the park is a topiary Big Bird, along with Ernie and Bert, whose hair is black Mondo grass. Jeff Salvesen, a college student on the grounds crew, confides: "What I really like is when the kids talk to them."

Pat Hammer also finds that instant topiary turns kids on. She shows me Longwood's Children's Garden, which has a leporine theme: a topiary rabbit sits on a stump, fishing in a pool. A topiary rabbit magician pulls a topiary baby rabbit from his topiary hat. Hammer points out something she has prepared for a Longwood chrysanthemum spectacular on an Alice's Wonderland theme. A life-size flamingo, its neck euonymus and its body pink *Crypthanthus,* a plant in the bromeliad family, is poised on delicate legs, which turn out to be concrete-reinforcing rods painted pink. Nearby, a White Rabbit the size of Michael Jordan lounges on a log, ready to check its pocket watch and hurry off. Next to the finished pieces, an unstuffed caterpillar frame looks abstract and inert.

Single-species geometric topiary takes years of pruning and manicuring to produce the desired effect.

Art or Engineering?

When it comes to stuffed topiary, clearly separating "art" from mere engineering is a tough call. Some old-style woody topiarists sniff at the new upstart. But we can't all be woody topiarists. It's too hard. My own front-door arborvitae topiary rocket (or whatever it is) has a midriff bulge that blocks light to the branches underneath, so that they are slowly dying. Hoping for hints on that problem, I sit in on a woody-topiary workshop given by the Strybing Arboretum's Bob Hyland. We students, mostly home gardeners, sit outdoors at picnic tables, staring at foot-high potted Japanese boxwood shrubs. We are to prune them into balls. Hyland shows us his favorite Swiss shears: "I've cut my hand with these, and they make a very nice, clean cut," he says, admonishing us to, at any cost, avoid anvil pruners, which crush stems. "We could spend two hours just on pruning, but the essence is to cut terminal buds and encourage lateral and latent buds, so you have fuller plants," he says.

We learn to slice twigs off at a 45-degree angle. If you miscut, you can get rot. Hyland also tells us, "If you have trouble trimming the bush into a globe freehand, hold a paper plate behind it as a guide." A bearded man wearing a San Diego Padres cap asks a question that has been troubling me, given my sick topiary at home: "If you prune a shrub into a ball, how do you get light to the bottom?" Hyland explains that trimming the trunk bare under the ball will in itself let in light underneath.

Woody topiaries can be created from a number of species. Yew, cypress, and boxwood are common. Privet is goofproof, regrowing even when cut to the bone. But professionals turn up their noses: "Privet is so common and easy that some real gardeners sneer at it," Pat Hammer tells me. For stuffed topiary, ivy is king. "As soon as Victorians started growing ivy as a houseplant," explains Hammer, "they had to find ways to control it." So topiary came naturally. Botanist Sabina Mueller Sulgrove, the American Ivy Society's research director, says that topiarists can choose among some 400 different types of ivies.

Wandering through the stuffed-topiary workshops at the conference, I find one group tamping sphagnum moss into wire frames shaped like rabbits. One of the tampers, Joann Clark, tells me why she opted against woody topiary: "I'm a little taken aback by the chance of ruining a huge, wonderful plant that's alive," she says. Kathleen Kinsey, an airlines worker, has a different reason for joining the stuffed-topiary workshop. "I'm a bunny person," she says.

Giraffes are among the most easily recognized animals in the world. Despite this renown, giraffe behavior is subject to many myths, misperceptions, and misinterpretations.

FAMILIAR STRANGERS

by Jane Stevens

A 14-foot (4.25-meter)-tall female giraffe strolls toward a whistling acacia to scrape delicate dark-green leaves from its branches. A robust, 15-foot (4.5-meter) male mimics her every step.

She suddenly snaps her head around to look at him. He stops short and gazes away.

She turns and steps across the long grass. The male lowers his head, then tiptoes in unison with her.

She stops and shoots him another long glance. He freezes, then raises his huge head to look in another direction, flicking his ears.

Paying no attention to me watching from the sidelines in Kenya's Nairobi National Park, the giraffes continue these antics for 20 minutes. Later I describe the scene to biologist David Pratt and collaborator Virginia Anderson, who are among the few people who have published observations of giraffe behavior in the wild during the past decade. The two chuckle knowingly. "That's typical of their games," says Pratt. "They seem very gentle, very affectionate," says Anderson.

Tall Tales

Gentle? Affectionate? What about the stories that these cud-chewing ungulates with the same four-chambered stomachs as cows are too dumb to have a personality? That they're boring? That they can't utter a sound?

All myths, say Pratt and Anderson.

"If you ask any three-year-olds to identify this animal, they always know it's a giraffe," says Pratt, professor emeritus at the University of Rhode Island. "Giraffes are one of the best-recognized animals in the world. Yet we really know little about them."

Perhaps because, as a species, giraffes aren't listed as threatened or endangered, they haven't been studied as extensively as other African animals, such as elephants. Or maybe, as Pratt puts it, part of this pronounced information gap comes from giraffes' normally low-key life-styles. "They don't have very outstanding social behavior. They don't have the territoriality of some other hoofed mammals."

Nevertheless, scientists have made some progress in understanding the physiology, feeding ecology, and behavior of the eight subspecies of giraffes, which range through sub-Saharan Africa. As a result, I discovered as I crisscrossed East Africa that we humans aren't entirely ignorant about giraffes. We know, for example, that they eat about 500 pounds (225 kilograms) of vegetation, mostly tiny, delicate acacia leaves, each week to fuel those massive bodies. We know that 280 feet (85 meters) of intestines twine around beside their stomach. And we know that they probably sleep about a half hour a day in five-minute giraffe naps.

Giants of History

Biologists still disagree on the significance of various giraffe behaviors, but then, myths, misperceptions, and misinterpretations have always surrounded these unusual mammals. In the 1200s, for instance, Arabian author Zakariya al-Qazwini declared that giraffes were conceived only during the dry season, when different animal species gathered at watering holes. If a male hyena and an Abyssinian camel produced a male offspring that mated with a wild cow, a giraffe would be born.

In reality, giraffes evolved from ancient creatures that looked like 10-foot (3-meter)-tall deer with two broad, flattened horns and one conical horn. They even roamed the land that is now Europe and Asia some 35 million to 50 million years ago.

Modern giraffes paraded through Rome in 46 B.C. Their generally pacific nature dis-

appointed Julius Caesar and his fellow Romans, who had expected ferocious beasts. Nevertheless, Caesar slaughtered them in the Forum with 40 lions and 40 elephants. The Chinese, on the other hand, dubbed giraffes the emblem of Perfect Virtue, Perfect Government, and Perfect Harmony in the Empire and the Universe, after ambassadors from Melinda, now Kenya, presented an awed emperor with a giraffe in the early 1400s. Egypt's pasha gave a giraffe as an annual offering to Constantinople's king for 1,500 years, the last time in 1822.

Today thousands—no accurate counts exist—of *Giraffa camelopardalis* range from Nigeria to Somalia, from Sudan to South Af-

With their excellent vision, giraffes can spot herd-mates over one-half mile away. Drooping ears (left) shoot up (right) when an adult female catches sight of another giraffe.

rica: "giant speckled flowers, floating over the plains," as *Out of Africa* author Isak Dinesen described them.

Necking Rituals

In Kenya's Lake Nakuru National Park, I watch as a group of 17 float along the edge of a yellow-barked acacia forest. From a distance, their heads high, their ears stick out horizontally like giant white bows pinned to the backs of their heads, the giraffes look

Giraffes generally live together in harmony, although some aggressiveness manifests itself during mating season. Fighting, usually in the form of "necking" (above), can sometimes be fatal.

other like wrecking balls on the end of cranes, often tangling their necks.

The young male rotates his head into the older giraffe, catching him in the side on the upswing. Both casually look around. Then the older one takes a halfhearted golf swing into the younger's side. They exchange a dozen hits, and the game stops when the older giraffe turns back to a tree and begins feeding.

Necking is not always so harmless. Just days before, in Kenya's Masai Mara Game Reserve, two adult males, set in battle, slugged each other so hard that sickening, bone-crunching "thuds" resounded across the plains, and the impacts lifted the 1.5-ton animals off their feet. Although sinus cavities fill a giraffe's head, an additional 20 pounds (9 kilograms) of bone form on the top of the skull throughout the animal's life to transform the head, with its rounded horns, into a titanic armored weapon. The battle ended when one male walked away. He forded a narrow, shallow river and dropped dead on the other bank, presumably from internal injuries.

Lazy Days

Giraffe life is rarely so dramatic. They spend most of their time—day and night—walking from tree to tree, eating or chewing their cud like a cow. They doze, standing in the afternoon shade. Occasionally they lie down to nap. I watch as two male giraffes who are settled in the middle of a small acacia grove chew their cuds, their heads on a level with mine as I sit on the roof of the Land Rover. In a curious, slow rhythm, each tilts forward slightly as a bolus of food from its first stomach flows up its neck and into its mouth. Each animal leans back and chews for about a minute before swallowing again. Meantime, tiny yellow-billed oxpeckers, which shriek warnings of approaching enemies when the giraffes nap, scramble over the placid animals' necks and heads to search for ticks.

Like a swan, one giraffe twists and drops his head onto his back to snooze. Although giraffes have no more vertebrae than humans, each of the seven bones is 1 foot (30.5 centimeters) long, and the whole neck

like a group of 10-year-old girls, straight-backed and demure, dressed in their Sunday finest. I move in closer, parked in the middle of the acacia grove where the giraffes are feeding, and let a typical African day slide past my Land Rover.

A young male and an older male stop eating to neck. Most of the time, giraffes go gently in the world. Occasionally, however, in a genetically prescribed formal duel, males fight over a female in estrus. Males often practice the dueling called necking or sparring. They stand side by side and swing their massive heads and necks into each is flexible enough for giraffes to scratch their back with their nose. Giraffes can get an extra 3 feet (1 meter) of stretch just by pointing their head and long tongue straight up to the tops of acacia trees.

Older nature books often say that giraffes don't drink water. They do, but probably only every three or four days on average, and less frequently if necessary. It's just as well. I watch a giraffe attempt to drink. It is an ordeal. She spreads her legs awkwardly. She lowers her head 8 feet (2.4 meters), about halfway to a small pond. She raises her head quickly to fake out any approaching lions. She clumsily spreads her legs further apart. She drops her head 14 feet (4 meters), close to the water. She raises it suddenly, again to fake out any potential attackers. She bobs her head several more times and never passes out.

If she had the circulatory system of a human, she would have fainted as the blood rushed to and from her head. But a special complex of blood vessels in her head swells, taking up the excess blood when it pours in fast. The upper part of her carotid artery is muscular, which controls the pressure. And her jugular vein, more than 1 inch (2.5 centimeters) in diameter, is lined with several valves that prevent her blood from flowing the wrong way.

Although a giraffe's heart is large, it is only 0.4 percent of the body weight, just like a cow's. But a giraffe heart beats 150 times a minute, more than twice that of a human's or a cow's. To distribute enough oxygen to

In order to drink water, a giraffe must assume a position that's both decidedly awkward and extremely vulnerable to attack. Ever alert to danger, the giraffe frequently interrupts its drinking to raise its head and scan about for potential predators. Fortunately, a specially adapted circulatory system prevents the giraffe from fainting as blood rushes to and from its head.

Height alone makes giraffes easy to spot on the open plain (above). Against a background of trees, however, the creature's characteristic markings act as camouflage. Red-billed oxpeckers (left) feed on parasites lodged in the giraffe's back.

power both body and brain, a giraffe has twice the density of red blood cells and eight times the lung capacity of humans.

Animal Watchtower

As I continue watching, a female giraffe strips a long piece of bark from a tree and munches. An olive baboon hops into the crook of the tree to lick off the sap bleeding from the trunk. The two animals, just inches apart, ignore each other. Other animals, such as zebras and certain gazelles, often graze near giraffes to rely on the larger animal's height and eyesight for spotting enemies, and, if danger approaches, for a warning snort. Little gets by giraffes—their huge eyes, the size of golf balls, bulge out and offer a 360-degree color view of the world. From their vantage point at second-story-window level, they can spot a cheetah 2 miles (3.2 kilometers) away.

Because of their ability to see threats at long distance, mothers allow their calves to wander afield. Mothers often leave their offspring with other females, who baby-sit several youngsters while the mothers search for water or food. Giraffe babies and adults can quickly accelerate to 35 miles (56 kilometers) per hour, but lack the stamina to maintain the speed as long as a horse can.

In Nairobi National Park, a mother giraffe and her three-week-old calf stroll at a distance from a main herd of 34 giraffes feeding across a couple of miles of savanna dotted with 8-foot (2.5-meter)-high acacia trees. The calf stays close to its mother's flank as she browses.

Born with Horns

Female giraffes become pregnant for the first time when they're four years old. Giraffe courtship is low-key, just like most of the life-style. A male follows a female around and determines her time of estrus by taking a sample of urine on his tongue, raising his head, and curling his lip back. If the female giraffe is ready, the male pushes his chest against her hindquarters and, if she stands still, mounts her. After 14.5 months of gestation, a giraffe gives birth to a 160-pound (72-kilogram), 6-foot (2-meter)-tall baby whose first real-life experience is a 5-foot (1.5-meter) drop to the ground.

Giraffes are one of the few ruminants that develop horns before birth. A pair of little cartilaginous spikes lie flat against the head at first, but straighten up during the calf's first week outside its mother. Giraffe calves grow 3 feet (1 meter) in the first six months, and double their height in a year. During this growth spurt, they spend 12 to 18 hours a day lying down.

Mother giraffes give birth in isolation, and keep their calves away from the herd for the first two weeks. When the mother and youngster rejoin the herd, members exhibit great interest in newborns, often standing over them for half an hour, touching and sniffing their head, mane, back, and rump.

Giraffe herds can number up to 100 or more (in the western Serengeti), but generally divide into smaller groups comprising females and juveniles. The groups' members come and go, seemingly at a whim. Biologist Bristol Foster, who studied giraffes in Nairobi National Park in the 1960s, followed one female that associated with 74 different giraffes as groups formed, dispersed, and re-formed. Adult males amble into a group to mate, but generally stroll alone or in pairs roaming over large ranges in search of females in estrus.

The calf in Nairobi Park approaches its mother several times to feed, but, in typical fashion, the mother moves away. Eventually she turns to stare at the youngster, which scampers to her and nurses for a couple of minutes before the mother moves on. Mother giraffes apparently don't believe in on-demand feeding. More often than not, they initiate nursing, and do so by staring at their calves until the babies get the message.

Staring seems to be a favorite form of giraffe communication. There are what look to human observers like hostile stares, come-hither stares, go-away stares, there's-an-enemy stares. When giraffes spot lions in the grass, a steadfast gaze alerts dozens of other giraffes that may be scattered over 1 square mile (2.6 square kilometers) of savanna. Mother giraffes stare at other adults to warn them away from calves. Males and females stare at each other in greetings,

Giraffes give birth standing, a practice which makes a baby giraffe's first experience a 5-foot drop to the ground. The calf generally weighs about 160 pounds.

games, and in enmity. Often duels fail to materialize because one of the two males loses a staring battle.

Domesticated Giraffes

Giraffes are not easy to catch, but, once caught, are easy to tame. In the 1970s Betty and the late Jock Leslie-Melville, in an effort to save the dwindling numbers of the Rothschild variety of giraffes, caught two calves and raised them on an estate outside Nairobi. The Leslie-Melvilles wrote *Raising Daisy Rothschild* about the experience, and collected funds to transfer to Lake Nakuru National Park some Rothschilds that were in danger of being killed by farmers. Today about 500 Rothschild giraffes live robust lives in six areas of Kenya. A small herd wanders off and onto Giraffe Manor, a bed-and-breakfast-with-giraffes (and education center) now run by Betty's son, Rick Anderson.

More than anything else, it is the size of the animals that stuns visitors to Giraffe Manor. As I stand in the hall, Anderson suddenly opens the front door, and the frame is filled by the head of Betty, an adult giraffe, and her 18-inch (45-centimeter) purple, prehensile tongue looping out in hopes of a few alfalfa pellets. I gulp in relief that she is relatively tame. I could have walked under

Giraffes can adeptly defend themselves with their powerful front hooves. Nonetheless, a hungry lion can sometimes best a young giraffe.

her belly and not scraped the top of my 5-foot, 4-inch (162-centimeter) frame.

As Anderson and I enter the sun room to have tea, two more giraffes stick their heads through the floor-to-ceiling windows, panhandling for pellets. I place some in the palm of my hand, and let the huge, soft tongues and lips pick off the treat. I feel a gentle tap on my left shoulder. I turn to go eye-to-eye with Betty, whose head is as big as most of my body. Yes, and a few more for me, please, say her dark eyes with their long, straight lashes. I oblige and stroke her thick, short, soft fur.

Studying Giraffe Behavior

Even though giraffes adjust to human presence, few scientists study giraffe behavior these days. Most research, concentrated in southern African countries, focuses on feeding ecology, including the impact of giraffes on their habitat. The studies reflect the current emphasis on sustainable management of Africa's wild animals as population pressures corral them into smaller, more-confined areas. Wildlife managers need to know how many giraffes a particular park can support before they must be culled.

But, as in most scientific research, giraffe studies have raised as many questions as they have answered. Some people wonder whether giraffes have another communication method besides staring: sound waves on frequencies that humans cannot hear. The idea has yet to be studied, but Pratt and Anderson have noticed giraffes stopping and waiting for others that were out of sight over a hill. The Leslie-Melvilles report that a young male giraffe tried to break through a brick wall, on the other side of which, far away in a field, stood his mother.

The last time I saw giraffes in Africa, a storm suddenly swept into Nairobi National Park. A group of 11 giraffes that had been feeding lifted their heads, turned their backs to the torrent, and, seemingly spellbound, stopped moving for the duration of the shower. I left the giraffes there, mottled sentinels in the savanna, all gazing steadily toward the same point in the distance where Mount Kilimanjaro hid in majestic solitude behind the clouds.

PARTY ANIMALS

by Jerry Dennis

Animal symbols have become an accepted feature in many areas of American business, politics, and government.

When Ross Perot announced he was dropping out of the presidential race in July 1992, Republican elephants trumpeted and Democratic donkeys brayed at the news; but a lot of voters were left scratching their heads in wonderment. Me, I was not surprised. Perot's problem from the beginning was that he did not ally himself with a recognizable animal symbol. He should have lifted an eagle from his celebrated collection of patriotic art or resurrected Theodore Roosevelt's bull moose. When it was suggested that

The editorial cartoonist Thomas Nast introduced the donkey as the symbol of the Democratic Party in 1870 (below). In 1874, another Nast cartoon established the elephant as a symbol for the Republican Party (left).

Perot scuttled his campaign because he could not bear the thought of losing, some pundit proposed that a good party symbol for Perot might be the chicken.

Elephants versus Donkeys

When faced with such big mysteries as the business of politics, I tend to focus on the little ones. On my desk are a couple of Arnold Roth cartoons depicting imaginary meetings of the Republican and Democratic national committees, at which party stalwarts are trying to decide on new animal symbols. The Democrats are at each other's throats, belting it out over a tangled heap of newsprint scribbled with graffiti and one goofy Pegasus-like horse with wings. In the other, the Republicans are sitting in staid order, raising their hands like bidders at an art auction, while elegantly framed paintings of chickens, snails, turtles, dinosaurs, a unicorn, and a dodo are held up for consideration. Not bad choices. A good symbol should have weight and meaning, something the existing party symbols lack.

It's fitting that the subject of party symbols be addressed in cartoons because the Republican elephant and the Democratic donkey were inventions of the 19th-century cartoonist Thomas Nast. In the November 7, 1874, issue of *Harper's Weekly,* there appeared a Nast cartoon of an elephant representing the Republican Party being

"A LIVE JACKASS KICKING A DEAD LION."

stampeded into a pit by a rabble of other animals. Leading the rabble was a donkey dressed in a lion's skin, representing the anti-Republican *New York Herald.* Nast's sympathies were obviously Republican, and perhaps he chose an elephant simply because that animal has such a formidable presence. His donkey was undoubtedly an ass—already a synonym for stupidity. The lion skin referred to one of Aesop's fables, in which an ass wears a lion's pelt to frighten a fox, but gives itself away when it cannot refrain from braying. That those two party symbols have survived for more than a century is

Other Beastly Symbols

• "Remember, only YOU can prevent forest fires," said an overweight black bear in baggy trousers and a park ranger's hat in 1947, and the world became a safer place for it. Smokey the Bear, who was actually born as a cartoon character on January 10, 1945, has appeared in thousands of posters, print ads, and television commercials, and now enjoys special trademark status bestowed by an act of Congress. Since 1950 the National Zoo in Washington, D.C., has kept a succession of bears rescued from forest fires as living Smokeys.

• Joe, the Arabian dromedary on Camel cigarette packages, is one of the most readily recognized symbols in the world. According to Hal Morgan in *Symbols of America*, R. J. Reynolds Tobacco Company brought out their first packaged cigarette in 1913, made from a blend of Turkish and American tobaccos, and named it Camel in honor of the Turkish ingredient. By coincidence the Barnum and Bailey circus was in Winston-Salem, North Carolina, at the same time graphic designers at R. J. Reynolds were looking for cover art. They sent a photographer to the circus; he snapped a shot of Old Joe, one of the stars of the troupe; and an artist sketched him in front of a background of palm trees and pyramids. The original art of the cigarette package hasn't changed, but, for advertising purposes, Old Joe has been made hip, complete with a bow tie, Italian suit, and debonair posture—an image that has sparked heated controversy because of its perceived appeal to young people.

• In 1867 a New York cigar company, Straighton and Storm, was struggling to come up with a catchy name for a new line of cigars. One night (according to legend), as Mr. Storm sat in his Long Island house pondering the problem, a snowy owl burst through the window.

Not one to ignore providence, he named the new line White Owl Cigars, and a snowy owl has appeared as its symbol ever since.

• There is hardly a community in America that doesn't support a local chapter of The Benevolent and Protective Order of Elks. Organized in 1868 as a drinking club for vaudeville actors, the Elks have grown to become one of the most successful and secretive of all fraternal organizations. Whatever it is Elks do in their closed-door meetings, it apparently has nothing to do with large ungulates. The name Elks was chosen only after Foxes was rejected (because they're "too cunning"), Beavers (because they're "destructive"), and Bears (because they were perceived as "coarse, brutal, and morose").

probably due more to Thomas Nast's influence than the appropriateness of the images. I'm certain the Democrats would like to think so.

Eagles versus Turkeys

In the world of politics and government, few symbols have been as successful as the bald eagle. Its meaning seems obvious to us now. It reminds us that we're brave, fierce, independent, and awesomely armed. Yet the eagle was not an automatic choice when the Founding Fathers set out to select an emblem for their new country. The July 4, 1776, meeting of the Continental Congress charged Thomas Jefferson, John Adams, and Benjamin Franklin with the task of designing a national seal, but they failed to reach a consensus. So did a second committee convened in 1780. An offering by one William Barton, a Philadelphian, in May 1782—a tiny eagle perched above a 13-striped shield—was initially rejected by the Congress. Finally, Charles Thomson, secretary of the Congress, modified Barton's idea and came up with the basic *E Pluribus Unum* design we use to this day (see the back of a $1 bill), and on June 20, 1782, the bald eagle officially became the national bird of the United States of America.

Not everyone was pleased. Benjamin Franklin wrote to his daughter to complain that the eagle was a poor choice because it was "a bird of bad moral character. . . . Like those among men who live by sharping and robbing, he is generally poor, and often very lousy. Besides he is a rank coward." In a passage that people frequently offer as evidence that Ben Franklin advocated the wild turkey as the American symbol, he wrote that compared to the eagle, the turkey is "a more respectable bird, and withal a true native original of America. He is, besides (though a little vain and silly, it is true, but not the worse emblem for that), a bird of courage."

It has never been determined whether Franklin was serious in his praise of the turkey, or whether he was indulging in a bit of ornithological whimsy. If he was serious, there was precedent. Old World cartogra-

The Great Seal of the United States.

phers had been using turkeys as symbols of the New World since at least 1555, and European silversmiths in the 16th and 17th centuries frequently used them as embossed emblems representing America (just as peacocks represented Europe; camels, Asia; and lions, Africa). Even as late as 1826, John James Audubon, the naturalist and illustrator of birds, was promoting the virtues of the wild turkey, making it the first plate in his *Birds of America,* and engraving its image above the words "America My Country" on the ring he used to seal his documents.

When we think of turkeys these days, most of us think of the notoriously stupid birds bred wholesale in poultry factories. Legends about domestic turkeys drowning during rainstorms may have no basis in fact, but we've nonetheless come to use the term *turkey* to mean anything foolish, stupid, and incompetent. The notion that wise old Ben Franklin would advocate such a creature for our national symbol is too attractive to resist. It was irresistible as well for the makers of Wild Turkey bourbon, who launched a nationwide magazine advertising campaign during the 1970s and 1980s that made liberal use of the Ben Franklin turkey story and featured a handsome painting of a wild turkey in flight above the headline, "Part of our National Heritage."

Hawks versus Doves

The success or failure of animal symbols usually depends on the appropriateness of the species. Eagles are practically fail-safe, as Anheuser-Busch and its subsidiary, Eagle Snacks, have discovered. When American Airlines tried to change its logo a few years ago by eliminating the eagle, company employees formed a "Save the Eagle" campaign. The bear markets of Wall Street are said to have originated in a folktale about a man who sold the skin of a bear before he'd actually killed the animal, leading to the expression "bearskin broker," for a person who speculated on falling market prices. A bull market, on the other hand, is one in which prices are rising, and a bullish speculator buys while prices are low.

Two age-old symbols that on the surface seem entirely appropriate are hawks for war and doves for peace. Never mind that hawks rarely battle among themselves, or that doves and pigeons are among the most quarrelsome of birds. The Roman poet Ovid wrote that hawks are hateful because they "always live in arms," a sentiment shared in early 19th-century America by critics of Republican expansionists, or "war hawks."

In Genesis a dove returns to Noah with an olive twig to show that the waters of the flood are receding, and in Christian art, white doves appear as symbols of the soul and the Holy Ghost. When Picasso set out to design a universal symbol of peace, he chose, naturally, a dove. During the war in Vietnam, it was an important schoolyard ritual to announce whether you were a hawk or a dove. Like many of my contemporaries, I lacked political backbone and declared myself a hawk early in the war and a dove later. Secretly, I remained a hawk in spirit, having always been fascinated by raptors. Besides,

The dove has long symbolized peace, despite its membership in the quarrelsome pigeon family.

I associated doves with pigeons, and was sure it would be more fun to spiral high into the sky than perch cooing on a statue's whitewashed head.

Psychologist Carl Jung believed animal symbols are important to us because we have animal natures as well as spiritual natures. To the Danish mythologist Sven Tito Achen, animals are the most important of all symbols, because in the eyes of flawed humans, animals appear perfect. "We would like to be like them," Achen claims. "We would like to *be* them!"

Which brings us back to donkeys and elephants. As far as I know, Democrats are not distinguished by their stubbornness, and Republicans don't have prodigious memories. If animals are significant symbols in the lives of humans, why would Republicans and Democrats allow themselves to be associated for more than 100 years with creatures that have no relation to party platforms? My advice to politicians: Instead of wasting time bashing opponents and telling voters what they want to hear, read *Walker's Mammals of the World* and *The Illustrated Encyclopedia of Birds,* and hope for inspiration.

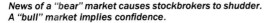
News of a "bear" market causes stockbrokers to shudder. A "bull" market implies confidence.

NO HONKING MATTER   *by Doug Stewart*

Norman Heifetz, facilities manager of the Bytex Corporation, remembers with a mixture of fondness and frustration the industrial park in Southboro, Massachusetts, where his electronics firm set up shop in the mid-1980s. The site was meticulously landscaped, with two ponds that invited restful contemplation.

When the ponds also proved an invitation to Canada geese, he recalls, "We thought, 'Isn't this beautiful?' People started feeding them. Then the geese had babies, and everyone thought they were cute. All of a sudden, we had an army of them. We had to hose down the sidewalks."

Heifetz tried mightily to make the geese leave. He banged pots and pans at them. He bought $350 worth of floating plastic swans, which he'd heard would terrify the geese (not true). He crept through underbrush, jumped up, and fired blanks at the birds with a starter's pistol. "The geese just looked at me," he says. Finally he bought a $500 radio-controlled speedboat to chase the birds. But even that worked only temporarily. Then, last year, much to Heifetz's relief, the company moved.

Once Nearly Extinct
In flight, Canada geese can be a thrilling sight. But on the ground the bird can be a walking, flapping, honking contradiction:

A family of Canada geese can add a regal beauty to any local pond. But in great numbers, the birds' quarrelsome manner, sloppy habits, and aggressive nature quickly make them a nuisance.

This symbol of the great outdoors, of crisp fall days and the mysteries of migration, is also a gluttonous, bad-tempered, loose-boweled trespasser. And though no hard data exist, experts agree the bird's numbers are exploding in this country.

The problem is most evident in the Atlantic Flyway. From northern Quebec to Florida, nonmigrating Canada geese—primarily the giant subspecies—may account for a quarter of the more than 600,000 birds in the midwinter population. Seventy years ago, in contrast, most Canada geese were migrants. And the giant subspecies was nearly extinct, a victim of overhunting. Then, as now, truly migratory Canadas nested almost exclusively in northern Canada. During the spring and fall migrations, the northeastern and upper midwestern states are just stopovers on the way to and from wintering grounds farther south.

Now, as giants multiply, migrant-geese numbers are shrinking. The winter population of all Canada geese in the U.S. part of the flyway dropped from 955,000 to 655,000 between 1981 and 1992. That suggests that the number of truly migrant geese has dropped more sharply than the total (in part, perhaps, due to weather extremes in their breeding grounds).

With a wingspan of 6 feet (2 meters) or more and a body weight that can exceed 20 pounds (9 kilograms), the giant Canada—the largest wild game bird—has long been prized as the big game of American waterfowl. Today's resident flocks in the United States are probably descendants of captive birds bred by gamesmen in northern states and used as decoys, a practice outlawed in 1935. (Although there has been some genetic mixing, residents and migrants have tended not to mingle.) Later, in the 1960s, the sedentary habits of giant Canadas made them the fowl of choice for stocking federal wildlife refuges: The birds are always around for nature lovers, and they aren't in danger of perishing elsewhere.

As the birds' numbers have surged, so have complaints from human neighbors—about droppings, all-night honking, crop damage, and aggressiveness. Most protests come from the grassy northeastern suburbs, where the resident fowl migrate only short distances, if at all. When the geese do take to the air, if they're low enough for you to hear their honking clearly, they're likely just moving from the park to the golf course.

Prime Goose Habitat

In feeding style, Canada geese are more cow than duck. Says Janet Sillings, a U.S. Department of Agriculture (USDA) biologist: "They like short-mown, fertilized grass, which describes most of northern New Jersey." Sillings's office, the animal damage control unit, receives more complaints about problem geese than about any other creature. One company complained about geese nesting on the roof and acting territorial on the pavement. "The gander was keeping people from getting to work," she says.

A Canada goose protecting its young will aggressively attack any person or animal that intrudes on its nesting area.

People have made the region into prime goose habitat. Says Sillings: "It has huge residential areas, lots of corporate parks with acres and acres of mown grass and ponds. Combine that with no predators to speak of, and the geese have no reason to move." And if they do, they probably won't go far. Migration patterns are not inborn; geese seem to learn them by example. If a goose grows up

In the eastern United States, Canada geese have all but taken over many of the ponds created by corporate landscapers (above). Unlike most waterfowl, the Canada goose tends not to migrate south for the winter (left), making it a year-round pest.

on an island in the middle of a water trap by the 15th hole, that's where it returns.

Because Canadas are large birds that eat virtually nonstop, damage to farm crops can be considerable. In addition to grains, "I've had farmers say they even go after broccoli and apples," says Laura Henze of the USDA, who is based in western Massachusetts. But most of all, the geese love succulent young shoots of lovingly tended grass.

Mead Park, a 40-acre (16-hectare) expanse of pastoral New Canaan, Connecticut, is a case in point. Says Steve Benko, the town's recreation director: "We have no grass around the pond at all now. We'll put down new grass seed, and as soon as the seeds germinate, the geese come along and eat them." Benko had wanted to plant wildflowers around the pond, which serves as a landing strip. (Geese prefer wide-open spaces with plenty of taxiing room and good sight lines in all directions; tall hedges and fences make them nervous.) "But we figured the geese would eat the flowers as soon as they started to come up," he says.

The Droppings Problem

The most common complaint about Canada geese has to do with what becomes of what they eat. An adult Canada goose can leave behind 1 pound (0.5 kilogram) of droppings a day. On the playing fields of New Canaan's junior high school, outfielders now dread long rollers through the infield gap.

The droppings may be more than just a nuisance. On Long Island Sound in recent summers, beach closures have become common after flocks of several hundred defecating Canada geese waddle onto beaches to pick up grit for their gizzards. Health officials are blaming geese feces for raised fecal-coliform bacteria levels in shellfish flats on Cape Cod and in inland lakes in suburban New York, although the evidence against the geese remains circumstantial. Officials also worry that geese might even taint reservoirs.

Once Canada geese settle into an area, they can be nearly impossible to uproot. In Belmont, Massachusetts, several hundred geese have become a fixture of the high school playing fields in the center of town.

The average Canada goose leaves behind an amazing 1 pound of droppings per day. Such a prodigious output makes even the best-planned picnic in goose territory a risky proposition.

"If there's a soccer game, they move to the football field. If there's a football game, they move to the soccer field," says Donna Saia, a secretary in the town's health department.

Belmont has been waging a futile harassment campaign against the geese. Highway workers have flapped towels and yelled at the birds at dawn. Policemen have fired firecrackers from shotguns. Considered and rejected: loud rock and roll, pink flamingos with spinning wings, and a one-day hunt, the meat to be donated to the homeless.

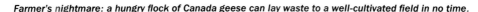

Farmer's nightmare: a hungry flock of Canada geese can lay waste to a well-cultivated field in no time.

"We heard that geese don't like garlic, so we considered loading garlic oil into the sprinkler system, but we haven't tried that yet," Saia says. Citizen dog-walkers were enlisted to come at specified times to let their dogs chase the geese "But," says Saia, "a dog has to be pretty darned big to scare off these geese." Actually handling or killing the geese is illegal unless during hunting season in an area where hunting is allowed.

Food First, Territory Second

Canada geese are fiercely territorial, particularly during the spring nesting season. A flapping, honking Canada goose recently chased one woman golfer in downstate New York right into the water. Another attacked the limousine of a corporate president at the business's front door.

The geese are programmed to claim and defend enough territory to feed themselves. "When you feed them," says Jay Hestbeck, a biologist with the U.S. Fish and Wildlife Service, "you overwhelm that traditional instinct. You cause the geese to bunch together, since their first instinct is to eat; defending territory is secondary. So you end up with higher concentrations of geese in an area, bigger flocks, and more droppings."

But giving $20 fines to goose feeders, as some Massachusetts towns are now threatening, won't get to the root of the Canada-goose problem. Neither will hunting, since most suburbs don't allow it, and hunters might not be able to distinguish residents from migrants. Some communities are taking out federal permits to oil, puncture, or otherwise damage goose eggs. This tactic works, however, only if the geese are breeding locally. Vasectomies, too, are an option, though expensive. The USDA is now testing a food additive used in bubble gum as a possible goose repellent: It's harmless to grass and humans, but geese apparently hate the taste of it.

For now, the only recourse is to scare away new populations before they become entrenched. And once the birds are in place, maybe there's a small consolation in the pleasures of noisily making a fool of oneself at dawn, spinning those flamingo wings or mastering the controls of a toy speedboat.

Canada geese appear to prefer a highly manicured environment, much to the dismay of country club owners.

PLANTS THAT EAT MEAT *by Michael Lipske*

All carnivorous plants derive some of their nutrients from insects. The Venus flytrap (top) has nectar-producing glands to lure a fly (middle) or other hapless prey to its leaves. When the fly lands, the leaf snaps shut (above), trapping the prey.

Cooking up a delightful movie comedy that stars a speaking vegetable cannot have been easy. So it is a little unfair to contemplate plot changes—especially at this late date—to director Roger Corman's wonderfully goofy 32-year-old film *The Little Shop of Horrors.*

You probably remember the story. It's sort of a "Tales from the Dark Side of the Victory Garden," with much of the gut-level emotion expressed by a giant, belching, flesh-eating plant, Audrey, Jr. Under the increasingly desperate care of Seymour Krelboing, klutzy clerk in a skid-row Los Angeles florist shop, the plant eats a railroad detective, a dentist, a burglar, and a prostitute. "Feeeeed meeeee!" the plant demands. Finally the pushy plant even eats poor Seymour.

Attracted by the boldly colored waterlily flowers, bee flies in search of food become a meal themselves when they are mired in sticky secretions (left). The flower is triggered closed around the animals (right) and they are digested.

Fierce Feeders

If Roger Corman ever schedules a sequel to his *Little Shop,* he should send scouts to America's real carnivore country. Here, on the Southeast's hot and humid coastal plain, glistening, sticky sundews, tall, tubular pitcher plants, and their vegetable brethren snare and digest insects and other hapless creatures by the millions. Botanically speaking, it is one of the strangest areas on Earth.

Although carnivorous plants grow in Canada, parts of the Northwest and Northeast, as well as California, nearly 90 percent of North American species can be found within a long arc stretching from Virginia to Texas, with some of the fiercest feeding occurring along the Gulf Coast. A skilled searcher working the zone between Florida's Apalachicola River and Louisiana's Tangipahoa can find as many as 13 species in a single bog. Growing in flat, grassy savannas pressed among the pinewoods, there are more kinds of carnivorous plants here than in any area of comparable size in the Western Hemisphere, and perhaps the world.

Quaker botanist William Bartram, exploring the Southeast in a time when "Bears, tygers, wolves, and wild cats [were] numerous enough," marveled at the hollow, liquid-bearing leaves of pitcher plants along the coastal plain. In his 1791 *Travels,* he was the first naturalist to accurately describe the inner structure of the leaves; he noted the stiff, downward-pointing hairs that prevent those insects "invited down to sip the mellifluous exudation" from crawling back up to freedom. Bartram wrongly believed that the pitcher leaves functioned as cisterns, providing water for the plants in time of drought, and he doubted that trapped insects were a source of nourishment. Less than 100 years later, Charles Darwin published proof that carnivorous plants not only snare insects, but also dissolve the insects' soft tissues in order to absorb liquid animal nutrients; carnivorous plants that were provided with such fare grew more vigorously in Darwin's experiments than did members of the same species deprived of animal meals. Scientists still marvel at the plants and their habitats.

"The areas are so fascinating they just kind of grab you and take you over," says George Folkerts, an Auburn University zoologist who has studied the ecology of what he calls "pitcher-plant bogs" in the Southeast since the early 1970s. Blooming with orchids, lilies, and other flowers, the bogs are among the richest plant communities anywhere, with as many as three species per square foot. But as many of these soggy sites have been drained and developed, this floral abundance is fading. Compared with the profusion of Bartram's time, only a pittance of pitcher-plant habitat remains: probably just 3 percent of the carnivorous kingdom that once covered thousands of acres in places along the Gulf Coast. "If something's not done," says Folkerts, "there will be, at the turn of the century, very, very few sites left where one can go and see pitcher plants." In the areas that do remain, over-harvesting—including poaching—threatens the future of the carnivores. North Carolina now levies fines of up to $2,000 for illegal taking of the Venus flytrap.

It is easy to understand the allure of these slightly creepy plants. "They turn the tables," explains Leo Song, Jr., manager of the greenhouse complex at California State University in Fullerton, and co-editor of the quarterly journal of the *International Carnivorous Plant Society*. "Usually plants get eaten by insects. And here's a plant that eats insects."

Chamber of Doom or Cozy Room?

Equally intriguing are cases where insects turn the tables yet again. A small moth belonging to the genus *Exyra* secures food and shelter from the deadly tubes of pitcher plants. The moth's larva chews a groove around the inside of the pitcher in which it has hatched, causing the upper part of the leaf to wilt and collapse. The remodeling prevents other insects from entering the tube, thus ending that leaf's ability to capture prey. A wasp, *Isodontia mexicana,* does the same when it lays an egg on the alternating layers of grass and paralyzed grasshoppers stuffed into the pitcher leaf.

One animal's chamber of doom is another's cozy room. Scientists have learned that an array of specialized arthropods maintain useful relations with pitcher plants. Some insects feed on tiny rotting carcasses or ply the plant's rich inner broth for other food. Larvae that profit from the supermarket provided by the plant may speed up the breakdown of dead prey, rendering nutrients more readily available to their host.

The strategies used by carnivorous plants are cunning indeed. The tiny trapping

The aquatic bladderworts get their name from their bulbous, bladderlike structures, which contain a suction trap that permits prey to enter but not to exit. An insect touching a trigger hair on the plant is carried into the bladder by an inrush of water. Unable to open the trapdoor (far right), the animal suffocates within a day and its decayed body is absorbed by the bladder lining.

mechanisms of bladderworts, a genus of rootless, often aquatic plants found throughout the world, sometimes are as small as a pinhead. Yet the traps exhibit some of the most sophisticated engineering of all carnivorous plants. Found on the plant's stems or leaves or both, the hollow traps have an entrance door that rests against a threshold sealed with mucilage-secreting cells. When the bladder trap is set for capture, its interior is almost free of water. A passing mosquito larva or other tiny creature that lightly brushes the trigger hairs on the outside of the door causes it to open inward; water shoots into the bladder, carrying the victim with it; the trap walls swell outward, the door reshuts, and the prey—free-swimming a second ago—is now imprisoned in a translucent sac, ready to be digested by enzymes secreted by glands within the bladder.

Like all green plants, the carnivores manufacture food from sunlight, water, and carbon dioxide. But they also possess the modified leaves that permit them to capture and consume animal food. Most of the carnivores grow in areas with highly acidic soils. That acidity ties up nutrients that otherwise would be available to plants. The ability to eat animals gives the plants a leg up in a difficult environment.

The Venus Flytrap

Probably the best-known insect catcher is the Venus flytrap. With curved spines projecting from the edges of its gaping red leaves, shaped like open clamshells, it even looks like a meat eater—a plant that Charles Darwin judged to be "from the rapidity and force of its movements . . . one of the most wonderful in the world." Darwin was only echoing the wonder of earlier observers. "Astonishing production!" declared Bartram. "See the incarnate lobes expanding, how gay and sportive they appear! Ready on the spring to intrap incautious deluded insects!" Nectar-secreting glands on each leaf lure the prey—most often ants and spiders rather than flies, despite the plant's name. At midpoint of each lobe of the leaf are trigger hairs. A potential victim must brush a single hair twice or two hairs in succession—within a span of no more than 20 seconds—to generate electrical signals that cause the trap to snap shut within another second. The projecting spines now function as bars on a cage, and the closed leaf becomes a stomach that begins to secrete digestive fluids. When the leaf trap opens a few days later, all that remains of the victim is a husk.

No less deadly are sticky snackers like the sundews, equipped with glittering, gluey filaments that snare insects. Several decades ago a British scientist described a dense, 2-acre (0.8-hectare) stand of sundews upon which a horde of migrating butterflies had settled. With as many as seven butterflies stuck to each plant, the number of doomed insects may have approached 6 million.

Hollow-leaved pitcher plants are passive killers. These include not only the 16 species scattered across North and South America, but also bizarre-looking members of the genus *Nepenthes*. These pitcher plants of tropical Asia and Madagascar often grow as climbing vines in the jungle. Their gourdlike dangling pitchers are shaped like bananas, cornucopias, or even toilets. Collecting cockroaches, centipedes, scorpions, and reportedly an occasional rat, *Nepenthes* pitchers can come in handy for travelers. Parched during his ascent of Mount Ophir, in peninsular Malaysia, the 19th-century naturalist Alfred Russel Wallace turned to the local pitcher plants. "Full of insects and otherwise uninviting," he wrote, the warm water inside the pitchers proved drinkable, and "we all quenched our thirst from these natural jugs."

Right This Way . . .

The pitcher plant's diabolical trapping technique could be narrated by Vincent Price. First comes cruel artifice. Often brightly colored, the tall, trumpet-shaped leaves of several North American species mimic flowers, and lure prey with good looks and sweet scent. Sometimes a trail of nectar-secreting glands starts at ground level and leads on up the outside of the leaf—a "right-this-way" summons to ants.

The hungry ant or other potential meal is lured to the mouth of the trumpet, so crowded with nectar glands it may be wet.

The sensitive, gland-tipped, green to reddish hairs of the sundew plant (above) are able to actively move an ensnared insect to the plant's center. The plant then wraps itself around the insect and digests it (right).

But below this mother lode of sugar, the interior of the pitcher tube is waxy and slick. This is the start of the plant's slippery slope, where victims lose their footing and slide into the increasingly narrow tube. Down inside, the inner wall of the leaf is lined with glands that secrete digestive enzymes, which trickle down and collect in the bottom of the trap. The insect slips lower, to where the surrounding wall is lined with downward-pointing hairs that discourage exit. In some species the bottom fluid contains an ingredient to stun the struggling captive. There may even be a wetting agent that helps soak and drown the victim.

The honeyed lure, the fatal slip, the plunge into a fatal pool—it's a clear case of life imitating Edgar Allan Poe. But the pitcher plant giveth, as well as taketh away. For those creatures that are ready and able to live in the belly of a vegetable beast, a bog ripe with pitchers is the ultimate happy hunting ground.

Cozy microhabitats, the hollow leaves of pitchers provide greater relative humidity, a more constant temperature, a bit of shade, even a food supply in the form of trapped, decomposing prey. Eateries, love nests, nurseries: the leaf tubes are prime real estate, attracting steady customers.

Inside its remodeled retreat, the *Exyra* larva feeds on the pitcher's inner skin, leaving the outer shell of the plant intact. The two species of *Exyra* that inhabit only pitchers with long, narrow tubes solve the problem with elbowlike spines, or lappets, on the sides of their body. The little bumpers pre-

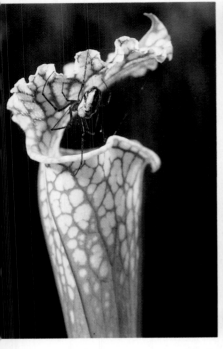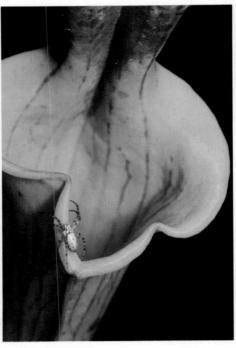

Pitcher plants contain tubelike leaves that are filled with digestive fluids (above left and center). Prey is attracted by glands on the rim of the pitcherlike traps (above, right).

vent the caterpillar from getting stuck if it falls into an especially narrow space.

Even after becoming a moth, *Exyra* spends most of its time inside pitchers. But where other insects might slide to their doom, "*Exyra* very calmly walks up and down the pitcher walls," says Debbie Folkerts, George's wife, an entomologist who spent a year and a half marking pitchers and tracking *Exyra* moths and caterpillars in an Alabama bog.

"The shape and size of their claws make it appear they can cling to the downward-pointing hairs," a feat other insects cannot manage, says Debbie. But the small moths also seem to rely on a behavioral adaptation to help them negotiate the treacherous walls. They unfailingly orient themselves with their head pointing upward, walking backward to go down inside the pitcher, and forward to go up. And while most other moths in their family copulate end to end, *Exyra* (which even mate inside pitchers) get

together at right angles, permitting each partner to avoid staring dangerously downward into the pitcher tube.

In all, 16 species of risk-loving insects or other small creatures live nowhere else but within North American pitcher plants. Leaves of the purple pitcher, which may snare hundreds of insects during the warm months, also provide a nurturing aquatic home for mosquito larvae belonging to the species *Wyeomyia smithii*. The wrigglers swim through the plant's indoor pool, grazing on microorganisms and suspended matter (the wrigglers' bodies are believed to produce a protective antienzyme that prevents their digestion by the plant). In the northern part of the pitcher's range, the liquid within its leaves freezes, and the tenant mosquito larvae spend the long, cold winter months inside a pitcher-sicle.

All this foraging activity within the pitcher leaves appears to be mutually beneficial, with the plants delivering meals that

Prey becomes enmeshed in wax from the pitcher plant's cuticle, causing it to lose its grip on the rim and fall into the plant's increasingly narrow tubal prison (above). Downward-pointing hairs prevent escape and digestive juices serve to drown and metabolize the struggling victim.

the tiny feeding animals digest and break down into nutrients that in turn are used by the plant. Such seems to be the role of mites found in virtually all pitcher plants. Tiny creatures with comblike mouthparts that slide back and forth, these pitcher-leaf dwellers spend their lives wandering slowly through the prey remains within the leaves.

Into this crowded world squeeze other opportunists, like squatters at an apartment complex. Several kinds of spiders work the pitchers, seizing dinners from the stream of insects attracted to the plants. One species even drops on silken lines to retrieve prey for itself. Tree frogs shelter in the mouths of some pitcher tubes, clinging to life with suction feet and, like the spiders, perhaps grabbing for themselves an occasional down-bound bug. On the island of Borneo off the coast of Southeast Asia, a species of tarsier, a small primate that eats mostly insects, visits pitchers of *Nepenthes* vines to scoop out and eat fresh prey.

Avian Water Fountains?

One early naturalist proposed that the liquid-containing leaves of North American pitcher plants served as drinking fountains for wild birds. Birds do visit the plants, but in a search for juicy *Exyra* larvae or pupae hidden inside. To one entomologist, writing more than half a century ago, the sight of caterpillar-sealed leaves "slit lengthwise by the beak of a searching bird" was yet more evidence of the role of the carnivorous pitchers "as focal points of contact between the plant and animal worlds."

Shedding light on such connections has occupied George Folkerts for almost two decades. "There is a real dearth of information on these sites," he says of the Gulf Coast pitcher-plant bogs. "We know something about the plants now," but next to nothing is known about microbial life in the bog soil, or how insects are affecting the plant community. "So we're probably missing a big piece of the puzzle."

Content as any *Exyra* caterpillar to spend his days among the nibbling vegetables, the tall, trim zoologist shows a visitor some of what Folkerts calls the "infinite number of interesting things" that occur in carnivorous-plant bogs. At a site outside Pensacola, Florida, Folkerts squishes across the wet ground in size-14 Nikes and mud-spattered chinos; bending down, he parts the coarse wire grass that hides a specimen of *Sarracenia psittacina,* the low-growing parrot pitcher with its beaklike leaf end. The scientist points out the raised flange, or wing, that runs lengthwise along each leaf and that serves, he says, as a "functional drift fence" for rerouting ants, centipedes, or other prey.

No Biting Back

To demonstrate the bad news about carnivore country, Folkerts drives up toward the Alabama hamlet of Perdido, on the corner of the Florida Panhandle. Parking alongside what as recently as 1982 was one of the Southeast's richest bogs ("You couldn't walk without your legs touching pitcher plants," he remembers), he points out how a combination of road building, pond making, and cattle pasturing has bisected, altered, tramped down, and otherwise brought low the 600-acre (240-hectare) privately owned complex.

The plants are unable to bite back at the two-legged omnivore that swallows pristine bogs and spits out pine plantations and shopping malls. For decades the plants have been losing the essentials of life—not bugs to eat, but a necessary mix of frequent fires and saturated soils that maintain their habitat. Natural fires, ignited by lightning strikes, once pruned away shrubs and pine trees that would otherwise shade out the carnivores. Pitcher plants, with their underground rhizomes, survived such blazes unharmed. But to protect towns and timber or to prevent smoke-caused smashups on interstate highways, natural fires that once raged for days in the coastal plain are now quickly snuffed out. As a consequence, the bogs have declined. Many carnivorous-plant bogs have been drained so they can be made into pine plantations. Others have been con-verted to cropland. Carnivorous plants growing on steep slopes had been safe until farmers began taking advantage of such topography to create farm ponds, obliterating the habitat in the process.

Then there is the brisk business in cut tubes of *Sarracenia leucophylla,* the white-topped pitcher plant that, with its colorful network of veins—sometimes green and sometimes a red that ranges from pink to crimson to maroon—is often declared the most beautiful member of its tribe. So striking are the yard-long pitchers that they are widely used in floral displays in this country and in Europe.

Estimates vary, but some authorities in the carnivorous-plant world say that *S. leucophylla* pitchers are being harvested by the millions, a situation a federal government botanist calls "out of control."

To George Folkerts, "simple logic" suggests that removing the pitchers is not good for *S. leucophylla* plants: "If you have a plant that supplements its resources by catching insects, and then you take off the structures it uses to do that, it's just like taking away its mouth, so to speak."

But to Richard Tarnok, a lanky 33-year-old nurseryman from Carriere, Mississippi, "the worst thing that's hurting these flowers is paper-wood companies plowing them up." Tarnok's family has founded part of its healthy cut-flower business on knowing where *S. leucophylla* grows, and on getting the pitchers out of rural bogs and off to wholesale buyers. Not surprisingly, Tarnok is displeased that Folkerts informed the International Paper Company that pitchers were being taken off their land in wholesale numbers, an act that resulted in the pickers being barred from certain boggy tracts they once leased.

In Tarnok's eyes, the scientist is misinformed. The family's *S. leucophylla* harvesting, even though it amounted to some 500,000 pitchers per year, was no more than harmless cropping, maintains the nurseryman. The real muggers of the plants, says Tarnok, are the big guys, the ones turning a bog into a "shopping center, homesite, cow field, farmland, or whatever. Whenever all of it turns to asphalt, they're through."

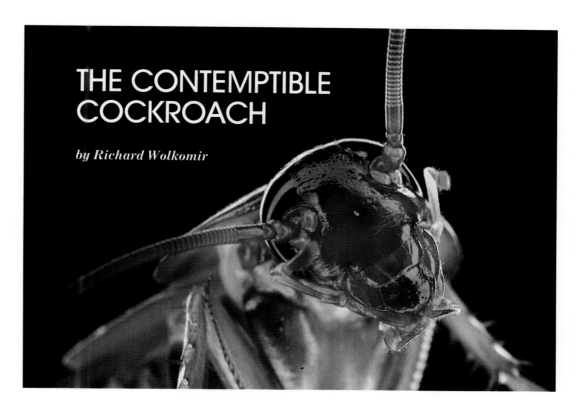

THE CONTEMPTIBLE COCKROACH

by Richard Wolkomir

Behind the U.S. Department of Agriculture (USDA) laboratory in Gainesville, Florida, sits a tiny house consisting of two kitchens and an attic. Shaded by oaks, it is an appealing, though somewhat strange, dwelling. Appealing, that is, for entomologist Richard Brenner and the cockroach tenants for which he built it.

The building, custom-designed by Brenner, offers its teeming inhabitants all the benefits of a simulated human occupant—including the all-important moisture from, say, cooking spaghetti or bathwater. In the floors, walls, ceilings, and roof are 120 sensors to measure humidity, air currents, temperature, and cockroach movement. The results go to a central computer, which Brenner, who works with the USDA Medical and Veterinary Entomology Research Laboratory, uses to study how the insects respond to changes in air, light, or availability of food.

Thousands of Species
This is serious work, with serious consequences: Cockroaches, researchers are finding, not only evoke a visceral "ick" from people, they also can threaten our health.

But whatever Brenner discovers about the lives of cockroaches in our homes, his findings will barely scratch the surface of what there is to learn about the insects. For only about 25 of the 4,000 known species worldwide of cockroaches ever intrude on the lives of people. They don't even *like* us—and have even been known to wash themselves after being touched by people.

Cockroaches come in sizes ranging from ants to mice, in colors ranging from brilliant green to translucent to mahogany. Most species inhabit tropical jungles; many also live in caves, deserts, temperate forests, and, of course, every manner of human dwelling—from Manhattan kitchens to the subterranean utility tunnels of Arctic air bases. The nearly flawless (by evolutionary criteria) creatures have been around for 350 million years, since long before the time of the dinosaurs. They can live for vast stretches of time with no food, and can withstand massive doses of radiation that would easily kill peo-

The cockroach, though much maligned by city dwellers, is nonetheless considered one of the great success stories of evolution.

Cockroaches are notorious for their appetite, feeding on virtually anything—from dead insects to leftover food.

ple. For scientists, then, nearly everything about these bugs is a lesson in survival that goes far deeper than their adaptation to mere human beings.

Eating Machines

Cockroaches have been so successful in large part because they are generalists, unlike animals that survive with exquisite adaptations to particular features of their environments. The newer creatures on Earth tend to be the specialists. For instance, lice, only about 75 million years old, evolved in tandem with mammals and birds, which they use as mobile restaurants. Koalas feed solely on certain types of eucalyptus leaves. And pandas eat primarily bamboo. Those strategies can work beautifully—but only as long as the food supply lasts.

Cockroaches, in contrast, are "so generic they can exploit a changing environment," says Richard Brenner. Like their relatives the mantises and grasshoppers, cockroaches have jack-of-all-trades mouths, with working parts evolved from legs. They can chew virtually anything, hard or soft—from algae to dead insects to the metabolite products of the digestion of microbes. Little goes to waste around a cockroach. They eat their own cast-off exoskeletons, and they nibble off each other's antennae and legs.

After a cockroach brood hatches, the adults apparently eat the embryonic membranes, the egg cases, and any unhatched eggs.

"In a home," says Brenner, "they'll eat salts in tennis shoes, bacteria or mold, grease spots on a wall from cooking, the starch on postage stamps and in wallpaper paste—just about anything looks good to them." When they must, cockroaches can survive on virtually nothing. Brenner's laboratory once kept a cockroach colony alive for two and a half years on a diet of no protein at all. Their secret is certain bacteria that live only inside a cockroach organ called the "fat body."

Most insects excrete uric acid. But cockroaches stockpile it for hard times. When a cockroach's diet is rich, the bacteria in its fat body transform the protein's nitrogen into uric acid crystals and store them. When food is scarce, the insect draws upon the stored uric acid for sustenance. When times are really bad, cockroaches sustain themselves on each other. "The strong eat the weak," says Phil Koehler, a University of Florida entomologist who works with the USDA lab. "They rip off the top of the abdomen and consume the fat body."

Some cockroaches dine inside caves where bats roost, feeding on the moist mounds of guano and bits of fruit and seed

the bats rain down, as well as on dead bats. Some cockroach species have dwelled in caves so long they have lost their wings and eyes. In the underground darkness, these bugs seem to set their biological activity clock by the bats' comings and goings, "telling time" by bat-wing-stirred breezes or by the warmth of bat bodies when the night hunters return at dawn.

Scientists speculate that caves were where the bugs first discovered the benefits of living with our ancestors. Later the insects stowed away aboard Roman and Phoenician galleys to Europe, rode

Spanish galleons and slave ships to the Americas, and took rides with traders and marching armies to go wherever humans have gone.

Experts believe that cockroaches evolved originally in tropical jungles, where most still live. When entomologist Coby Schal of Rutgers University studied Costa Rican rain-forest cockroaches, he found that every evening, large numbers of adults of some species fly or crawl several meters up into the bushes. At dawn the bugs sink back into the leaf litter. Schal theorizes the cockroaches ascend to avoid spiders and other nighttime predators of the forest floor, and descend in the morning to escape lizards and birds.

Cockroaches range in color from the dull brown of the Australian roach (left) to the brilliant green of the Cuban cockroach (above). They can be as small as an ant or as large as the mousesized Madagascar hissing roach (below).

Proof of the cockroaches' versatility is that these insects of humid jungles even colonize deserts, finding moisture in fungi coating desert shrub roots. The bugs' Achilles' heel is the speed with which they dry out. So, to survive the searing desert day, they move into the moist burrows of tortoises and rodents like the kangaroo rat. (In a range of climates, they also move into nests of bees, wasps, termites, ants, and birds.)

Many Babies

If cockroaches are the eating machines—the turbines, even—of diners, they are the turbojets of reproducers. The Gainesville lab has determined that in one year a single Asian cockroach (particularly prevalent in Florida) whose young all reproduce will cause 10 million new cockroaches to come into the world. In that same time, a German cockroach (the most ubiquitous urban dweller) can produce more than 500,000 descendants. And many species live as long as four years. "The strategy is to live long and reproduce often," says University of Florida's Phil Koehler.

As far as scientists know, most cockroaches have a few mating tactics in common (with the exception of a North American strain of the species *Pycnoscelus surinamensis,* which consists only of females, and reproduces essentially by cloning). In several species, males perch higher than females. That way, males can smell females' pheromones, or sexual perfume, wafting upward. A tiny gust of unscented air from below—possibly stirred by a predator—is likely to send a male upward. But at a whiff of female pheromone, the male climbs down.

Since the bug's skeleton is external (a hardened carapace called a "cuticle"), mating rituals can appear—to people, at least—like undertaking an assignation while clanking around in full armor. Even the bugs' genitalia are an array of hooks and grapples to assure proper positioning.

Consider *Gromphadorhina portentosa,* the Madagascar hissing cockroach, big as a mouse, and aptly named for a noise it makes by pumping air through its breathing holes. The ritual starts with mutual antennae touching (cockroaches taste and smell through chemical sensors on their antennae). Then the male circles the female, hissing and posturing in something of a dance,

The cockroach's large size and ease of rearing make it an ideal experimental animal. Scientists hope to learn more about its unusual survival skills and its incredible adaptability to changing environmental conditions.

at the same time releasing a chemical attractant. Stage three, copulation, is an act of engineering. In many species the male secretes a tasty fluid from his back. When the female clambers up to eat the attractant, she is positioned for the male to attach his grappling hooks. Then the two pivot rear to rear for the transferral of sperm.

The all-important chemicals in these machinations also rule other elements in cockroach lives. An "aggregation" pheromone, for example, draws together German cockroaches, which crowd convivially in dark crevices. The densities and age groups, in turn, are determined by chemicals that broadcast the reproductive state of the females (whether they are pregnant, carrying egg cases, or ready to mate).

Controlling Cockroaches

So much for the more esoteric details of cockroach life. What about the ways these bugs bug us? And how might we rid ourselves of them? First the bad news: Far from being benign, they are a serious cause of human allergies ranging from tearing eyes to severe wheezing. As much as 60 percent of the 11.5 million Americans who suffer from asthma are allergic to the creatures (whether the insects are alive or long dead—even decades old). In rare cases, some species can also cause vertigo and even potentially fatal anaphylactic shock.

And despite all our human ingenuity, we have not yet found any foolproof ways to control cockroaches. These masters of the art of survival are sure to outlive our own species; we are, after all, a mere blip on their time scale. Cockroaches quickly develop defenses against almost anything we poison them with, probably because they've coped for eons with chemical warfare from plants. "Plants have been putting out anti-insect chemicals since Day One," says biologist Jim Moss, who studies insecticide resistance at the Florida lab.

The good news is that everything scientists are learning about cockroaches holds clues for future ways to deter them. Take the reason desert-dwelling cockroaches seek the kangaroo rat's burrow (for moisture). Richard Brenner has found in his Gainesville labora-

Cockroaches tend to be elusive in their habits, but they will enter human habitats to scavenge any food left exposed. Little evidence supports cockroaches as vectors of disease, although they are a serious cause of human allergies.

tory that cockroaches typically spend most of the day backed into a crevice, antennae sticking out, sensitive for air currents. The reason? They must be just as careful of drying out in your home as they are in the desert. One food-deprived cockroach colony in the Gainesville lab died only when a technician forgot to water it over a weekend. "Deny them water, and you've hit them hard," says Brenner. So, too, he hopes, will the drying action of a ventilation system he is testing.

The lab's sensors have revealed that cockroaches scurry for cover at the slightest air movement. "We used to think they avoided the light, and that's true, but they also hide away to avoid moving air," says Brenner. (As for avoiding light, anyone who's experienced an invasion of Asian cockroaches can attest that at least one bold species loves illumination—even the glare of a television screen underfoot.)

Until air blowers designed to dry out cockroaches come along, we may be able to thwart the critters with fans at key locations. The experts also recommend general cleanliness and an absence of clutter that can harbor the bugs. As for poisons, anything that will kill a cockroach may also hurt people. "The problem is that we are trying to eliminate something that intimately affects our environment," says Brenner. "The point to our research is learning to very carefully manage the cockroaches' environments without harming people."

A case in point is boric acid (to which cockroaches can't seem to develop resis-

tance—yet!). In powder form, it sticks to the insects, which then ingest it as they clean themselves. Boric acid is best used spread in small amounts where the bugs are likely to tread, but not where it might poison humans and risk toxifying our own environment. Also effective, and far less potentially toxic to people, are baits—which often come as small black plastic rectangles. The cockroaches actually feed on the poison and then die later.

Those of us who are alarmed by cockroaches may find some perverse consolation in the notion that people provide only a small fraction of the bugs' habitats. The insects might prefer, say, a tree hole to a house. Recalls population biologist Donald Strong of Florida State University in Tallahassee: "The first time I saw a hollow oak cut down, I thought there was a fire inside because the tree hit the ground, split—and black fumes seemed to issue out." The "smoke," of course, was cockroaches.

If that image is somehow not soothing, we might learn something from entomologist Harley Rose of Sydney, Australia. Rose maintains that Australia's giant burrowing cockroaches, the length of an adult finger, are ideal pets. And he's not alone. For two years, he has sold about 10 mating pairs a month (at $50 a pair). Like the cockroaches that may be benefiting from the steam wafting off your dinner plate, these "cockies" are maintenance-free. "You can go away on holidays without any problems about looking after them," says Rose.

Botany

PEP PILLS FOR STRESSED-OUT PLANTS

The concept is so simple, it's a wonder no one figured it out earlier. An entomologist at the University of Wisconsin at Madison has found that plants withstand stress better if they're given minute doses of vitamins—notably vitamins C and E. Dale M. Norris and his research team observed that plasma membranes surrounding plant cells contain a protein that monitors stress. When the protein senses danger—perhaps from a nibbling insect or an invading fungus—it triggers production of natural substances that defend the plant.

Unfortunately, these sensor proteins can be destroyed by oxidation—the same process that causes cars to rust or paper to turn yellow. Norris decided to test if antioxidants such as vitamins C and E would protect the sensor protein. He found that plants treated with these vitamins stayed healthier than their untreated counterparts.

This vitamin therapy has the added benefit of helping plants fight off stress from most sources—insects, diseases, drought, air pollution, heat, herbicides, and mechanical injury. So gardeners wanting perkier plants won't have to figure out which type of stress their plant has, then sort through the various plant-related products on the market to find an appropriate cure.

Another plus is that the vitamins continue working long after they are applied. Even plants that grew from seeds soaked in a vitamin bath showed increased tolerance to stress. The research suggests that the effects of fat-soluble vitamin E are longer-lasting than those of water-soluble vitamin C.

The drawback, at least for now, is that the vitamins must be applied in parts-per-million dosages, a quantity too small for the measuring equipment most home gardeners have on hand. And dumping on a large amount doesn't work, be-

The first bioengineered food to hit grocery store shelves will be the Flavr Savr tomato. Scientists have deactivated the enzyme that causes the tomato to rot, giving the tomato twice the life span of normal tomatoes.

cause excessive doses can be toxic to plants. Fortunately, Norris anticipates that plant vitamin products in the proper dosage should be available commercially by 1994.

THE FLAVR SAVR FLAP

It's a controversy that plays on two of the most basic human drives: the need to feel that our food is harmless, and the desire to buy tomatoes that taste like they ripened in the backyard.

The vegetable (or is it a fruit?) at issue is the Flavr Savr, a genetically engineered tomato that, once picked, doesn't get soft as fast as regular tomatoes do. This genetic tweaking means that Flavr Savrs can ripen on the vine, developing all the tasty sugars and acids that say "home-grown," then be harvested and shipped to supermarkets without turning to mush. Compare this to unaltered tomatoes, which must be picked and packed green and flavorless to withstand the rigors of making it to market.

Despite its appeal to the palate, the Flavr Savr's debut as the first genetically altered consumer food has aroused controversy. The Food and Drug Administration (FDA) decided that the Flavr Savr need not be labeled to reveal its biotech origins. The FDA reasoned that splicing new genes into a tomato or other crop is no different from the well-accepted practice of hybridizing, in which two plants are crossed to give their offspring desirable traits. Splicing longevity into the Flavr Savr, the FDA reasoned, is the genetic equivalent of, say, breeding extra sweetness into sweet corn. And since the FDA requires no special labels for hybridized foods, it would require none for gene-spliced foods.

The FDA's decision in general, and the Flavr Savr in particular, have met heated opposition from both consumer and environmental groups. Having long feared that foods with manipulated genes might pose unforeseen risks, the groups were especially alarmed that altered foods could be sold with no warning to consumers who might choose to avoid them.

One such organization is the Foundation on Economic Trends, which has been campaigning to get grocery stores to refuse to sell the Flavr Savr, and restaurant chefs to refuse to serve it. Establishments participating in the foundation's Pure Food Campaign indicate so by posting an identifying symbol.

The Foundation on Economic Trends is just one of many groups who want the FDA to test and label genetically altered foods. Should the FDA change its position, testing requirements are likely to slow commercial release of hundreds of gene-spliced crops already in laboratory and field trials. And, given the way marketing works, we might someday see warning stickers on our produce saying, "Try me! I'm genetically altered!"

EXPLORING PLANTS' PAST

While genetic engineers take plants into the future, archaeologists and botanists are uncovering more about their history. Botanists at the University of Georgia at Athens may have found the origin of the endosperm, the starchy part of a seed that provides nutrients for the seedling as it sprouts.

The researchers studied the seeds of a desert shrub in the *Ephedra* genus. In these primitive seeds, both nuclei of a pollen cell fertilize an egg, creating, in effect, a set of fraternal twins. As the seed develops, one of the fertilized eggs dies, and the other draws nourishment from the dead sibling's embryo sac.

The researchers hypothesize that this twin arrangement in *Ephedra* shows the

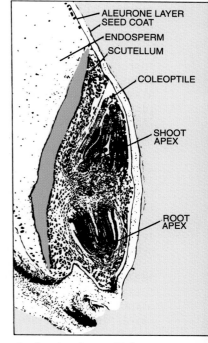

ALEURONE LAYER
SEED COAT
ENDOSPERM
SCUTELLUM
COLEOPTILE
SHOOT APEX
ROOT APEX

Paleobotanists may have found the origin of the endosperm, the food supply of higher plants. In wheat (right), 80 percent of the bulk of the wheat kernel and 70 to 75 percent of its protein are stored in the endosperm.

endosperm's genetic beginnings. Paleobotanists believe that before this double fertilization evolved, mother plants had to put great energy into storing nutrients in single embryo sacs before fertilization —a wasted effort if the egg wasn't fertilized, or if the embryo died.

The Georgia team speculates that once plants evolved the ability to create a food supply at the time of fertilization, plants were able to devote energy to evolving more complex life cycles, including flowering.

In another look into the past, archaeologists from Southern Methodist University in Dallas, Texas, have found evidence suggesting that crops may have been domesticated in Africa as far back as 8,000 years ago. This challenges the long-accepted theory that African agriculture began in about 4400 B.C. with the domestication of wheat and barley in northern Egypt.

While exploring an early-Neolithic site in southern Egypt near the Sudanese border, archaeologist Fred Wendorf discovered thousands of seeds of more than 40 different plant foods—including grains, mustards, and legumes—near shallow cooking holes in four houses. The seeds, which carbon dating indicates are about 8,000 years old, show that the prehistoric inhabitants harvested great quantities of many wild plants. Harvesting, Wendorf says, is the first step toward planting and protecting plants, which leads to the genetic changes that result from domestication.

Wendorf suspects that one of the grains found at the site—sorghum—may in fact have been domesticated. Infrared spectrography shows that the fat composition of the sorghum harvested at the site more closely resembled that of today's domesticated sorghums than of existing wild sorghums.

UPDATE: FINDING ANOTHER YEW

Cancer researchers have found a substance that may reduce the demand for taxol, the cancer drug derived from the rare Pacific yew *(Taxus brevifolia).*

A team at the University of Texas M. D. Anderson Cancer Center in Houston, Texas, is testing taxotere, which is derived from the English yew *(T. baccata),* for its effectiveness in treating some types of cancer. A small study found an antitumor effect in 6 out of 10 women with ovarian cancer.

This research not only shows promise for potential cancer treatment, but it may provide a solution to the controversial harvesting of the endangered Pacific yew for taxol. Because the English yew is far more common, taxotere should be more readily available if it does prove to be an effective cancer treatment.

Erin Hynes

Endangered Species

Controversy, hostility, standoffs, confusion. A whirlwind of ill ease surrounds the Endangered Species Act (ESA), which is due for congressional reauthorization this year. The focus of the furor is the battle to balance economic growth and jobs with the rights of endangered wildlife to survive.

Caught in the midst of this complex debate, perhaps more than any other creature, is the threatened northern spotted owl, found in the old-growth forests of the Pacific Northwest. The owl's fate pits loggers against conservationists in an emotional fight for dominance now locked in a legal stalemate.

The northern spotted owl's case is not unique. A plan to reintroduce the gray wolf into Yellowstone Park has galvanized ranchers who fear livestock losses. Loggers are up in arms over timber-cutting restrictions geared to protect the red-cockaded woodpecker's habitat in Florida's panhandle. And most recently, a songbird called the California gnatcatcher, which may soon be listed as endangered or threatened, has snagged developers' plans to build on prime Southern California real estate.

The problems loom so large that there's growing momentum for a complete revision of the ESA. Despite all these obstacles, there have been notable strides made in the recovery of several species.

WHOOPING CRANE

The endangered whooping crane continues to slowly rebound, and, with establishment of a wild flock in Florida in 1993, the outlook for its long-term survival seems a bit brighter.

This elegant, long-legged bird with snowy-white plumage and black-tipped wings never existed in large numbers. Estimates in the late 1800s range from 500 to 1,400 birds. They historically nested from Illinois to southern Canada, and wintered in areas from the Carolinas to Mexico. As agriculture and other human intrusion swallowed up more and more of their habitat, their numbers plummeted. By 1941 the one remaining flock had fewer than 20 birds.

Over the years, various conservation measures have aided this showy bird, perhaps starting with the creation in 1937 of the Aransas National Wildlife Refuge in Texas. This site is where the only self-sustaining natural wild population winters. Other actions include: A captive-breeding program begun in 1967; public education about not disturbing this shy bird in the marshes and waterways it inhabits; further habitat protection; radio tracking whoopers' movements; and marking power lines along migration routes to prevent collisions.

To date, 240 whooping cranes exist in three wild populations and five captive locations. The Aransas flock has 136 birds, a small group reared by wild sandhill cranes in the Rocky Mountains has eight whoopers, and nine birds form a new flock released in January 1993 in the Kissimmee Prairie of Florida. If all goes well in Florida, more whoopers will be released there in late 1994.

Recovery leader Jim Lewis of the Fish and Wildlife Service in Albuquerque, New Mexico, feels optimistic about the continued recovery of the Aransas flock and the ability to establish two more self-sustaining flocks. But he warns, "We think it will be the year 2020 before we would have those three populations well established and can downlist them from the endangered category to threatened."

GRIZZLY BEARS

Gory stories abound about the powerful grizzly bear, but for all its prowess, the grizzly's survival is not assured. The threatened bear now occupies less than 2 percent of its historic territory in the United States, yet recently its numbers appear to be stabilizing, and may be increasing in one area where it lives.

The grizzly once roamed from the mid-Plains to the California coast and south into Texas and Mexico. Between 1800 and 1975, its numbers dropped from 50,000 to less than 1,000. This decline came from habitat loss, hunting, trapping, and protection of human life. Grizzlies inspired fear and were killed on a wide-

Habitat loss and hunting have combined to greatly reduce the population of the grizzly bear in the United States. Efforts are under way to improve the grizzly's threatened status by creating areas of safe habitats.

spread basis. Now an estimated 800 to 1,000 grizzlies live in six or perhaps seven wilderness areas, mountainous regions, and national parks in Washington, Idaho, Montana, and Wyoming.

Efforts go on to ensure that the bear has adequate habitat, the most crucial element in grizzly recovery. Grizzlies' home ranges can cover as much as 1,000 to 1,500 square miles (2,600 to 3,900 square kilometers). In places where the bear doesn't have enough undisturbed land, it comes into contact with people, and problems arise.

Concern over fragmentation of natural habitat and the lack of genetic diversity remaining in small island populations prompted a five-year study now under way. The study will determine the viability of linking with one another the separate areas the grizzly occupies.

Even if areas can be joined together, the grizzly recovery is "going to be a struggle," says Ann Vandehey, a wildlife biologist with the Fish and Wildlife Service. "If you could stop time right now, I think they'd have a very good chance," she adds. But continued development brings added pressure on the bear's remaining ecosystems. Only time will tell if people will leave room for the grizzly.

CALIFORNIA CONDOR

California condors have soared the skies since prehistoric times, but modern humans brought them to the brink of extinction. Recently it appears that humans may have taken the bird out of immediate danger as well.

California condors, this country's largest bird, weighing as much as 25 pounds (11 kilograms), with a 9-foot (2.7-meter) wingspan, once numbered in the thousands. These vultures ranged all along the coast from British Columbia to Baja California, west to Florida and north to New York. Loss of habitat, lead poisoning from eating carrion killed with lead shot, and falling prey to hunters and collectors dropped their numbers to a perilous low. By 1982 only an estimated 21 to 24 condors remained in the wild. A con-

troversial decision was made to remove all the condors from the wild and to breed them in captivity. The last condor was captured in the wild in 1987.

A year later the first captive-bred chick hatched. The breeding program at the Los Angeles Zoo and the San Diego Wild Animal Park "has been successful beyond anybody's wildest dreams," says Marc Weitzel, project leader. There are now 64 birds in captivity, and 12 eggs ready to hatch.

The California condor has been saved from the brink of extinction by a novel zoo breeding program. Some have already been released into the wild (above).

The breeding program has gone so well that in January 1992, two young condors were released on cliffs high in Los Padres National Forest. Sadly, one of the pair died in October of that year from kidney failure after eating antifreeze left alongside a road. But on December 1, 1992, six more condors got their taste of freedom at that same site, and are adapting wonderfully.

The ultimate goal is to create three separate populations, one in Los Padres National Forest, one in the zoo (as a buffer against natural disasters taking the wild birds), and one perhaps in the Grand Canyon. "I'm very optimistic that the recovery efforts will be successful," says Weitzel. "This shows that man's intervention can have a positive impact, too."

Linda J. Brown

Zoology

SEA-LION ESPIONAGE
The U.S. military has long employed marine mammals for reconnaissance missions with military instruments and weapons on the ocean floor. Now scientists are recruiting sea lions for a spying

mission that should reveal previously top-secret information about the underwater antics of other marine mammals, as well as about sea lions themselves. Researchers at the Long Marine Laboratory at the University of California at Santa Cruz are training sea lions to carry sophisticated video equipment on their backs to take underwater movies of whales, dolphins, and other marine species that typically defy careful study.

By teaching sea lions to respond to complex diving and swimming instructions, scientists hope to direct them to perform such specific duties as swimming beside or under a whale, or swimming beside a dolphin's eye or mouth, all the while shooting video footage. Whereas whales or dolphins behave differently in the presence of human divers or submersible vehicles, they are quite accustomed to the presence of sea lions. So their secretive underwater behavior, such as nursing calves or mating, can be observed unobtrusively via the video footage taken by the spying sea lion.

An added advantage to the espionage training of sea lions is that the amiable creatures can then be asked to simulate their natural diving and foraging patterns in the ocean while scientists precisely track the metabolic processes that occur in the animal's body from one dive to the next. Scientists hope to use this information to learn about the remarkable adaptations displayed by marine mammals that, among other advantages, permit them to avoid decompression sickness (also called the bends)—a painful, life-threatening condition that humans develop when they ascend too quickly from an underwater dive. In addition, scientists hope to gain insights into how marine animals are affected by weather patterns, human fishing practices, and other events.

PLAYING AS TRAINING FOR ADULTHOOD
Scientists have often observed that young animals expend an inordinate amount of energy on simply playing. This seemingly random, purposeless activity—which may include giddy leaping, jousting,

pouncing, or chasing tails—also frequently places the youthful animal at risk, as by falling during dangerous maneuvers, or through unnecessary exposure to predators. Scientists are now beginning to suspect that this rigorous, playful behavior is more than just the reckless excesses of youth, and actually may have serious purpose.

Recent studies have found that young animals are at their most playful at the point when their brain cells are developing synaptic connections to transmit electrochemical messages from one area of the brain to the next. Scientists theorize that the sensory and physical stimulation that accompanies play activity is necessary for the growth of these brain synapses and the development of motor coordination. Furthermore, playful movements help muscle tissue mature to produce the proper combination of fast-twitch muscle fibers needed for quick movement, and slow-twitch fibers necessary for aerobic activity.

In addition to the physiological reasons for play, there is a definite behavioral aspect that cannot be discounted. Scientists have observed that animal play is a rehearsal for many adult behaviors, including stalking prey, mating, and socializing. For instance, young cats, wolves, and hyenas pretend to capture prey by stalking, pouncing, biting, and swiping at objects with claws extended. Gordon M. Burghardt, Ph.D., of the University of Tennessee in Knoxville has observed that hatchling sea turtles play at shaking a front foot in a playmate's face, a gesture used years later during courtship.

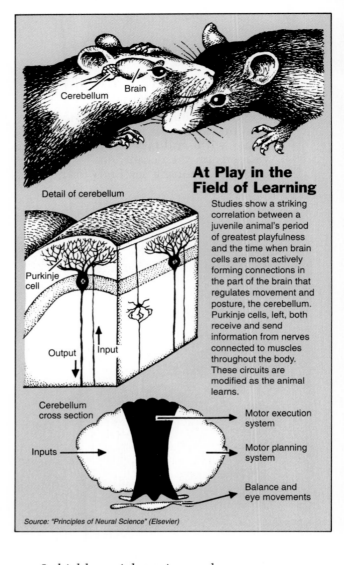

Detail of cerebellum

Purkinje cell

Output Input

Cerebellum cross section

Inputs

Motor execution system

Motor planning system

Balance and eye movements

Source: "Principles of Neural Science" (Elsevier)

At Play in the Field of Learning

Studies show a striking correlation between a juvenile animal's period of greatest playfulness and the time when brain cells are most actively forming connections in the part of the brain that regulates movement and posture, the cerebellum. Purkinje cells, left, both receive and send information from nerves connected to muscles throughout the body. These circuits are modified as the animal learns.

In highly social species, such as monkeys and even humans, playing helps young animals to curb their selfish or ag-

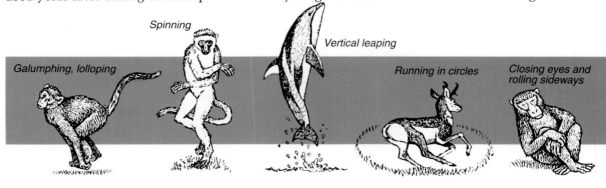

Galumphing, lolloping   *Spinning*   *Vertical leaping*   *Running in circles*   *Closing eyes and rolling sideways*

gressive behaviors to better fit in with a troop as an adult. In an interview with *The New York Times*, Stephen J. Suomi, chief of the laboratory of comparative ethology at the National Institute of Child Health and Human Development, observed, "Through play bouts, an animal's aggressive tendencies are socialized and brought under control. The animal learns when to submit and when to pursue, and it will learn how to fight gracefully."

CANNIBALISM MAKES GOOD SENSE

Scientists used to believe that cannibalism, the eating of an individual's own species, was a behavioral aberration forced upon animals that were overcrowded or starving. But new insights discussed in *Cannibalism, Ecology and Evolution Among Diverse Taxa* (Oxford University Press, 1992) suggest that the reasons for cannibalism are more complex. The behavior is actually a shrewd survival mechanism that is so advantageous, scientists are puzzled why they don't observe it more often in the animal kingdom.

Perhaps the most basic advantage of cannibalism is simply good nutrition. Studies with mosquito fish and tiger salamanders show that species members may make the most nutritionally balanced meals. Cannibalism also produces a balance between food supply and population. The notorious undertakers for the insect world, burying beetles, for example, will fight for control of the corpse of a bird or mouse. The war ends when one male and one female are left to claim the corpse, which they prepare for eating and then bury. The victorious couple mate, and when new beetles are hatched, the parents begin eating the brood until the family is pared down to a size that will be supported by the available food supply. Experts theorize that the beetles are improving the chances that the remaining young will grow into healthy adults.

Another sort of "head-start" program seen in the animal kingdom occurs when animals stock the nest or womb with extra offspring. This guarantees that the strongest and fastest-growing of the young will have a life-supporting food supply and will grow up strong, healthy, and better adapted for survival. Embryonic sharks, for instance, feed on their own siblings while in the womb. A Midas-cichlid-fish family will also breakfast on each other when young.

Perhaps the most lurid example of cannibalism occurs during mating. Praying-mantis females have been observed noshing on their partners even before mating is completed. Some animals have developed elaborate tricks to avoid becoming a postmating tidbit. The male orb-weaver spider, for instance, tries to mate with females immediately after they've molted, and are therefore too limp to instigate a deadly attack.

But there are some animals that virtually jump at the opportunity to be eaten alive. During mating, Australia's male redback spider will leap into the jaws of his mate. Some researchers propose that the spider is gallantly offering himself to nourish the mother of his future offspring.

Lisa Holland

Chasing tail

Bucking

Somersaulting

Pirouetting

Hanging upside down

TECHNOLOGY

CONTENTS

RAPID RAILS

by Liz Crutcher

I n New York City's cavernous Penn Station, passengers mill around a large board suspended from the ceiling. With a noisy click-click-click, the board flips up the departure track for the Metroliner from New York to Washington, D.C., and the crowd rushes to the gate. Travelers funnel onto the platform and with controlled frenzy select a car. Greeted by plush seats and a spacious interior almost befitting the Concorde, the passengers take their seats and discover other creature comforts: glass-enclosed conference areas, seatside stereo jacks, cocktail service, and other luxuries that promise to transform the 224-mile (360-kilometer) journey into a first-class trip.

Ironically, the luxury of the European-style compartments steals the travelers' attention away from the more important aspects of the journey. Indeed, with its tilting cars, pivoting axles, and on-board computers, this train —a Swedish import on loan to Amtrak—represents a 40-year leap in conventional train technology for the United States. Remarkably, a trip on the Swedish "tilt train" bears witness to just a small part of the technology that promises to revolutionize the U.S. rail system —and perhaps someday renew America's sense of romance about train travel.

From the ordinary commuter trains of New York City to the flying trains of Florida, Americans have begun to get a taste of the

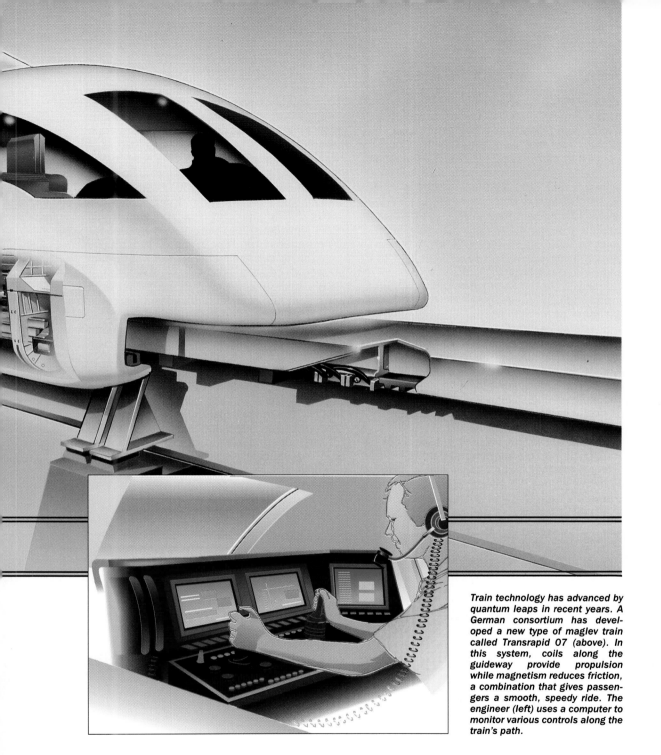

Train technology has advanced by quantum leaps in recent years. A German consortium has developed a new type of maglev train called Transrapid 07 (above). In this system, coils along the guideway provide propulsion while magnetism reduces friction, a combination that gives passengers a smooth, speedy ride. The engineer (left) uses a computer to monitor various controls along the train's path.

efficient, powerful rail service that has long been commonplace overseas. Overshadowed by the speed of airlines and the convenience of automobiles, railroads in the United States fell into neglect after World War II. Fifty years later, air pollution, traffic congestion, and the increasing hassles of airline travel have taken their toll. Finally investment in new transit systems has become an attractive option for federal, state, and local governments.

Move with the Groove

Today Amtrak is trying to strip away those World War II cobwebs by testing out the X2000, just one in a family of high-speed trains that can move 30 percent faster than the speediest Metroliner, currently Amtrak's fastest train in the Northeast Corridor. Built by a Swedish-Swiss joint venture, Asea Brown Boveri Company (ABB), the train has cut the grueling 4 1/2-hour trip from Stockholm to Göteborg, Sweden, down to 3 hours by achieving speeds up to 125 miles (201 kilometers) per hour.

In the United States, the Swedish bullet train winds up and down the Northeast Corridor route at an astonishing 135 miles (217 kilometers) per hour. Amtrak's fastest Metroliner travels at 125 miles (201 kilometers) per hour, but only on the straightaways. "A regular train could go faster around curves, but you wouldn't want to because you'd be knocking people over and at least spilling drinks," explains Amtrak project engineer Ed Lombardi.

What allows the X2000 to take a curve at 100 miles (161 kilometers) per hour is an axle system that lets a train's wheels "move with the groove." Called radial steering, it allows the train to take curves safely at 30 to 50 percent faster than do conventional trains, whose axles keep the wheels straight, producing more friction and force as a train rounds a curve. The radial steering feature of the independent axles allows the suspension of the X2000 train to flex and follow the curve of the tracks and drive right through turns. "There are enough of these curves between Washington and New York that if you don't slow down for them, it can save about 20 minutes in trip time," says Lombardi.

Taking corners faster is great, but doing so would mean a pretty jerky ride without ABB's hydraulic tilting system. "The train senses immediately that it's entering a curve and, through the software, the master computer, and the individual computers on each coach, tells the tilting system how to operate," says Joe Silien of ABB.

Passengers may not actually feel the floor rise on one side as the train approaches a curve. Looking out the window, the horizon tilts as you round the curve. But only 70 percent of the force is compensated by the tilt system. If the train adjusted 100 percent, passengers would experience a train-sickness of sorts, because the difference between what the eye sees and what the body feels causes sensory confusion.

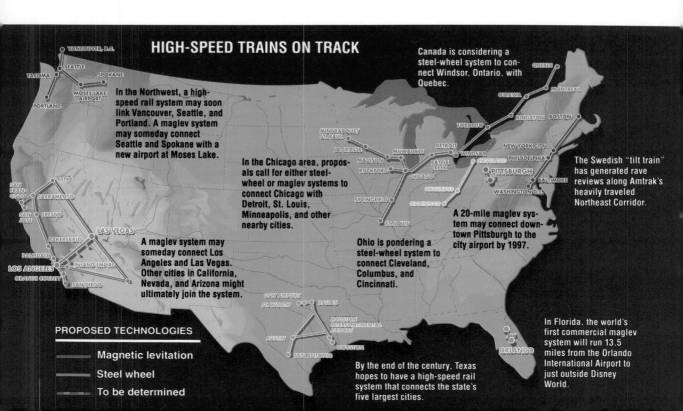

HIGH-SPEED TRAINS ON TRACK

Canada is considering a steel-wheel system to connect Windsor, Ontario, with Quebec.

In the Northwest, a high-speed rail system may soon link Vancouver, Seattle, and Portland. A maglev system may someday connect Seattle and Spokane with a new airport at Moses Lake.

In the Chicago area, proposals call for either steel-wheel or maglev systems to connect Chicago with Detroit, St. Louis, Minneapolis, and other nearby cities.

A maglev system may someday connect Los Angeles and Las Vegas. Other cities in California, Nevada, and Arizona might ultimately join the system.

Ohio is pondering a steel-wheel system to connect Cleveland, Columbus, and Cincinnati.

A 20-mile maglev system may connect downtown Pittsburgh to the city airport by 1997.

The Swedish "tilt train" has generated rave reviews along Amtrak's heavily traveled Northeast Corridor.

In Florida, the world's first commercial maglev system will run 13.5 miles from the Orlando International Airport to just outside Disney World.

By the end of the century, Texas hopes to have a high-speed rail system that connects the state's five largest cities.

PROPOSED TECHNOLOGIES

— Magnetic levitation
— Steel wheel
— To be determined

The smoother, more powerful acceleration for those fast curves comes from the four axle-mounted motors' alternating-current (AC) propulsion. Most trains in the United States run on DC, or direct current. Because DC motors use brushes, they need frequent inspection and maintenance. Engineers have to remove the motors and clean the copper relays that make up the brushes.

AC propulsion "improves on DC motors by eliminating many of the moving parts," says Silien. If you compare AC to advances in automobile engines, it holds all the advantages of fuel injection over carburetors, essentially by eliminating some very complicated machinery. "Over time, more railway and transit cars will have AC propulsion," Silien predicts.

AC traction motors and the speed and comfort produced by the axles and tilting mechanism are what makes the X2000 such an advanced train. "Putting together the AC technology, the radial steering, and a reliable active tilting system is unique to the X2000," explains Silien.

The X2000's AC system is controlled by an onboard master computer. "When the engineer operates the throttle, he's really operating the computer that tells the traction system to go faster or slow down." This electronic center also works some of the simpler creature comforts of the train, like air-conditioning and door operation. Surprisingly, "most of the trains we ride in the U.S. today don't have any computers on board," says Silien.

The impact of all these technical marvels is lost on most passengers until they reach their destination 20 minutes sooner. "With our fastest Metroliner, we can run a 2-hour, 40-minute schedule between Washington, D.C., and New York. Now, with our X2000, we can run a 2-hour, 19-minute schedule," explains Amtrak's Lombardi.

But the X2000 is just one option Amtrak is testing to speed patrons between cities. The carrier will also test out the latest German high-speed-train technology with an Intercity Express, or ICE train. With the power of two locomotives and 13,000 horse-

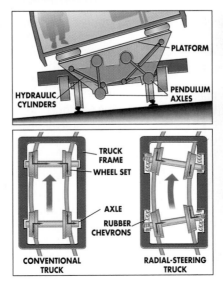

Amtrak's "tilt train" (above) has been a hit in the Northeast Corridor. On conventional trains, steel wheels are mounted rigidly (1A) and remain parallel, even on curves. The train uses a self-steering assembly (1B) that enables the axles to follow curves instead of remaining rigid. To avoid passenger discomfort when the train rounds a curve, sensors "tilt" the train a maximum of 8 degrees (B). Pneumatic suspension (C) is another way to make a train ride smoother.

PLATFORM
HYDRAULIC CYLINDERS
PENDULUM AXLES

TRUCK FRAME
WHEEL SET
AXLE
RUBBER CHEVRONS
CONVENTIONAL TRUCK
RADIAL-STEERING TRUCK

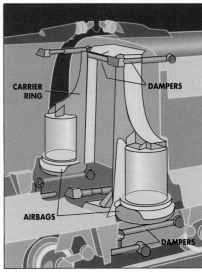

CARRIER RING
DAMPERS
AIRBAGS
DAMPERS

power, the ICE uses what Lombardi terms the "brute force" approach. "It slows down for curves, but then takes off like a shot," he states. The train also has a lower center of gravity and uses lighter-weight materials than do conventional trains.

In Europe and Japan, however, these brutes don't have to contend with the meandering curves of U.S. railways. Tracks laid down in a straight line from city to city allow trains like the ICE and the French TGV version, or *Trains à Grande Vitesse,* to accelerate up to 186 miles (300 kilometers) per hour in regular passenger service. On some test runs, the trains have even exceeded 320 miles (515 kilometers) per hour.

Such a route would be impractical for the Northeast Corridor, where the railroads have to use existing tracks. But officials in Texas, Pennsylvania, and other states are looking into this straight-shot approach.

In the end, Amtrak may order a fleet that uses a combination of the two approaches. Whatever is decided, the goal is the same. "We are aggressively trying to get those people out of the Delta shuttle," says Lombardi.

Techno-perks for Commuters

While Amtrak tries to lure passengers from the sky, the transit authority in Los Angeles hopes to coax commuters out of their cars with a brand-new system. The newest section of the network is a 4.4-mile (7-kilometer) Red Line route that will form the backbone of a planned 23-mile (37-kilometer) subway system for Los Angeles County. Far from the squeal and jolt so familiar to riders of older subways, these sleek new trains have adopted many of the techno-perks that high-speed passenger trains have enjoyed for years.

Using air-bag suspension, regenerative brakes, and microprocessor-controlled braking and propulsion, the trains are a dream of efficiency and reliability to rail managers, and a smooth ride for passengers.

When a subway arrives at a station and passengers wait for the doors to open, a noise akin to that of a balloon inflating is heard. That puff sound is the air-bag suspension. Like an automatic leveling device, it inflates the cars to compensate for a shift in weight to one side of the train. Before this type of automatic-leveling system, the train would actually be below the platform, and passengers would have to step up.

Computerized braking on the Red Line is controlled by a microprocessor that senses all the inputs from the train. "The speed of every axle, the weight of the train, every function that is going on in the train has an input to this microprocessor," explains Dave Kalasnik, a rail vehicle maintenance manager at Southern California's Rapid Transit District. This microprocessor improves upon the older systems, in which the braking response was slower and not as accurate.

To keep the trains running smoothly and to save wear and tear on the wheels, the subway also has a state-of-the-art lubricating device that even many freight and passenger cars lack. In the old days, as a train traveled around a curve or through a switch, the train wheels would trip an arm, and an oil pump would shoot lubricant at the wheels and axles—a messy, inefficient procedure. In the new subways, the wheel-flange lubricator travels on the train and has the consistency of a candle. When it senses a change in heat or pressure, it feeds the lubricant directly onto the wheel.

Such automation means that the modern subway will require little more than an attendant to open and close the doors at the station. "Leave the driving to us," is the motto of a central control station, which sends commands to the track. Instructions are then picked up by the train with a small antenna. "The computer will do all the commands that an operator normally would do," explains Kalasnik.

While the Red Line trains have an operator to work the doors and make announcements, the new Green Line will eventually be fully automated. Scheduled to open in 1994, the light-rail line will run about 20 miles (32 kilometers) east-west across Los Angeles. The person in the cab is on board only to operate the train manually if a system fails. Ultimately, among the Blue, Green, and Red Line trains, Los Angeles County plans to lay down some 300 miles (480 kilometers) of track.

Metro-North, a major commuter rail service in the New York area, is developing a "Super Tower" control room (above) to better monitor its trains. The commuter trains in Los Angeles County (left) have received rave reviews from passengers and engineers alike.

New York's Challenge

As Los Angeles builds up an all-new train network around its sprawling metropolis, New York City tries to grapple with a crumbling infrastructure that has been all but forgotten since World War II. Adding together subway, intercity service, and commuter lines, about 6 million riders step on board trains for transportation to or from Manhat- tan. About 200,000 of these train travelers enter the Big Apple via the Metro-North commuter lines from upstate New York and Connecticut. Once considered the "nursing home" of the freight industry, these commuter lines have been busy revamping their operations since they came under the domain of the transit authority in 1983.

Commuting to Manhattan used to be a grueling trip exacerbated by broken-down trains and interrupted service. Incredibly, in just the past 10 years, the incremental improvements have produced a commute that can almost be described as pleasant. "It's better than a heart transplant," claims Don Muyskens, assistant chief mechanical officer at the commuter agency. Muyskens can claim first-hand knowledge of those improvements: He rides from upstate New York every day into Metro-North's headquarters at Grand Central Terminal.

Muyskens's latest project is the introduction of AC propulsion to dual-mode locomotives—locomotives that work on both diesel and electric power. For the passenger, such a system translates into smoother acceleration and braking. For the transit authority, it translates into increased reliability and reduced maintenance.

ABB Traction of Elmira, New York, has been working with Metro-North to rebuild its trains from the inside out. "We're putting brand new technology into 40-year-old locomotives and it's been a challenge," says ABB Traction's Karen Breinlinger. "Every component except the car body and trucks is being replaced, and Metro-North has essentially new locomotives with state-of-the-art drive systems. They are the first locomotives in North America to be fully dual-powered *and* utilize AC traction," claims Breinlinger.

Directing Traffic

More-efficient engines would mean little without the parallel transformation in traffic control, such as the one taking place for getting trains into Manhattan.

For Metro-North, the huge merge of trains that takes place during the morning rush hour is like musical chairs on a grand scale. When the music stops, however, there are only four tracks to accommodate the more than 120 trains that converge on Grand Central between 6:00 and 10:00 A.M.

Directing traffic for the 4 miles (6.4 kilometers) of track that connect the main lines at the terminal works much like it has since 1913: a train director, receiving information from an open telephone, shouts instructions to operators, who then throw switches to align tracks for incoming and outgoing trains.

Such manually operated traffic control will be replaced by the "Super Tower"—a computerized mission-control center. As train dispatchers follow train movements indicated by blinking lights along a giant board, they will relay directions to train engineers using a computer keyboard that triggers green, yellow, or combination lights to appear on the console of the engineer's train. Traffic safety will be maximized by the use of these onboard controls and by an elaborate system of wayside signals, which work much like traffic lights on roads.

Futuristic Mass Transit

As the United States enters the 21st century, it is only natural that futuristic trains with magnetic-levitation technology will roll onto the scene. It's no surprise that these trains will make their debut at the headquarters of fantasy in the United States: Orlando, Florida, where millions of tourists each year will be transported by magnetic levitation, or maglev, trains from the city's airport to a tourist center near Disney World.

Even though the idea of a flying train was first conceived by an American researcher while caught in a traffic jam, it is Germany and Japan who now lead in maglev technology. This technology uses the push-pull characteristics of magnets to create a magnetic field that acts as a cushion of air. The German Transrapid consortium and the Japanese High Speed Surface Transport (HSST) system use electromagnetic attraction. Magnets in the wings of the vehicle hug an elevated track. The draw of powerful magnets in the stationary guideway elevates the train by about three-eighths of an inch (9.5 millimeters).

Repulsion technology, designed by the Japanese National Railways, works on the principle that magnets with different polarity resist coming together. As superconducting magnets line a guideway, they push on magnets embedded in the train just enough to raise the vehicle about 6 inches (15.24 centimeters) off the guideway. Because the train and guideway don't touch, there is little more than air molecules to get in the way.

While the designs differ in their approach, the effect of eliminating the steel-wheel-on-steel-rail friction is the same—no clickety-clack of the wheels, only an aerodynamic whoosh. Because the train is powered by a magnetic field, there is no engine. The train is so quiet that it can even streak through a pasture of cows unnoticed!

After intense competition, Maglev Inc., the private interest heading up Florida's Disney-Orlando link, chose Transrapid to build a 250-mile (400-kilometer)-per-hour maglev using magnetic-attraction technology. Scheduled to open in 1996, the Florida project will be only the first of numerous high-speed links that transportation planners are cooking up around the nation. Ohio, Texas, and Pennsylvania, just to name a few, are making tracks to build links that will bring German, Swedish, French, or Japanese ingenuity to hundreds of miles of track.

FIGHTING TODAY'S FIRES

by James R. Chiles

Firefighting was long an occupation bound by tradition, and fearless to the point of recklessness. For the nation's 1.6 million modern firefighters, however, the fire service now shows a keen interest in keeping up with the dangers of technology when it catches fire, explodes, or collapses.

The list of dangers keeps getting longer. A candle-size flame catching hold in synthetic upholstery can turn a room into an inferno of flame and poisonous gas in three to five minutes. A formula resembling rocket fuel has been concocted by a small band of arsonists in the Pacific Northwest: it can ignite unstoppable building fires hot enough to vaporize steel. Hoppers, pipes, and tanks in sprawling industrial complexes can become traps for workers and the firefighters who try to rescue them. New, lightweight, truss-type roofs and floors make homes affordable, but also crumple quickly in a fire, risking the lives of residents, let alone the firefighters.

I interviewed fire officials around the country, and spent time with departments in Los Angeles and Colorado to see what the service is doing to stay ahead of the fire king. All fire departments are making do with less money, even as demands increase. Fire Chief Richard Marinucci of Farmington Hills, Michigan, says that some departments send out engines with only two firefighters aboard. "Our busiest engine company had 300 calls a year 25 years ago," says Seattle Deputy Fire Chief Stewart Rose. "Now it's up to 3,000!"

In a typical day, a firefighter might battle a roaring house fire (top) or rescue a trapped child. Modern training techniques and high-tech equipment ensure that today's firefighters carry out their duty in the safest and most efficient way possible.

The Oakland holocaust of October 1991, at a cost of 25 lives and $1.6 billion in damage, is a reminder of the terrible havoc that a single fire can cause. Nationally, fires still claim more than 5,500 Americans a year. It's a fearsome toll, but compare it with 1971, when nearly twice that number died. The reasons for the drop are better prevention and containment, and improved firefighting and rescue techniques.

Prevention

Begin with the all-important prevention side. It's lunchtime, and I'm on the 42nd floor of the Southern California Gas Company Tower, a new skyscraper in downtown Los Angeles. Followed by eight representatives of the builder and manager, city fire inspector Jesse Franco takes a whirlwind tour of the upper floors where construction has just been completed. Franco has the magnetic pull of a celebrity, because a bad report from him can derail any fast-track occupancy schedule. No new commercial building in Los Angeles higher than 75 feet (23 meters) opens without a say-so from the fire department's High-Rise Unit. And once open, every tower will receive an annual follow-up visit from the unit, which inspects more than 700 buildings every year.

Franco nods toward a corner at the end of a hallway. A technician uses a rod to trigger a smoke detector in the ceiling; seconds later a warning signal sounds through recessed speakers, and a steel shutter slams down, closing off a utility area. Franco nods, checks off an item, and leads his entourage off in a rush.

Without fire codes, the potential for high-rise disaster in the nation's cities would be awesome. If stairwells were not designed to exclude smoke and flame, a fire on the lower floors of a big skyscraper could trap thousands of office workers above. The city's rules now call for sprinklers, water outlets, and smoke alarms on every floor; fire-rated doors and walls; helicopter landing pads on the roof; insulation sprayed on every exposed steel beam; and hundreds of other details.

I notice that every fifth window in the Gas Company Tower has a little decal marking it as "tempered," meaning that when broken, it will crumble into tiny pieces. These windows would have been helpful during the 1988 First Interstate Bank fire in Los Angeles, among others. The bank tower's high-strength windows, when smashed, broke into long shards that sailed 100 feet (30 meters) or more from the building, cutting hoses and threatening firefighters like flying guillotine blades. One shard punched

Like all skyscraper fires, the 1988 blaze in a Los Angeles high-rise (right) presented many obstacles to firefighters. After the World Trade Center bombing in February 1993 (facing page), thousands of office workers had to evacuate the 110-story towers via smoke-filled stairwells.

through the roof of a car. In the building that is being inspected today, firefighters would know which windows they could break safely to let smoke out.

All of these precautions in skyscrapers are useful and even vital, but if they had to, firefighters would trade away every requirement but one: a good sprinkler system. When a fire takes hold in a skyscraper without sprinklers, says Kevin Mellott of the fire-safety consulting firm Erase Enterprises, "you can just kiss it good-bye. A big high-rise fire is capable of beating the [expletive] out of 1,000 firefighters and coming back for more." An all-out effort by the Philadelphia Fire Department couldn't stop the One Meridian Plaza fire of 1991, which burned out eight floors and killed three firefighters, but sprinklers on the 30th floor did.

"Ninety-six percent of the time, when the sprinkler system was operational in a modern building," Mellott adds, "the fire was extinguished or held in check until the fire department got there." Many communities require sprinklers in all new commercial and apartment buildings, but few require them in new houses. Los Angeles County requires sprinklers in all rural housing not connected to water mains, at a minimum cost of about $2,000. Firefighters credit smoke alarms and sprinklers for the drop in fire deaths. "Prevention is gaining momentum," Mellott says. "Chiefs no longer have the budget for massive firepower."

In the new era of fire and disaster services, special problems and solutions are found on the waterfront. One of the earliest hazardous materials was something we all take for granted: creosote-treated wood. Firefighters at wharf fires complained that smoke from burning creosoted pilings caused lung problems and severe skin irritation. The fires usually lay hidden under pavement, and firefighters spent hours blundering through the choking fumes, chopping random holes until they located the fire. In

the early 1960s, departments began using scuba divers equipped with floating nozzles supplied by hose from a fireboat. Using the back-thrust from the nozzles, the divers steered themselves deep into the wharves, between the pilings, then doused the fires from underneath.

Though some such wharves remain, they're diminishing as a fire hazard. "Our tactics are continually changing along with technology," Los Angeles city pilot Mike Corcoran tells me aboard Fireboat 2. "We've tended away from wharf attacks and more toward shipboard firefighting.'' Typical of marine fires today is a smoky, concealed blaze on board a containership; firefighters go aboard and hunt among thousands of 40-foot containers for the offending box, and douse the fire with carbon dioxide or water. Infrared detectors aid the search.

Corcoran's red-and-white fireboat—though built in 1925 and designed to handle huge wharf, petroleum, and shipboard fires—has been adapted to meet new firefighting demands. The department added a hydraulic lift arm with a basket on the end, allowing the boat to put firefighters on ships with high decks. But its original bronze pumps can still pump 18,000 gallons (68,000 liters) of seawater a minute, and launch it far enough to reach a ship 500 feet (150 meters) away.

Fireboat 2 is perhaps an extreme example of longevity, but all fire departments plan on long lives for their machinery. In Los Angeles County, a fire engine bought today is likely to stay in daily service through 2012, and then will go on the reserve list through 2017.

Hook and Ladders

The basic pumper is still a good investment because large quantities of water remain a cheap and highly effective tool. Chemicals have earned a place, but are too expensive or problematic to replace water for regular use against building fires. For example, chemicals in the halon family snuff fires in electronic equipment, but they are expensive, and their use contributes to damage of the Earth's stratospheric ozone layer.

During the 1988 Yellowstone National Park fire, firefighters saved many buildings with a new type of water-based foam employing an additive similar to liquid soap. Los Angeles County fire stations began experimenting with a similar foam two years later, and found it so effective that the county plans to equip all pumpers with it.

To get a feeling for how the classic hook-and-ladder station is adapting, late one evening I visit city-operated Fire Station 11, near Los Angeles's MacArthur Park. With more than 20,000 calls in 1990, Fire Station 11 is one of the nation's busiest. Its calls have been driven up by a dense and

Most major ports maintain a small fleet of fireboats to help contain fires on oil tankers (below) and other ships. The newest fireboats are completely computerized and have a host of high-tech features.

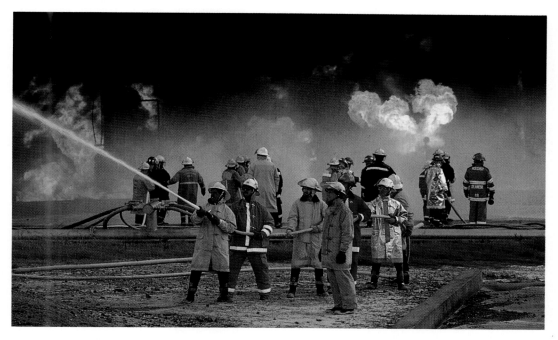

Many modern firefighters learn their trade at special academies. Working in small groups (above), future firefighters *practice the standard containment procedures used to battle a chemical fire.*

transient population, drug use, and gang activity. There are so many rescue calls that fire is almost an afterthought. "We only spend 3 to 5 percent of our time at fires," firefighter Rick Erquiaga tells me. The rest of the time is consumed by assists for medical calls, fire inspections, training sessions, and station and equipment maintenance. An alarm sounds as we talk, and two paramedics prepare to answer the medical call. They don dark-blue vests before adding their bright-yellow rescue coats. Bulletproof vests, says Erquiaga. "We always wear them on family disputes, assaults, and shootings." I am told that gangs usually hold their gunfire when fire engines and paramedics arrive, though occasionally, while paramedics are still working, gunmen will circle back to ensure that their victim has died.

Along with new problems, the station has new solutions. A computer display mounted in the fire engine's cab shows the driver the truck's exact position and direction on a detailed street map, along with all the water hydrants. In the near future, automatic radio links will transmit the truck's location to the city's new communications center. That way a dispatching computer will know the unit nearest to the address of a 911 call, and be able to shave seconds off the response time—seconds that might well mean someone's life. The same radio network already can save a firefighter's life: if trapped in a building, he or she can press a red emergency button on a personal radio set; this triggers an alarm at headquarters, where a computer tells dispatchers the identity of the firefighter whose radio sent the call for help. Dispatchers then relay the emergency to an officer at the scene. The station's firefighters also carry a small orange device that blares if it remains stationary for a half-minute—usually meaning the wearer has been overcome or injured and needs immediate help. Erquiaga activates one and sets it on a ledge; 30 seconds later, it squawks for help.

At the Colorado Fire Fighters Academy, a training conference held annually in Durango, I see a demonstration of the fire service's high-temperature "bunker" clothing. Even though this type of clothing was available in the mid-1970s, very few firefighters started using the equipment until the mid-1980s. Now these composite suits are almost universal.

Feeling Too Mummified

Some firefighters have mixed feelings about this protection, which costs close to $3,000 for clothes and breathing gear. Made of Kevlar, Nomex, PBI, and other space-age fabrics, the suits shield them from heat, and the air gear protects their lungs from the toxic fumes of today's fires. But they complain of being so mummified that they have no way of sensing when a room is overheated and ready to flash into flame. Traditionally, explains Tom Smith of the U.S. Fire Administration, firefighters used their exposed ears as heat detectors.

"Our new bunker gear lets us get into deep-seated fires that we have no business being in," fire officer Hubert Albrandt of Longmont, Colorado, tells a class at the academy. He warns the group that modern fire-resistant clothing, if pushed too far, gives way suddenly by melting onto the firefighters. "If it gets that hot," he says, "any citizens in that room aren't going to be alive anyway." Albrandt goes on to describe the danger signals that should alert fully suited firefighters that their surroundings are getting too hot: the air from their tanks turns warm, and vapors begin rising from upholstery and carpets. "If fabrics are steaming," he says, "it's almost time for a flashover."

Concerns about overprotection notwithstanding, there's no going back to the old days, when firemen went into smoke holding only wet hankies over their face. In fact, the future holds even more protection. The National Aeronautics and Space Administration (NASA) has developed lightweight liquid-air packs capable of furnishing hours of air. A British firm sells an astronaut-type helmet that seals around a firefighter's face, hooks directly into a self-contained breathing apparatus, and offers the option of a complete communications system with built-in radio headset and microphone. The deluxe model is mounted with a miniature thermal-imaging camera and a tiny display module so the wearer can locate survivors through dense smoke.

To give us an idea of how well modern protective gear works, the Colorado academy suits up photographer Layne Kennedy and me and allows us inside a house trailer during a "live-fire" training exercise. Remember the hat, high rubber boots, and overcoat of the traditional fireman? They are gone; the gear now is half astronaut, half Indy race-car driver. We end up in heavy yellow Nomex pants and overcoat; fire-rated helmet and visor; treated leather gloves and boots; and Nomex pullover hood covering all but the face, leaving room for the helmet and a tight-fitting mask to receive fresh air from a scubalike tank on our backs. I feel like a potato wrapped in foil, ready for baking.

A few minutes later, I change my mind about the discomfort. As smoke from the burning interior thickens, filling the trailer with an ugly brown layer that hugs the ceiling, the air from the tank suddenly feels fresh and cool. Tom Kaufman, our safety officer, estimates that the temperature at the ceiling is close to 800° F (425° C).

On a third trip into the trailer, Kennedy has the opportunity to see how well the gear performs under fire. The smoke ignites in a flashover, filling the trailer with 1,200° F (650° C) flame. Even the best fire-rated clothing cannot tolerate such heat for more than a few seconds, so the eight men in the trailer make an orderly exit after spray from a fire hose pushes back the flame. "If you had been in there without that gear," one firefighter tells Kennedy, "you'd be dead now." It's almost a badge of honor among some firefighters to sport smoke-stained bunker gear, with edges of the reflective tape melted to a blackened crisp. On the other hand, instructor Hector Lagomasino, who was tempered by urban firefighting experience in Dade County, Florida, before moving to Colorado, tells me that he regularly cleans his bunker gear with a high-pressure water hose at a car wash.

Overhauling

I tag along as Lagomasino inspects the charred walls in the trailer. Every building fire, whether trailer or skyscraper, is followed by such a process, called "overhaul." Overhaul is the part of the firefighter's job that news cameras rarely record; it's also the only way to prevent the rekindling of a fire hours later. Firefighters drag out furniture and pull down ceilings to soak any lingering embers.

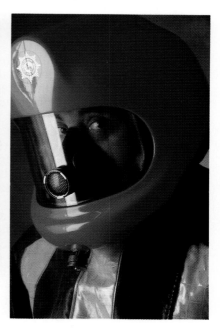

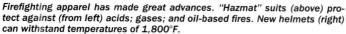
Firefighting apparel has made great advances. "Hazmat" suits (above) protect against (from left) acids; gases; and oil-based fires. New helmets (right) can withstand temperatures of 1,800°F.

Increasingly they are being told to wear breathing gear during these hours of hard work, but some still don't. Captain Richard Anderson of Minneapolis tells me that he wonders whether this exposure will cause a new spike in lung-cancer rates in future years. "The number one cause of death for firefighters is cancer," says consultant Kevin Mellott. "The word on the street is that synthetic materials are responsible. While this is probably right, there's never been a good study of the baseline for cancer."

Fires and accidents claim many victims, and their rescue usually falls to the fire service. The possibility of saving a life is the main reason a firefighter runs into a burning building when everybody else is heading in the other direction. When no lives are at stake and a fire is out of control, standard procedure is to protect nearby buildings and get water into the structure: "surround and drown," in fire lingo. Crouching low or crawling, firefighters search each room—by feel, when the smoke is thick. In stressful times like these, a firefighter can use up a 30-minute tank of air in 15 minutes.

One way that fire services train their people for this work is to send them through a dark "confined-space maze," wearing full bunker and breathing gear. Maze training has been spurred by alerts from the National

Institute for Occupational Safety and Health (NIOSH), which found that of all the deaths recorded in confined-space emergencies, two out of three were of would-be rescuers.

Crawling Blind
In Los Angeles County, the maze, which simulates an earthquake collapse, is a multilevel plywood labyrinth with cramped stairs, ledges, and a short ladder, barely big enough to crawl through. Virginia's Fairfax County Fire Department uses concrete-manufacturing plants and sewage-treatment facilities for its rescue exercises. But the training can be as simple as sending a participant crawling through an unfamiliar living room with his or her mask taped over. With all senses cut off or muted, rookies find that the most difficult part is getting out. They learn to place beacons by doorways and to follow a fire hose back to safety. Most important, they learn how to control their fear.

"Research on the body shows that stress-filled situations like these trigger the release of more than 700 different chemicals," Captain Wayne Ibers tells me. Ibers is an urban search-and-rescue specialist for Los Angeles County. "The maze is antipanic training for firefighters. We give them information to help reduce the stress, to know the difference between anxiety and panic."

That's one way to describe how I feel during another Los Angeles County Fire Department exercise, with water-rescue instructor Roger Wilhelm. Employed as a firefighter specialist in one of the county's hazardous-materials squads, Wilhelm is qualified to teach open-water rescue by virtue of his military training. During the Vietnam era, he was a member of two elite U.S. Navy teams—Underwater Demolition and the SEALs.

Wilhelm has me up in a helicopter, dressed in a wet suit, standing on a skid, ready to jump into the Castaic Lake reservoir. I am participating in the tail end of a session to train 15 firefighters for ocean rescue, in the event of an airline crash in the waters off Los Angeles International Airport. The copter has already made several passes, dropping rescuers in pairs, who have inflated a big yellow raft.

A thumbs-up sign from the jumpmaster, and Wilhelm and I drop 10 feet (3 meters) into the water. Wallowing around in the water trying to get my flippers on, with the copter hovering above, I can understand how disoriented a frightened victim might feel in such a situation. The helicopter's rotors are deafening and blow water in my face like a waterspout. But the copter has its advantages; it edges closer after Wilhelm and I are aboard the raft, and we begin plowing through the water like a slow motorboat. The pilot is using his rotor wash to propel the raft toward a dock.

Ocean rescues are just one part of the bigger role that firefighters find themselves training for these days. Even fire-department helicopters have a split personality now: besides being good for dropping 10 firefighters at a time into the drink, Los Angeles County Fireships 16 and 17 are each equipped with a hoist and spotlight for cliff rescues and a 360-gallon (1,360-liter) tank for dumping water on wildfires.

The evening after my simulated ocean rescue, I have the opportunity to watch a real helicopter rescue by the Los Angeles City Fire Department. A 55-year-old man has tumbled down a seaside cliff at Point Fermin. Firefighters climbing across the rocky headland find him unconscious but alive. As the Sun sets over the Pacific, two helicopters arrive. The smaller one takes up a position at the side and throws a spotlight. The other lowers a paramedic and then a rescue cradle called a Stokes basket. Within five minutes the injured man is aboard. Onlookers clap as the helicopter arcs into the sky and heads for the hospital.

Such helicopters represent one extreme of fire-department medical equipment. For the average person at the other end of a fire call, the medical help arrives via the common fire engine. In Los Angeles County, all firefighters have received basic medical-emergency training, including cardiopulmonary resuscitation (CPR). Some injuries and medical conditions are beyond them, though, and the minutes before a paramedic unit arrives can mean life or death. The county is filling the gap by equipping fire engines with a new machine called a semi-automatic defibrillator. About the size of a laptop computer, the unit reads the patient's heart rhythm through electrodes attached by the firefighters. It analyzes the pattern to check for certain kinds of arrhythmia, charges itself, and applies the proper voltage of electric shock. The county estimates that the machine has saved the lives of 25 percent of all the people it has been hooked to.

With fire prevention, equipment, and rescue tactics better than ever, many more people are being saved today than two decades ago. But seasoned firefighters, who know that nothing can ever make their world safe, warn rookies against placing too much faith in their high-tech gear. The fire equipment is so good, they say, that complacency can set in.

I know what they're talking about. While taking part in a Los Angeles County training exercise on how to deal with a pool of burning gasoline, I had used up my extinguisher and turned to walk away. Al Harris, a safety supervisor from a Texaco refinery, grabbed me by the shoulder. "Don't ever turn your back on it," Harris told me. "Fire is your enemy. Don't ever let it get behind you." Like some ancient king, quick to take offense, fire will be ready with a harsh penalty for those who—even for a moment—forget where the real power lies.

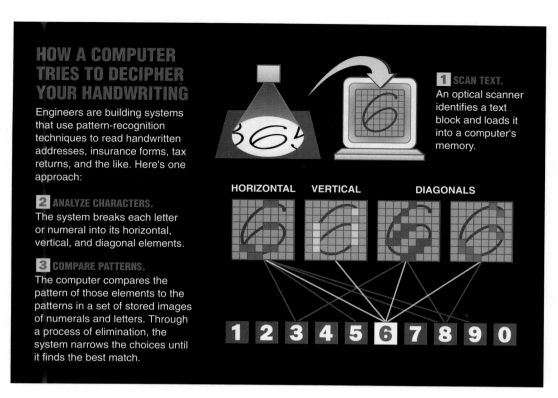

HOW A COMPUTER TRIES TO DECIPHER YOUR HANDWRITING

Engineers are building systems that use pattern-recognition techniques to read handwritten addresses, insurance forms, tax returns, and the like. Here's one approach:

1 SCAN TEXT.
An optical scanner identifies a text block and loads it into a computer's memory.

2 ANALYZE CHARACTERS.
The system breaks each letter or numeral into its horizontal, vertical, and diagonal elements.

3 COMPARE PATTERNS.
The computer compares the pattern of those elements to the patterns in a set of stored images of numerals and letters. Through a process of elimination, the system narrows the choices until it finds the best match.

HORIZONTAL VERTICAL DIAGONALS

1 2 3 4 5 6 7 8 9 0

TEACHING A COMPUTER TO TELL a "G" from a "C" *by Mark Lewyn*

For years, clerks at American Express Company laboriously typed the hand-written numbers on every charge slip into data-processing equipment. Then, six months ago [in June 1992], the financial-services giant finished installing optical character readers (OCRs) that can decipher and process 60 percent of the 900,000 slips that each day pass through AmEx's two processing centers, in Phoenix and in Fort Lauderdale. The remainder—which even humans have trouble figuring out—are handled by clerks. The system, installed by TRW Inc., cost more than $10 million, but it is expected to pay for itself within four years. Says Cliff Dodd, an AmEx senior vice president: "It is critically important to us."

As pleased as AmEx may be with its state-of-the art equipment, computer-aided handwriting recognition is a technology that is just learning to crawl. A variety of writing styles, combined with sloppy or incomplete penmanship, continues to stymie scientists' efforts to devise machines that can decode every scrawl. "It's really one of the more difficult problems in computers," says Gregory K. Myers, a manager involved with SRI International's research into the area. "People don't appreciate just how difficult this is."

Huge Savings
If they can advance the technology, however, a wide range of companies—from AT&T to Hughes Missiles Systems—expect to reach a huge new market. Banks, insurance companies, and government agencies are sure to be interested in handling their paperwork more efficiently. The technology is also important for pen-based computers, which have to decipher words written with an electronic stylus instead of a keyboard. "The market could be in the billions," says Charles L. Wilson, a manager at the Commerce Department's National Institute of Standards & Technology, which is conducting research in the field.

Already the Internal Revenue Service (IRS) uses a system that reads handwritten 1040EZ forms. And in 1994 the IRS plans to award a $1 billion contract for technology that will enable it to scan all tax forms by computer by the year 2000. But the biggest potential customer of the technology is the U.S. Postal Service, which handles 555 million pieces of mail a day, 20 percent of it with handwritten addresses. The Postal Service foresees huge savings: It costs $3 for automatic processing of 1,000 letters, compared with $35 by hand.

Today's most advanced commercial systems, like the one at AmEx, are best at reading legible handwritten numbers and letters in predefined forms—such as charge slips. That gives the computer an edge: It knows exactly where to look. Coping with a handwritten address is a bigger challenge. Though the Postal Service has equipment that can read most typed or printed envelopes, the systems are often stumped by handwritten street names, numbers, and ZIP codes. And when they can decipher the characters, the systems work at the painfully slow rate of about one address per second. The Postal Service wants to reach the speed of 12 complete addresses per second.

ZIP Code

Making computers into speed readers involves a variety of technological strategies. At the State University of New York at Buffalo, researchers are working on computer programs that place a character in context. Given an address in Washington, D.C., the machine might have trouble if the first two zeros of the ZIP code 20005 are looped together, making it hard to tell if they should be 00 or 06. But the machine may see that the word before the ZIP code begins with a "D," and recognize that this designates the state. Using a database of addresses and ZIP codes, the system would find that there are no codes with a 6 in the middle position in the District of Columbia or in states beginning with D. Hence, it concludes the number is 00.

So far, this system can read an address and ZIP code only 30 percent of the time, though it can decipher the ZIP code alone 75 percent of the time. The Postal Service, which is helping fund the work, might begin implementing the technology once the success rate for addresses gets up to 50 percent. Sargur N. Srihari, a SUNY Buffalo computer-science professor, expects to hit 50 percent in the next two years.

At the IRS, a computer operator (left) loads a program tape which will allow the computer to read a handwritten 1040EZ tax form. As the computer scans, the form can also be read by the operator on the computer screen (below).

At the U.S. Postal Service, optical readers have the capacity to "read" all lines of an address containing a five-digit zip code. The computer can then apply a nine-digit bar code to each piece, and thus ensure the delivery of mail to the correct address.

At AT&T Bell Laboratories, nearly two dozen researchers are trying a different technique: "neural networks," a series of small computer programs, or algorithms, that work together to try to mimic the brain. In the first stage, the computer is shown different characters and told what they are. The network then sets up parameters that define those characters. To read addresses, it breaks letters or digits into parts. The number "6" would be broken down into horizontal, vertical, and diagonal strokes. If these components fall within the specific parameters the system has learned, it concludes the number is 6.

One benefit of the neural-network approach is that the system keeps learning. It continually adds new variations to its store of parameters for each letter or numeral. For example, it quickly learns to recognize a European-style numeral 7 even though it was never taught that a 7 might contain a horizontal slash. "We have a long way to go before we approach the performance of a human being," says Bell Labs researcher Charles E. Stenard, "but we're light-years ahead of where we were a year or two ago." Indeed, the lab is working to apply the technology to a related problem: deciphering license plates on cars whizzing by at high speeds. That would enable the police to find a particular car by checking the plates on all passing cars. Stenard says highway agencies are interested in testing the system.

Pen Pal

At Hughes, engineers have gone far afield to solve the handwriting challenge—to the pattern-recognition technology smart bombs use to find targets. Instead of trying to isolate and recognize individual characters, the system is designed to analyze and identify larger patterns, shapes, lines, or combinations of lines. That helps it ignore scratched-out words, for example, and makes it more accurate than existing systems that try to recognize individual letters, says Carol D. Campbell, who heads Hughes's efforts to adapt military technologies for commercial markets.

The challenge of decoding human handwriting is important for new pen-based computers, too, but these machines have an advantage. The computer is "watching" as the person writes—so it can pick up clues from the sequence of the strokes and identify where the writer lifts the stylus between words. And pen computers don't have to be as speedy as mail-sorting machines. Still, they haven't yet been able to master reading even careful penmanship.

Handwriting-recognition machines may never be able to decipher every doctor's prescription. But if researchers maintain their current pace of progress, sometime later this decade, perhaps, people may be able to buy the high-tech equivalent of pharmacists—computers that can read all but the most illegible scrawls accurately and quickly.

TROUBLE IN
CYBERSPACE

by Christopher King

A large automobile manufacturer is hit by a computer-based foreign-exchange fraud at a cost of some $260 million. Members of a West German hackers club face charges for breaking into U.S. computers in search of technical secrets to retail to their contacts in the Soviet intelligence community. A viruslike "worm" program, running rampant through the nationwide network known as Internet, temporarily disables upwards of 6,500 computers, causing damages in man-hours and lost productivity estimated in the hundreds of millions of dollars. A teenage hacker, penetrating Defense Department computers, nearly manages to start World War III.

That last incident, of course, is pure fiction, depicted in the 1983 movie *War Games*. The other events, however, are quite factual. All occurred at the end of the 1980s—the personal computer's first decade of explosive growth. And all underscored the continuing vulnerability of computer systems to intrusion and crime. While no experts fear nuclear Armageddon at the hands of a young computer hacker, some are concerned that hackers could do real damage. In 1990 the National Academy of Sciences (NAS) sounded a warning note in a report titled *Computers at Risk*. "So far," began the report, "the nation has been remarkably lucky in escaping any successful systematic attempts to subvert critical computing systems. Unfortunately, there is reason to believe that our luck may soon run out. . . ." In addition to that chilling prospect, experts are also concerned by the more mundane matters of thievery, fraud, and malicious destruction of data —all by computer.

Estimates about the extent of computer crime vary considerably. One rough calculation, for example, places the cost of such crime at $3 billion to $5 billion a year. Another figure, supposedly based on Federal Bureau of Investigation (FBI) estimates, puts annual losses to computer crime at anywhere from $500,000 to $5 billion. However, according to Donn Parker, a computer-security expert at SRI International in Menlo Park, California, such figures are so speculative or unreliable as to be meaningless. "There are no valid, representative statistics on computer crime," he says; "therefore, we do not know how big the problem is or whether it is increasing or decreasing in any quantitative way."

Parker and others, however, have noticed a disturbing trend in recent years: an influx of new hackers, motivated not by the sense of adventure or intellectual enterprise that characterized earlier generations, but by profit. "These new hackers have the old ones very mad at them," says Parker, "because they've violated the old ethic of not hacking for money." *Forbes* magazine offered an even less flattering description of the new breed: "Hacker hoods."

A New Electronic Landscape

The explosion in computers and telecommunications technology has created a vast new electronic realm that has been dubbed "cyberspace." Voice communication, faxes, and modem-equipped computers all add to the digital jumble in this new frontier. Selfstyled "cyberpunks," named after a recent subgenre in the literature of science fiction, have staked a claim to the international reaches of cyberspace. Most, of course, are satisfied to ply the "data highways" in a manner that is entirely legitimate, accessing commercial databases such as CompuServe or Prodigy, or logging onto smaller computer bulletin boards to exchange messages or chat electronically.

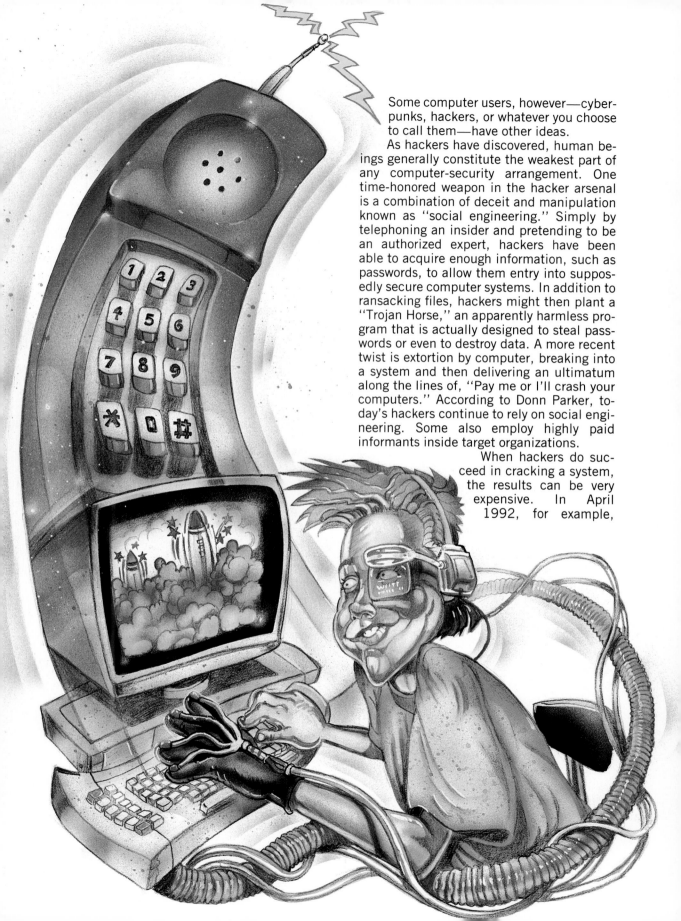

Some computer users, however—cyber-punks, hackers, or whatever you choose to call them—have other ideas.

As hackers have discovered, human beings generally constitute the weakest part of any computer-security arrangement. One time-honored weapon in the hacker arsenal is a combination of deceit and manipulation known as "social engineering." Simply by telephoning an insider and pretending to be an authorized expert, hackers have been able to acquire enough information, such as passwords, to allow them entry into supposedly secure computer systems. In addition to ransacking files, hackers might then plant a "Trojan Horse," an apparently harmless program that is actually designed to steal passwords or even to destroy data. A more recent twist is extortion by computer, breaking into a system and then delivering an ultimatum along the lines of, "Pay me or I'll crash your computers." According to Donn Parker, today's hackers continue to rely on social engineering. Some also employ highly paid informants inside target organizations.

When hackers do succeed in cracking a system, the results can be very expensive. In April 1992, for example,

police in San Diego busted a ring of young hackers who had managed to gain access to the computers of national credit-card companies and credit-reporting agencies. Members of the group stole card numbers and other account information and were able to make millions of dollars in fraudulent credit-card purchases.

Recently hackers have also turned their attention to the computers that handle voice-mail systems for corporations. Once they've barged onto such a system, hackers commandeer mailboxes and begin leaving messages for one another, exchanging illegal information such as stolen access codes for long-distance telephone service, or filched credit-card numbers. Even when their activities are discovered by system operators, hackers do not always retreat. In fact, hackers have been known to threaten to crash an entire system unless they are granted their own space on voice-mail systems. Such voice-mail scams became particularly popular after law-enforcement officials began cracking down on the "pirate" bulletin boards on which hackers had traditionally traded illicit information.

Not-So-Funny Phone Calls

Voice-mail intrusion represents merely one of the latest quirks in the long hacker history of bedeviling the telecommunications industry. Although the era may have passed in which hackers fashioned "blue boxes"— home-built devices that mimicked phone-company switching frequencies, allowing hackers to use long-distance lines without being billed—the practice of so-called "phone phreaking" remains very much alive. When it comes to roaming the long-distance airways of cyberspace, hackers remain determined to let others pay for the fun.

Hackers have been able to penetrate the computers of long-distance services, such as MCI, in order to steal account numbers and access codes. Also emerging as popular targets are the sophisticated telephone systems used by large corporations—systems known as public branch exchanges, or PBXs. By cracking a PBX, hackers gain access to long-distance lines, letting the company get stuck with the bill for the resulting calls. Often this illicit phone service is also retailed on the street in operations known as "callsell." Among the best customers are illegal immigrants, who are compelled to maintain a low profile in society, and who have a very hard time getting phone service. "I understand there are illegal immigrants in this country who literally have no idea how to make a legitimate long-distance call," says Bruce Sterling, a noted cyberpunk novelist who has covered computer crime. "As far as they know, when you want to call home, you go talk to the guy in the trenchcoat who hangs out by the public phone."

The widening use of mobile cellular phones has also been profitable for computer criminals. With a personal computer and some custom software, such phones can be "cloned" with a mobile identification number (MIN) and electronic serial number (ESN) stolen from a legitimate customer. Cloned phones are then leased or sold, primarily to drug dealers. As *Forbes* magazine reported, dealers will pay up to $3,200 for a cloned phone to keep an eye on operations in Bolivia or Colombia—and the legitimate customer gets the bill. One New Mexico customer received a $20,000 surprise.

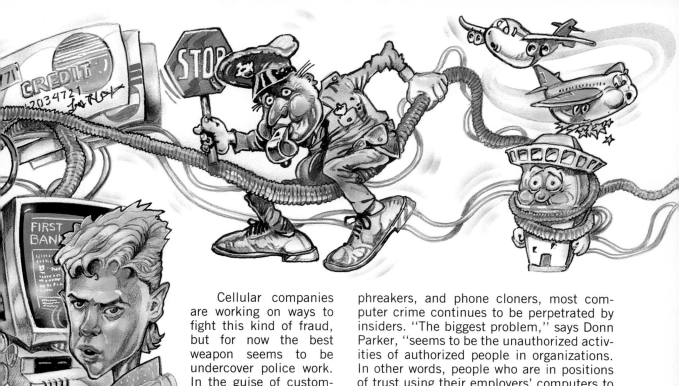

Cellular companies are working on ways to fight this kind of fraud, but for now the best weapon seems to be undercover police work. In the guise of customers, officers arrange to buy cloned phones from the shady suppliers. Once they've obtained the evidence, the officers return to make arrests, using search and seizure to grab up phones and the computers and software used to clone them. "Hopefully, in the near future, we'll be seizing the rest of the computers out on the street that are doing this," says Officer Donald Delaney of the New York State Police, who's in charge of computer and telecommunications fraud in the New York metropolitan area. "And we'll thereby slow these people down in their quest to bankrupt all the telephone companies."

Inside Jobs

Despite the intrusive misdeeds of outside hackers, telephone phreakers, and phone cloners, most computer crime continues to be perpetrated by insiders. "The biggest problem," says Donn Parker, "seems to be the unauthorized activities of authorized people in organizations. In other words, people who are in positions of trust using their employers' computers to engage in misuse or fraud of some kind."

An insider might create a program to unobtrusively channel funds to a specified account. Such programs are designed to subsequently erase themselves and any trace of the transactions. In one so-called "salami" attack at a large New York bank, an insider skimmed a penny or so from each account until he had amassed over $200,000 .

Companies do seem to have gotten smarter about computer security, from inside as well as outside assault. Some countermeasures are as simple as regularly changing passwords, controlling physical access to computers, and using "dialback" equipment to make sure that all incoming calls are received from authorized sources. Digital "smart cards" containing microprocessors are also used to control access. "We are getting much better at putting controls into computers," says Parker. "There's better authentication of users through secret passwords, the use of smart-card tokens, and controls that limit how much each individual can see inside the computer. Methods of detecting unusual activity, relying on very sophisticated statistical techniques, have also been developed. These systems report activity to a security station, so that action can be taken against computer crime."

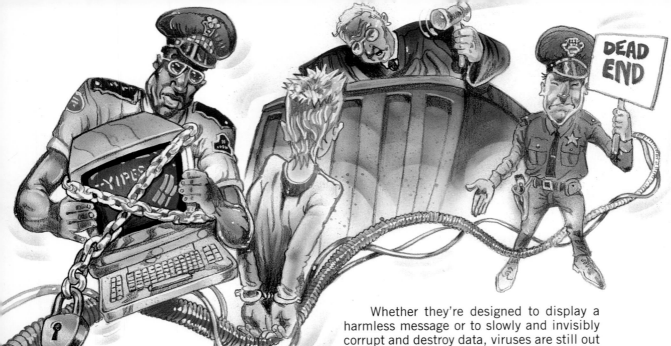

Another weapon is encryption—the algorithmic alteration of passwords and other data into scrambled code. This technology is becoming more widely available. Unfortunately, hackers have also learned how to encrypt, making it harder to detect and counteract their intrusions. This is particularly true in the case of computer viruses.

The Continuing Plague

By all accounts, the problem of viruses seems to be growing worse. Matching the diversity of their biological counterparts, computer viruses can come in hundreds of different varieties, and insidiously replicate themselves from program to program, computer to computer. Some of the better-known viruses constitute a roll call of infamy: the "Lehigh" virus, named for the Bethlehem, Pennsylvania, university where it attacked computers and destroyed data in 1987; the "Jerusalem" virus, which slows computers and erases files, typically attacking on a Friday the 13th; the "Michelangelo" virus, which caused worldwide headlines and much consternation (but little actual damage) on the artist's birthday in March 1992; and the notorious "Internet Worm" mentioned at the outset of this story, which afflicted more than 6,000 computers in 1988.

Whether they're designed to display a harmless message or to slowly and invisibly corrupt and destroy data, viruses are still out there. For organizations using 400 or more personal computers, some estimates have placed the risk at more than 50-50 that a virus infection may occur in any given month. So-called "shrink-wrap viruses" have even corrupted commercial software before shipment from the manufacturer.

One popular measure to combat viruses is the use of antiviral software, such as "scanning" programs that recognize virus patterns and automatically expunge them from a disk. Maintaining backups of all programs and data is an important element in combating viruses—as is avoiding public-domain software obtained from bulletin boards. "The effect that viruses have had on people's willingness to exchange information has been quite profound," says Bruce Sterling, who finds it difficult to comprehend why anyone would create a virus. "As far as I know, no one has ever made a dime from writing a virus program," he says. "It's purely some combination of intellectual curiosity and vandalism. But it's like throwing dead cats into a pure stream."

Crackdown

On January 15, 1990, Martin Luther King Day, AT&T's long-distance phone system crashed. Although the problem turned out to be a software glitch, many people in the telephone industry and in law enforcement were convinced it was the work of hackers. A crackdown was at hand, and in 1990 it was carried out with a vengeance. In New York State, in Chicago, and in a series of raids known as "Operation Sundevil" in Arizona, authorities seized computers running pirate bulletin boards. Operation Sundevil alone netted 40 computers and 23,000 floppy disks. The message was clear, as Bruce Sterling writes in his book *The Hacker Crackdown,* a nonfiction account of the law-enforcement backlash: Hackers were put on notice that "state and federal cops were actively patrolling the beat in cyberspace."

The message was driven home by highly publicized cases in the early 1990s. The prosecution of Robert Tappan Morris, creator of the Internet Worm, attracted wide interest (see sidebar at right). And members of an Atlanta hackers club known as the Legion of Doom drew prison terms for stealing information from the BellSouth phone company.

But was law enforcement going too far? In some instances, police had confiscated computers and other electronic equipment from hackers, keeping the gear for months or even years without filing formal charges. Disturbed by the implications of the crackdown, several computer professionals formed the Electronic Frontier Foundation, an organization devoted to safeguarding the civil rights of law-abiding hackers.

A Dead End

It seems inevitable that lawbreaking by computer will continue. Some experts, such as Donn Parker, are concerned that terrorist hackers could do real harm—interfering with computers that handle air-traffic control, for example, or crashing phone systems in times of emergency. Parker is currently exploring ways of keeping young people out of the hacker culture.

Meanwhile, in New York, Donald Delaney—a cop whose experience has included homicide investigations—seems less optimistic about human nature. "Every time we've made a technological advance to slow hackers down and stop them, they've managed to do something else," he says. "I can't imagine what they're going to do next. But by making arrests and getting publicity for what we're doing, we're keeping a lot of other people honest."

Computers and the Law

"Computer crime is a nascent field that as yet has no solid definition," write legal scholars Cynthia K. Nicholson and Robert Cunningham. Indeed, the law has had a hard time keeping up with the rapid changes in computer and information technology. The first federal computer-crime law was passed in 1984: the Counterfeit Access Device and Computer Fraud and Abuse Act. This act was subsequently replaced by the Computer Fraud and Abuse Act of 1986, directed at intentional, unauthorized access to government computers.

All 50 states now have legislation aimed at combating computer crime. Such cases, however, are difficult to prosecute. Laws can be ambiguous in their definitions of such matters as "access" and "authorization." Experts agree that more-stringent and technically specific laws are needed.

The world of cyberspace offers other prickly legal dilemmas. Should computer bulletin boards—even pirate boards that may harbor illegal information—be protected under the First Amendment? What exactly is the nature of "property rights" and "theft" when so much data is being flashed about on the electronic frontier?

Still more thorny issues lie ahead. For example, new digital-telecommunications technology will soon make it physically impossible for law-enforcement agencies to tap phone lines for their investigations, thus robbing the police of one of their most powerful weapons against organized crime. If wiretaps are to continue, the capacity to tap lines will have to be, in effect, engineered into the new systems. The implications of this—and of possible abuse—have provoked a debate among telephone professionals, civil libertarians, and law-enforcement agencies.

HIGH-TECH OLYMPIANS

by David Bjerklie

The thrill of victory and the agony of defeat may be unchanged since the first Olympics, but competition sports equipment undergoes constant technological improvement. Whether bicycles, kayaks, javelins, barbells, archery gear, or fencing swords, just to cite a few, the tools of the athlete's trade are now lighter, stronger, and better designed than even a few years ago.

In fact, sports equipment is so good that it raises some troublesome questions. Should attempts to advance the technology of sports equipment be a wide-open race to "build the best?" Should such equipment be allowed to drive world records? Or must rigid standards

be imposed to ensure fair competition? "In theory," says Chester Kyle, a designer of bicycles and clothing for the 1984 U.S. Olympic cycling team, "you want a contest between athletes, not machines; but in practice, it doesn't work that way."

From Boards to Bars
Many of the most dramatic improvements in sports equipment came on the heels of World War II. Diving springboards, for example, were transformed by way of the aircraft industry when Norman Buck introduced a design in 1948 that used 300 interlocking pieces of war-surplus square aluminum tub-

Equipment innovations may be turning the Olympics into a technology tourney. The hull and oar designs used in the 1992 kayaking events (left) would have been an overwhelming advantage in 1968 (above).

ing. The "Buckboard" soon gave way to designs that used even stronger and lighter aluminum alloys. By the 1960s the enhanced springiness of diving boards made possible dives at the 1-meter (3.2-foot) height that had previously been performed only at 3 meters (9.8 feet).

Today's boards are springier still, and provide 15 percent more lift than those of the 1960s. They also feature superior non-skid surfaces, and the light weight and resiliency of the board tip minimize injuries, especially to the head. In fact, the 2-foot (0.6-meter) section at the end of the springboard now weighs less than 10 pounds (4.5 kilograms).

Pole-vaulting also got a boost from improved materials. In 1960 Herb Jencks, manufacturer of fiberglass fishing poles, had just built a new deep-sea fishing rod 10 feet (3 meters) long and more than 1 inch (2.5 centimeters) in diameter. Jencks's son, a junior high school vaulter, borrowed one of his father's poles for a practice vault, and, much to his surprise and delight, surpassed his personal best by half a foot.

Until then, pole-vaulters used bamboo poles and landed in sawdust pits, and the world record was 15 feet (4.6 meters). Today the world record is just over 20 feet (6 meters), thanks to carbon composite poles that are custom-built to the vaulter's weight, takeoff speed, and hold technique. Foam-rubber pits that cushion falls more effectively also allow vaulters to be ever more fearless in their quest for height.

Safety has clearly been the force behind improvements in fencing technology. Epée and foil weapons must now be made from maraging steel, a jet-fighter alloy that is

Compared to those used in the 1960s Olympics, the boards used by American Kent Ferguson (above) and other Olympic divers during the 1992 games provide 15 percent more lift, better traction, and more resiliency.

stronger and less brittle than conventional carbon steel, according to Dan DeChaine, armorer for the 1992 U.S. Olympic team and member of the technical commission for the International Fencing Federation. The most dangerous situations in fencing arise when blades break in the heat of competition and puncture the fencers' protective garments. While 50 to 100 carbon blades typically snap in a world-class meet, only three maraging-steel blades broke in the last world championship.

Safety standards for clothing have also been made more stringent. Fencing jackets and bibs—now made of Kevlar, a flexible synthetic fiber that is stronger than steel— are essentially bulletproof. Partly in response to the Soviet fencer who was killed in 1982 when his mask was pierced by a broken blade, masks are now constructed of stainless steel with a thicker mesh and denser weave that can withstand twice the force of the standard puncture test.

Archery bows made from new materials are a far cry from the sleek wooden relics that were still standard less than 30 years ago. In the mid-1980s, Hoyt Archery introduced the latest in a series of materials innovations, a take-apart bow with a core made

In terms of safety, fencing equipment has made great advances since the 1960 Olympics (below). By 1992 (left), stringent standards for blades and clothing had reduced fencing injuries considerably.

of syntactic foam—a material composed of tiny glass beads embedded in a rigid foam matrix—and wrapped in layers of carbon fibers and fiberglass.

The Hoyt bow is lighter and more stable than wood-core bows, and is impervious to temperature changes, says archer and equipment expert Donald Rabska of Easton, the sports equipment company that now owns Hoyt. "With wood, you eventually get flex fatigue in the fibers," he says, "and no matter what you put on it, wood absorbs moisture and is affected by temperature changes." This means that wood bows shoot fast in cold weather, but get mushy when the temperature rises. Rabska says the lightweight synthetic material also provides a speed boost: "Foam core returns more of the release energy stored in the bow when the archer pulls back on the string."

Bowstrings and arrows have also been transformed. Originally made of linen, strings are now made of a lightweight and low-stretch polyethylene that provides a velocity gain to arrows of several feet per second. The latest arrows—made from hollow aluminum with walls only 0.006 inch (0.015 centimeter) thick—are the lightest, and therefore fastest, yet. In fact, "they are about 20 feet [6 meters] per second faster than the all-aluminum arrows," says Rabska.

Such design changes have sent winning scores soaring. In a competition round, archers shoot a total of 144 arrows, aiming for a perfect score of 1,440. Thirty years ago, winning scores in major international archery tournaments hovered in the 1,100s. To win today an archer must shoot around 1,350.

Even the lowly discus, traditionally made of wood with its weight concentrated in a metal hockey-puck center, has been re-engineered. In the 1960s manufacturers began to experiment with different materials and weight distributions, hollowing out the

wood rim and then replacing the hollow wood with a plastic shell. Designers then lined the edges with lead weights, distributing the weight nearer to the perimeter. "The theory," says Jay Silvester, chair of the U.S. discus development committee and former world-record holder in the sport, "is that weighting the edge stabilizes the discus in flight, which translates into longer throws." Certainly the numbers don't argue against the change. While Silvester set the world mark at 60.56 meters (198.6 feet) in 1961,

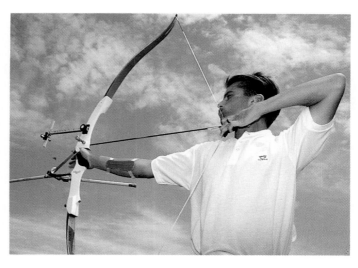

By 1992, hollow aluminum arrows, used with bows and strings made from high-tech materials, had become the norm in Olympic archery.

the current record, set in 1986 by Jurgen Schult of East Germany, stands at 74.08 meters (243.06 feet).

Today the newest barbells are made from "clock-spring steel" that is so flexible it can be bent nearly into a U and still resume its shape. "Lifters can take advantage of this," explains Lyn Jones, a coach of the U.S. Olympic weight-lifting team, "by timing the 'whip' of the barbells. Because the ends of the bar are momentarily lower when the center is jerked to chest height, the lifter can wait until the ends swing back upward to complete the lift."

Another technological innovation is a nylon sleeve that fits around the bar. Because the bar can rotate freely within the

Today's Olympic pole-vaulters use springy carbon-composite poles (above) to lift them to heights never dreamed of by the medal-winning pole-vaulters in the 1960 Olympics (below), held in Rome, Italy.

sleeve, the lifter can maintain balance and grip more easily while snapping the wrists under the tremendous strain.

Even the weights are better today. Since the 1972 Olympics, weights have been encased in rubber so that lifters could drop the barbell to the floor without causing damage. Having to lower it carefully wastes precious reserves of strength needed for the next lift.

Wavering Committees

In the early 1980s, a study by the U.S. Olympic Committee (USOC) found that American athletes earned more medals in low-tech sports such as running and swimming than in high-tech sports such as kayaking and cycling. "It seemed paradoxical," says Eric Haught, chair of the sports science and technology commission of the U.S. canoe and kayak team, "that one of the most technologically advanced nations in the world performed worst in just those events in which it would be expected to excel." In an effort to close the technology gap, the USOC formed the Sports Equipment and Technology Committee (SETC) and enlisted researchers throughout the country to improve the equipment available to U.S. athletes.

For example, Haught, an engineer who specializes in fluid flow, and Edward Van Dusen, president of Composite Engineering, received SETC grants in 1985 to perform extensive hydrodynamic testing of "slender-body" forms such as kayaks and racing shells. From that research, they designed a kayak with a more rigid body, which Greg Barton and Norm Bellingham paddled to gold medals at the 1987 world championships and Pan American Games, and then used to win two gold medals at the 1988 Olympics in Seoul.

Competition kayaks, once made of mahogany veneers, are now constructed of a combination of carbon-fiber cloth, Kevlar, and high-temperature epoxies—a stiffer design that minimizes the amount of energy wasted in flexing as the hull moves through the water. The hulls are also much lighter, but because there are minimum-weight requirements in competition kayaks, the weight saving is applied to new features such as reinforced foot braces and seat supports.

These enable the athlete to sit in a position that is ideal for both mechanics and comfort, explains Haught.

Paddles, once made of solid poplar or birch, are now constructed of lightweight high-tech composites. Even more significant is a new, spoon-shaped "wing" paddle that pulls more efficiently and achieves lift, thereby increasing speed. This design has required changes in paddling techniques, but the results have been so dramatic that all world-class kayakers now use it. "The wing was a revolution," says Haught, who estimates that the paddle has trimmed several seconds off world-record times.

from competitors' eyes, so they specify that boats have to pass inspection at the race. A team might decide to surprise the competition with a new design, he says, but the committee may throw the boat out of the contest right at the meet.

When it comes to run-ins with the rules, bicycle racing is in a class by itself. The sport has a long history of technological improvements, and an equally long history of ambivalence over them. For example, in 1911 the French designer Etaine Boneau Varilla patented an enclosed bike that resembled a mini-dirigible on wheels, which indeed did "fly" past competitors. But by

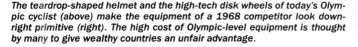

The teardrop-shaped helmet and the high-tech disk wheels of today's Olympic cyclist (above) make the equipment of a 1968 competitor look downright primitive (right). The high cost of Olympic-level equipment is thought by many to give wealthy countries an unfair advantage.

Studies continue on the hydrodynamics of slender bodies. "We basically run races on the computer and see which design performs better," says Norman Doelling, assistant director of the Sea Grant Program at MIT.

But even when the innovations of science and engineering come together in a dazzling new boat design, there is no guarantee that it will be embraced by the rules committees. What's worse, explains Haught, is that you can't prequalify a boat.

Kayaking officials believe it wouldn't be fair to approve a new design in secrecy, away

1914 the International Cycling Federation decided that it was in the best interest of the sport to ban features added to a bike solely for streamlining. Bicycle design has been a game of hopscotch between engineers and Cycling Federation officials ever since.

In 1932 French engineer Charles Mochet designed the recumbent bike, a low-slung affair in which the cyclist reclines and pedals as if in a Barcalounger. The bike used a rider's leg and lower-body strength more effectively than did an upright bike. "Even when piloted by second-rate cyclists," says bike designer Chester Kyle, "Mochet's bike beat every cyclist in Europe, and also promptly set a new world record for the distance covered in an hour." Just as promptly, the authorities outlawed the revolutionary new bike from all official competition. The recumbent, the ruling body declared, was not a bike at all.

Kyle's own designs for cycling equipment have been outlawed eight times. "But the federation seems to waver back and forth in its interpretation of the rules," he maintains—a feature that is declared illegal one year may become the legal standard the following year.

For example, in 1984 Francesco Moser, an Italian bike racer, broke what wheel designer Steven Hed refers to as "the Babe Ruth home-run record of cycling." Moser covered more than 31 miles (50 kilometers) in one hour, surpassing the mark set in 1972 by Belgian cycling great Eddy Merckx, the most successful professional racing cyclist of the modern era. Moser accomplished this feat riding a custom bike—crafted by Antonio Delmonte, designer and physiologist for the Italian national team—featuring novel wheels in which the spokes had been replaced with a thin but rigid disk that supported the tire rim.

Disk wheels were suddenly hot. Moser's were priced at $6,000, but European manufacturers were soon selling them at a discounted but still stiff $1,200. In 1984 both the U.S. and Italian teams battled the International Cycling Federation, arguing that disk wheels should be allowed into Olympic competition. When the cycling authorities eventually relented, Italy and the United States cleaned up: the United States earned nine medals, its first since 1907.

Disk wheels do have a drawback: they can be particularly tough to control in crosswinds. The best solution often proves to be a bike with a rear disk wheel and a conventional front wheel, especially since designers like Hed, whose wheels rolled bikes from 10 countries at the Barcelona Olympics in the summer of 1992, have further refined the performance of the front wheel. The designers have cut drag by reducing the number of spokes and flattening the remaining few into thin blades that slice through the air. The rim has also been made more aerodynamic with a new, egg-shaped cross section.

An even more significant innovation is the aero handlebar. Because it extends far over the front wheel with the hand grips close together, the aero bar forces the rider's body into a low forward tuck with arms extended almost prayerlike. "It's one of the biggest technological innovations since the beginning of the bicycle," Hed claims. The aero bar was developed by former U.S. ski-team coach Boone Lennon, who brought downhill-skiing and wind-tunnel experience to his enthusiasm for bike racing. The bars were first used by triathletes, but really caught on when Greg LeMond used them to win the 1989 Tour de France. "In a 25-mile [40-kilometer] time trial," says Hed, "aero bars probably make a three-minute difference. In these events, if you don't have aero bars, you don't have a chance."

Bicycle engineering innovations don't end with hardware. Helmets are teardrop-shaped, visored, and vented to eliminate turbulence, and are form-fitting, thanks to an inflatable bladder that can be pumped up to desired firmness. And new suits made of breathable nylon with silicone ribbing on the back and shoulders smooth out wind turbulence by helping air follow the natural contours of the cyclist's body.

Air resistance, in fact, is the barrier against which cyclists expend 90 percent of their energy. "What's most important is the turbulence created in the rider's wake—it's like an invisible hand holding the rider back," says John Sipay, former manager of team support for the U.S. Cycling Federation. Installing a fairing, or smooth wrap-around structure that reduces air resistance, behind the seat could clean up 20 percent of that swirling turbulence, he says. However, what's clearly illegal, says Chester Kyle, "is when things are done for streamlining purposes only." Everything on the frame has to be a functional part of that frame. The dodge, therefore, is to try to make the fairing part of the structure, not an add-on. In fact, already on the drawing board for the 1996 Olympics in Atlanta are designs that make a seat fairing "an integral part of the bicycle."

Sipay, for one, believes that the sky should be the limit when it comes to technical improvements in cycling. "In the world of auto racing, no one believes we should still be racing Model T's, do they?"

Selecting for Athletes

The actual consensus, at least in principle, is that sport should remain a contest between athletes, not machines. But Haught stresses that changes in the rules governing equipment can determine who will excel. "Why do elite rowers stand 6 feet 4 inches to 6 feet 6 inches (1.93 to 1.98 meters) tall and weigh about 210 pounds (78.3 kilograms)," he asks, "while elite kayakers stand 5 feet 10 inches to 6 feet (1.78 to 1.83 meters) tall and weigh 180 pounds (67.1 kilograms)?" The reason: a rowing shell can be as long as desired, whereas the length of competition kayaks is limited.

In a rowing shell, stronger is better, even at the expense of added weight, because longer boats can distribute more weight over a larger area without riding low in the water, and thus creating excessive drag. Performance in a kayak, however, says Haught, is more dependent on an ideal "aerobic strength-to-weight ratio." At a certain point, added bulk is a disadvantage despite any increase in strength that a heavier paddler can bring to the race.

This important consideration applies to nearly every sport: change the equipment—make it lighter, faster, stronger—and you change the athlete. Or, more accurately, you select for athletes who possess the traits that are able to maximize the performance of the new and improved equipment.

Take the javelin. When it was redesigned to be lighter and possess a more aerodynamic taper and center of gravity, athletes accordingly honed their technique to make the javelin sail. "If thrown at exactly the right angle, velocity, and orientation," explains Mont Hubbard, a professor of mechanical engineering at the University of California-Davis who applies computer modeling to sports, "the aerodynamic javelin would achieve a powerful lift, floating with

Between the 1976 Olympics (above) and the games in 1992 (left), a new javelin design was introduced that was so highly aerodynamic that athletes were shattering existing records almost daily—and by meters. In 1986, the design was banned and the new records voided.

its nose up even as it passed its zenith and began to descend back to Earth. If all went right, the javelin would dip nose downward only at the very end of its trajectory."

The technique required a great deal of finesse, but those who mastered it staggered the competition. In 1983 American Tom Petranoff broke the world record by 10 feet (3 meters) with a stunning 99.72-meter (327.18-foot) toss. Then, just days before the 1984 Olympics, Uwe Hohn of East Germany smashed Petranoff's world record by 16.4 feet (5 meters) with a soaring 104.80-meter (343.85-foot) throw.

Unfortunately, this match of exquisitely refined spear and perfect technique was rendering the world's sports stadiums obsolete —not to mention endangering spectators and other athletes warming up on the far side of the track. Javelins were being thrown out of the field of competition, and what's more, says Hubbard, they had a nasty habit of veering sharply to either side if misthrown. The sport's rules committee responded in 1986 by banning the new design. With the new deengineered model, the world record dropped precipitously by 60 feet (20 meters). And as the finesse throwers suddenly faded into the second ranks, the strong arms of power throwers once again prevailed.

Popularizing the Sport

Today the highest priority among rules-committee officials is to keep their sport affordable to many athletes. "It's not fair to compete on one-of-a-kind designs," says John Tarbert, technical director and effectively keeper of the rules for the U.S. Cycling Federation. For instance, disk wheels were initially banned from Olympic competition because they were so expensive they really couldn't be considered available to most cyclists, he says.

Andy Toro, a former bronze medalist who now sits on the board of the International Canoe Federation (which oversees both canoe and kayak competition), wrestled with this very issue at a committee meeting in October 1992 in Madrid following the Summer Olympics. On the agenda were several proposals that called for all canoes and kayaks to conform to a single hull design.

Toro supported the concept in theory. "Our sport is considered expensive, and we run the risk of pricing ourselves out of existence if we let the technological race continue wide open," he says, pointing out that new designs using exotic materials cost some $3,000 each. Costs must be reduced before the sport can be expanded internationally, he notes; for example, only three African countries are currently competing.

But Toro spoke out against the specific proposals because of concerns that the owner of the chosen design would suddenly have a monopoly. His proposed solution was to allow technological improvements, but at a slower pace: "I think we can establish a standard that we can review every four years and change if necessary to accommodate development."

The Federation of International Target Archery devised a similar approach—also to introduce competition archery into less affluent countries. Four years ago the federation established a "standard round" in international archery competition, in which the bows, strings, and arrows that can be used are all carefully specified. "It's very good equipment, but not the top end of cost and technology," explains Easton's Donald Rabska.

Part of the reason for the success of the standard round is that "there are a lot of archers who just don't want to futz with the cost and the latest technology," Rabska says, but at the same time, an open competition challenges competitors who shoot the latest equipment, such as bows that use pulleys to amplify the archer's pull strength. By allowing competition at different equipment levels, archery resembles car racing, which has managed to fuel competition along a broad spectrum of equipment ranging from stock cars to Formula One racers.

Some observers argue that the goal should be to keep a sport alive for athletes and fans—not to mention manufacturers—by encouraging rapid technical innovation. But when push comes to shove, rules committees seem to place popularizing the sport first. As Haught says, "The more children you can put into boats, the more chances you have for an Olympic champion."

QUIRKY KEYBOARDS

by Barnaby J. Feder

The computer keyboards of tomorrow will have a decidedly unusual but nonetheless ergonomic design intended both to reduce repetitive stress injuries and to make typing more efficient.

The quest to design a radically new keyboard, once the realm of tinkerers rarely taken seriously, is beginning to look a lot less quixotic. The main reason is concern about mounting reports of muscle and nerve injuries among clerical workers, journalists, and other office workers who use computers extensively. Doctors are saying part of the blame lies with the design of the traditional keyboard.

These ailments are usually called repetitive stress injuries, or RSI. Faced with lost productivity, rising insurance and worker's-compensation claims, and, more recently, a surge in lawsuits against keyboard manufacturers, business is finally taking seriously the decades-old claims of inventors that a sensible keyboard would look a lot different than those now used.

Standard Arrangements

"The worldwide keyboard market needs a major updating," says Stanley Hiller, Jr., the new chief executive of Key Tronic Inc., the largest independent keyboard manufacturer, based in Spokane, Washington. Mr. Hiller says that since his arrival at the company in March 1992, Key Tronic has begun work on

a number of new designs, but none is ready for introduction yet.

Getting from here to there will be tricky. Some major manufacturers—like Lexmark, the maker of IBM keyboards—are reluctant even to discuss the possibility of significant redesigns of keyboards for fear that it will look like an admission that current designs are defective, encouraging lawsuits.

The consensus among ergonomic experts is that design, posture, and workload all affect who gets RSI and how seriously they are injured. Most everyone agrees that the typical keyboard forces users to strain muscles from the shoulders to the fingertips in order to keep wrists horizontal to the keyboard and reach all the keys. What is not known is how much new designs would help.

"It's reasonable to assume that some new keyboard design would reduce injuries, but I don't think anybody has done statistically significant medical research to speak of on the new keyboards," says Peter Amadio, M.D., a hand surgeon and professor of orthopedic surgery at the Mayo Clinic in Rochester, Minnesota.

Standard keyboards in mass production can cost less than $20 apiece and retail for

The shallow V-shape and center peak of this MyKey Keyboard permits a more natural hand-wrist position than do conventional keyboards. MyKey's 12 function keys are arranged in a circular pattern to resemble a clock face, enhancing "touch" typing ability.

under $100. Most of the new units will cost far more. Their selling claim: they will pay for themselves quickly in reduced health-insurance costs and higher productivity.

Most of the new designs keep the standard arrangement of letters and numbers, known as the QWERTY design, after the six letters running from left to right in the top row of letters on a keyboard. The assumption is that corporate customers will not be interested in a product that takes employees more than a few days to learn, particularly if they frequently use temporary help.

Unusual Designs

The Kinesis Corporation in Bellevue, Washington, shipped the first units of its $690 keyboard for IBM Personal Computers (PCs) and PC clones in August 1992. The keyboard relocates the keys from a standard board into two shallow wells just over 6 inches (15 centimeters) apart. The aim is to let the arms and wrists run straight from the shoulder instead of bending the forearm in and the wrist out. The wells allow the shorter outer fingers to travel no farther than the middle fingers to strike a key. A number of control keys are placed in a slightly raised position that is comfortable for the thumb to reach; the raised plastic shelf in front of the pads provides support for the palm.

Another new model, the Tony!, developed by Anthony Hodges of Mountain View, California, is hinged in the middle, so the center of the keyboard can be raised to move

the hands into a more natural, thumbs-up position. Hinges also allow the halves to pivot forward or backward, and individual keys can be moved to fit the user.

The Comfort keyboard, developed by the Health Care Keyboard Company of Menomonee Falls, Wisconsin, is also highly adjustable. Health Care has broken the standard keyboard into three units, each adjustable in height, pitch, and direction. The $590 board also has features for people with disabilities that impede typing.

The simplest tack is that taken by Dr. Alan Grant, an optometrist in Chevy Chase, Maryland, whose modified QWERTY keyboard rises gently to a ridge in the middle, has a trackball to control the cursor with the thumbs in the front, and has the 12 function keys in a circle like a clock face to the left. The unit should retail for under $200.

Johan Ullman, M.D., a Swedish anesthesiologist and inventor, has a design that raises the center of the keyboard to a fixed height like Dr. Grant's, separates the hands like Kinesis's, and supports the lower arm. His company, Medinova A.B., has the backing of Volvo Data, several government bodies, and Swedish unions. The BackCare Corporation, an ergonomics consulting firm in Chicago, Illinois, has been demonstrating the Ullman design in the United States.

Industrial Innovations of Scottsdale, Arizona, is gambling on a more radical change. Its Datahand design, which has cost $1.5 million to develop, has padded hand rests with individual finger wells. Each finger can type four different characters by moving left, right, forward, or back in its well. The thumbs, by pressing down softly or harder on the thumb key, can create different "modes" that give new meanings to any other finger's stroke. The thumbs are also able to reach a mouse control.

You don't have to move your hands at all, which makes true touch-typing possible, says Dale Retter, the company's chief executive. It costs close to $53,000 to make a single unit but Retter thinks Datahands could retail for $300 if mass-produced.

Chordal Systems

Health concerns are not the only force encouraging new keyboard designs. Perhaps the most radical new products are keyboards in which each character is produced by hitting a different combination of keys. For example, users of a device made by Infogrip Incorporated of Baton Rouge, Louisiana, make the letter "a" by pressing the thumb, index-, middle-, and ring-finger keys; "b" by hitting the index- and ring-finger keys; "c" by hitting the thumb and middle-finger keys; and so on. Such systems are called chordal because they resemble the way a pianist produces a chord by hitting several notes.

Using this approach, typists can perform all the functions of a normal keyboard with one hand and as few as seven keys. That has attracted computer users who have only one good hand to use, and others, particularly graphic designers, who like to work with one hand typing on keys, and the other controlling the cursor with a mouse or trackball device.

More important, the chordal systems could be much more compact and portable than traditional keyboards. Tinkerers have already used such systems to build beeper-sized note-takers.

The Macintosh version of the Infogrip product, on sale since March 1992, costs $295 for one hand and $495 for two, which allows for faster typing. Infogrip is now focusing on a new IBM-compatible model.

BackCare is working with Northwestern University to set up comparative tests of several designs. Major computer users are expected to run many similar tests privately. As in other fields, though, the best design may well prove to be a commercial also-ran in the long run.

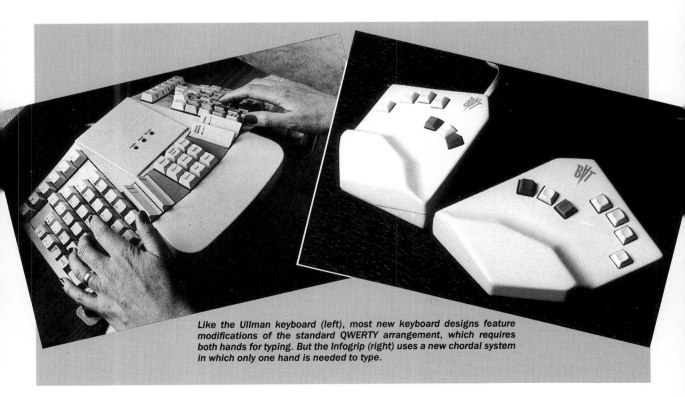

Like the Ullman keyboard (left), most new keyboard designs feature modifications of the standard QWERTY arrangement, which requires both hands for typing. But the Infogrip (right) uses a new chordal system in which only one hand is needed to type.

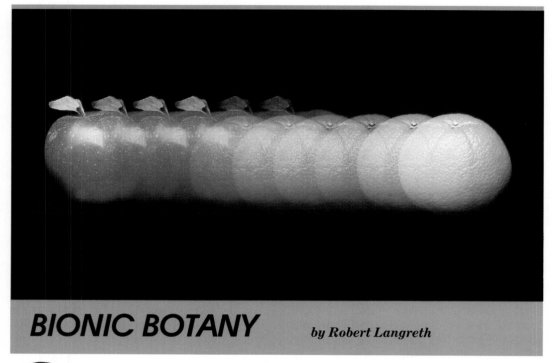

BIONIC BOTANY

by Robert Langreth

Growing in six small fields scattered across Florida and California are some extraordinary tomato plants. Not that you would ever know it by looking at them: Both the plants and the tomatoes themselves look like garden-variety crops.

The unique feature of these crops is hidden inside their DNA—the genetic "blueprint" that determines all their physical and chemical properties. In addition to the thousands of genes that make up the plants' DNA, scientists have inserted a single gene not present naturally. The extra gene slows down a series of chemical reactions that cause ripe vegetables to rot.

Most commercial tomatoes spoil so quickly that they must be picked while green and later artificially ripened with a chemical called ethylene, says William Hiatt, a researcher at Calgene Fresh. A subsidiary of Calgene, Incorporated, a biotechnology company in Davis, California, Calgene Fresh is producing the new tomato. The result of forced ripening is flavorless fruit. In contrast, Calgene's Flavr Savr tomato can stay on the vine a full week longer than regular tomatoes—allowing the necessary sugars and acids to develop—and still be shipped across the country. "Because it can go to color on the vine," explains Hiatt, "it's a better-tasting tomato."

Next year the company plans to start selling the Flavr Savr to retailers. In all likelihood, it will be the first genetically engineered fruit or vegetable available to the public.

Calgene's tomato, however, is just the beginning of what will be a many-splendored high-tech harvest. During the past decade, scientists have transplanted genes with all sorts of new traits into more than 50 different plants—from asparagus to sugar beets. Now, spurred by a U.S. Government decision to streamline the regulation of genetically engineered foods, biotechnology companies are rushing an array of these new crops to the market.

During the next several years, consumers will see "a steady flow" of high-tech produce, says Richard Godown, president of the Industrial Biotechnology Association, based in Washington, D.C. Most of the initial products will be vegetables genetically engineered for resistance to pests, viruses, and blights that have dogged farmers for centuries. Agritech companies are also focusing on traits that will appeal to consumers. Some likely products: tomatoes that are bruise-resistant or that will freeze without becoming mushy when thawed; extra-nutritious potatoes; and vegetable oils with lower saturated fats.

Genetic engineering "will eventually enter the production of all our food crops," predicts Godown. "It's going to be pervasive. It's going to be worldwide. It's the single best hope of being able to feed all the people on the planet."

Poor Image

To live up to this grand vision, biotechnology will first have to overcome its poor public image. Environmental groups, in particular, argue that directly tampering with plant genetic material poses unknown risks.

"Genetically engineering produce is [similar to] adding chemicals to food," says Rebecca Goldburg, a senior scientist at the Environmental Defense Fund. "Although most chemicals . . . have proven safe, every now and then something comes along that's not so safe. Think of some of the food dyes that have been taken off the market" because of cancer and other health risks.

To Goldburg's dismay, a new Food and Drug Administration (FDA) policy does not require government safety approval for most genetically engineered foods, as it does for food additives. Introduced in May 1992, the policy lets companies perform their own safety testing, and recommends they consult the FDA if they find anything unexpected.

Reflecting public fears, headlines have labeled genetically engineered produce "Frankenfood" and "Brave New Food" (referring to *Brave New World*, Aldous Huxley's novel about technology run amok). A *New Yorker* cartoon perhaps captured these misgivings best. It depicted a crazed scientist clenching his fists and cackling triumphantly at his giant-jawed plants: "There's splendid news from the FDA, my pretties!"

A History of Genetic Tampering

What many consumers probably don't realize is that farmers have been tampering *indirectly* with plant genes almost since the be-

ginning of agriculture. Explains Ganesh Kishore, a plant researcher at the St. Louis-based Monsanto Company: "Traditional plant breeding is not very different from [genetic engineering]. The end result of both is that people are reorganizing plant genes to introduce desirable properties."

Take crossbreeding. In many species, pollen from one plant fertilizes eggs from that same plant. In crossbreeding, pollen from one plant is used to fertilize a second plant to create a hybrid that combines the traits of both. For instance, one might try to mate a plant that produces large, bland strawberries with one that produces small, sweet berries, to get a fruit that's ideally both large and sweet.

This technique has produced hundreds of new crops, such as broccoflower, a hybrid of cauliflower and broccoli. But it's somewhat unpredictable, often resulting in unwanted traits along with the desired ones. Moreover, it works only with plants from the same or closely related species.

In 1973, for the first time, Stanford University biologists combined DNA from two unrelated organisms. Common to all living beings (except for some viruses), DNA is a long molecule shaped like a spiral staircase. Each "step" in the staircase is composed of one of four different chemicals called bases; the sequence of these chemicals determines what properties an organism will have. A gene, in fact, is simply a group of steps responsible for determining individual chemical components of cells. Genes are grouped into a fixed number of chromosomes (46 for humans, 20 for corn, for example), which are responsible for complex traits like color or size of an organism.

Like film editors splicing together sections of different videotapes, the Stanford scientists cut open a strand of bacterial DNA using a kind of chemical scissors called a restriction enzyme, then inserted a gene from a second, unrelated bacterium into the opening.

Microbe Carrier

This accomplishment had enormous implications for plant breeders. At least in theory, scientists could now take genes from any organism, plant or animal, and add them to any other.

Furthermore, instead of adding hundreds of new genes in crossbreeding—some desirable and some not—scientists could add exactly the genes they needed.

In practice, however, plants proved more difficult to address than bacteria. Unlike bacterial DNA, plant DNA is protected by a thick cell wall. Since scientists didn't know how to insert materials through this wall without damaging the cell, they looked for a microbe that could carry new genes into the plants for them.

In the early 1980s, scientists from Monsanto and the Max Planck Institute in Cologne, Germany, showed that *agrobacterium*—a common type of bacterium that infects tomatoes, potatoes, and many other crops—does the trick quite effectively. Over millions of years of evolution, *agrobacterium* has learned how to penetrate cell walls and attach disease-causing genes to the plant's DNA. The research teams first removed these harmful genes from the bacteria, then replaced them with genes they wanted to add to the plant. Next they mixed the altered bacteria together with plant cells and let the microbes do the rest.

It worked. And researchers had already figured out how to regenerate whole plants from individual plant cells. The race to develop genetically engineered foods was on.

Making Agriculture Environmentally Friendly

The first fruits and vegetables of biotechnology will hit the supermarkets in three "waves," according to Robert Fraley, the head of agricultural research at Monsanto, which is developing several high-tech vegetables. Calgene's tomato being the exception, the first wave of gene-spliced produce will be mostly pest- and disease-resistant.

Monsanto and several other companies, for instance, have added genes from the caterpillar-attacking bacteria *Bacillus thuringiensis* into cotton, tomato, and potato plants. Once inside the plants, the genes produce proteins that kill crop-eating caterpillar larvae, but are harmless to other animals and plants. Other companies have added new genes to cantaloupe, squash, and potato plants to make them resistant to deadly viruses. In yet a third development, Monsanto and others have genetically altered cotton plants to make them tolerant

The cotton industry stands to reap the benefits of genetic engineering. At right, the cotton plant that has been genetically engineered to repel insects (right) has much more luxuriant foliage than its unadulterated counterpart (left). Likewise, the genetically engineered plant produces bigger cotton bolls (above) than a normal cotton plant.

Bruised tomatoes (above) spoil very quickly. Biologists are now working to extract genetic material from injured tomatoes (left) in hope of finding a genetic way to enhance bruise-resistance.

to glyphosate, an environmentally safe weed killer. (Many environmentalists, however, complain that such crops would promote chemical use. Some worry that insects could eventually become resistant to the pest-proof plants, resulting in widespread crop failures. Conversely, others speculate that human-made plants may crowd out naturally occurring foliage.)

Within five years, more than a dozen pest- and pestilence-proof crops could hit the marketplace, say industry leaders. In Third World countries, insect-repelling vegetables could help alleviate serious food shortages. "Conventional agricultural techniques will not provide the more than doubling of food output that the world will need" in the next 40 years, argues Fraley. "Genetically modified crops are one key" to doing so. In the more affluent United States, bug- and virus-proof crops "are going to be a tremendously positive step toward making agriculture more environmentally friendly," Fraley says. Besides reducing pesticide use up to 60 percent for some crops, he notes, these crops may make it possible for farmers to plow their fields less often, slowing soil erosion. Currently farmers must till their soil often to get rid of weeds and bugs.

"Coming quickly on the heels of [insect- and virus-resistant crops] are bioengineered vegetables that are targeted toward food processors or provide unique consumer traits," says Fraley. Much of this second wave of products will involve tomatoes, for two reasons: Tomato plants are easy to work with in the laboratory, and tomatoes command a relatively high price.

Besides Calgene's Flavr Savr, both Calgene and Monsanto are working on a more advanced way of controlling the ripening of tomatoes and other fruits. Their efforts are based on the work of Department of Agriculture researcher Athanasios Theologis, who recently confirmed the long-held theory that naturally occurring ethylene is the driving force behind the ripening of fruits. By "turning off" the gene that causes ethylene production, fruits could be made to last many weeks before spoiling. (Flavr Savr delays spoiling only about a week.)

"Today 50 percent of the world food crop often goes down the drain because of spoilage," explains Peter Quail, assistant director of the Plant Gene Expression Center in Albany, California. "Theologis' contribution could make a widespread improvement."

Other ideas are downright weird: Plant geneticists at DNA Plant Technology Corporation of Cinnaminson, New Jersey, are trying to insert a synthetic version of arctic-fish genes into tomatoes, so they can be frozen and thawed without becoming mushy. If the company succeeds, the tomato-fish combo could arrive at your fruit stand in 1996.

Beyond Tomatoes

Not everything is coming out tomatoes. Monsanto has perfected a genetically engineered potato with higher levels of starch than usual. This will make the fried potatoes tastier and more nutritious (because they will absorb less oil during cooking), says Kishore, the lead researcher on the project. "Food scientists have been trying to do this for 70 or 80 years, without much success," he boasts. "We did it in two."

Buoyed by this feat, researchers are trying to alter the chemical content of vegetables in other ways. Scientists at DNA Plant Technology hope to make sugar snap peas and red peppers that stay sweeter longer. At Calgene, researchers have engineered canola (a plant with a seed similar to soybean) to produce oils with potentially healthier saturated fats, for use in margarine and other foods. It may even be possible to create a caffeine-free coffee bean or, according to experts, antiflatulence kidney beans.

"In the future, biotechnology companies are going to concentrate on changing the taste and flavor of food," says Vic Knauf, the head of research at Calgene.

In Fraley's third wave of bioengineering, "manufacturers will use genetically engineered plants as factories" for making drugs, industrial chemicals, and even fuels. The basic idea is this: Plants naturally produce an incredible variety of oils, waxes, rubbers, and other substances. Why not use genetic engineering to tailor plant products to human needs?

BIOTECHNOLOGY: A Third World Savior?

For the past 10 years, researchers have hailed genetic engineering as a technology that will reduce hunger and poverty in developing countries. Starting about a year from now in Mexico, they will get their first chance to prove it.

The Mexican Government, using seeds and expert assistance donated by St. Louis-based Monsanto Company, is adapting the company's virus-resistant potato plant for Mexican varieties of spuds. By the beginning of 1994, Mexico plans to start providing farmers with seeds for these plants.

Every year, between 10 and 60 percent of Mexico's potato harvest is destroyed by insect-borne viruses, according to Luis Herrera-Estrella, a Mexican researcher who is coordinating the project. For small farmers who can't afford pesticides, the virus can wipe out entire crops.

"For many farmers, this genetically engineered plant could mean survival," Herrera-Estrella says. "Overall, the project could reduce crop losses in Mexico by 15 percent."

Strangely enough, the source of the virus resistance turns out to be the virus itself. To reproduce, the potato virus must first remove a chemical "coating" that normally covers it. The genetically engineered potato plants, however, contain the viral genes that allow them to make this coat. "Every time the virus gets ready for sex by taking off its coat," the plant simply covers it with another coat, says Herrera-Estrella. Even after a month of constant exposure to the virus, the plants are hardly affected.

Monsanto is also collaborating with African countries. In giving away its technology, however, the biotech giant isn't simply exhibiting corporate altruism. Besides generating some rare positive publicity, says Monsanto research chief Robert Fraley, the company hopes that gene-altered vegetables may someday make Third World farmers rich enough to become paying customers.

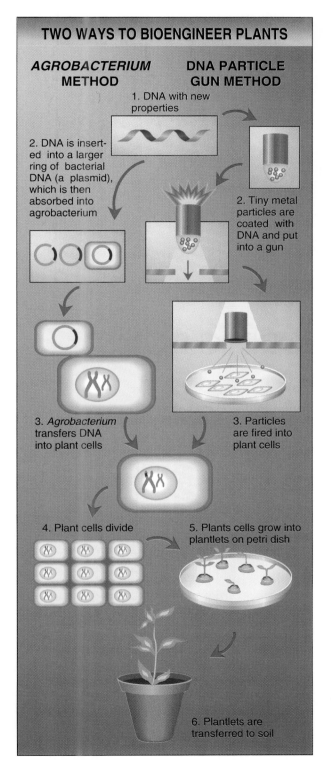

TWO WAYS TO BIOENGINEER PLANTS

AGROBACTERIUM METHOD   DNA PARTICLE GUN METHOD

1. DNA with new properties

2. DNA is inserted into a larger ring of bacterial DNA (a plasmid), which is then absorbed into agrobacterium

2. Tiny metal particles are coated with DNA and put into a gun

3. *Agrobacterium* transfers DNA into plant cells

3. Particles are fired into plant cells

4. Plant cells divide

5. Plants cells grow into plantlets on petri dish

6. Plantlets are transferred to soil

Indeed, genetic engineers have already altered canola plants to produce oils for laundry detergents, and modified tobacco plants to grow an enzyme used by bread manufacturers. In the most heralded accomplishment of this sort, botanists at Michigan State University recently engineered a relative of the rape plant to produce tiny granules of biodegradable plastic throughout the plant. If the technology can be improved to make larger quantities of the material, the researchers say, farm-grown plastic could eventually be used in plastic containers and plastic wrap.

Future Challenges

There are still many limitations to gene-splicing technologies. Only in the past year or so have biologists succeeded in genetically altering major grain crops such as wheat. Unlike most vegetables, these plants are resistant to *agrobacterium,* so scientists have had to develop other ways of getting genetic material inside plant cells. The most successful technique so far has been to literally shoot new genes through the cell walls, using a blank cartridge of a .22-caliber gun. However, this method has not proven as reliable as using the bacteria.

Researchers also can't add more than three to five genes into a plant at once—a procedure required for controlling more-complex properties. For example, several years ago, many university professors spoke excitedly about being able to engineer corn for so-called nitrogen fixing. This would allow the corn to convert atmospheric nitrogen directly into critical nutrients—eliminating the need for expensive fertilizers. But nitrogen fixing involves about 20 genes, and biologists aren't close to succeeding.

Nevertheless, the biggest question mark hanging over high-tech plants isn't the technology itself. ''What will happen won't be determined by what genetic-engineering technology is capable of,'' says Calgene's Knauf, ''but by what consumers want.''

Right now consumers aren't sure whether they want gene-altered foods. But if scientists can overcome the public ambivalence, gene-splicing gurus like Monsanto's Kishore foresee a bountiful future.

The Return of the GEODESIC DOME

by Gene Knauer

Geodesic domes are coming back. These spherical structures—most identified with the 1960s and engineer/futurist R. Buckminster Fuller—are in growing demand throughout the United States, Canada, and in Asia, Western Europe, and the Middle East.

The staggering cost of energy—both in economic and environmental terms—is forcing individuals and nations to look for ways to minimize their consumption. Energy efficiency is the main reason for the renewed interest in geodesic domes.

"We're seeing the beginnings of a new interest in domes," says Cynthia Kerstiens, executive editor of *Dome,* an industry magazine with a readership of several thousand. "They fit the 'do more with less' mentality of today."

The Home in the Year 2000
A large Southern California developer recently conducted a survey titled "The Home in the Year 2000" to gauge future expectations within and outside the building industry. In response to the question "What will

homes in the 21st century look like?", both homeowners and builders predicted that homes would become more spherical in design. Energy efficiency, structural integrity, and economy of building materials were all cited as major factors pushing this trend.

Dome structures were first introduced into the popular culture in the 1960s. Back then, domes celebrated the same ideals of simplicity and efficiency embodied in the Volkswagen Beetle. The dome ideal in the 1960s did not become an unblemished reality, however. Eager but unqualified builders with utopian visions fashioned domes in the back woods and for communes, using everything from papier-maché to tin cans. When it rained, these do-it-yourself domes leaked profusely, since all those triangular joints weren't fitted snugly enough together. The absence of skylights made most of those early domes dark, and insulation was either nonexistent or mediocre.

Today's dome manufacturers, using precision-cut materials and professionally designed dome shells and interior floor plans, say they have perfected the structures. Growing numbers of buyers are also attracted to domes for their architectural merits and their ability to withstand virtually every natural disaster except fire and flood, in addition to their low energy consumption. One Florida-based dome manufacturer offers full replacement if any dome they sell is damaged by a hurricane, earthquake, or tornado. The company has in fact concentrated its marketing efforts in areas hard hit by recent natural disasters, helping to rebuild hurricane-ravaged areas in Florida, the Carolinas, Virgin Islands, California, and Alabama with dome homes.

The Development of Domes
Domelike living structures can be traced back to the days of the Romans and the domed huts of ancient Tunisians. Eskimo igloos, the spherical yurts of nomadic Mongols, and certain domed tepees of American Indians are all predecessors of the geodesic dome.

Modern domes have been built primarily according to designs by Buckminster Fuller, who conceived of constructing domes as a network of intersecting triangles, the triangle being the strongest geometric shape. One of Fuller's first dome-home designs called for a home supported by a pole, with space below for a car and a jet-propelled, wingless aircraft. Called the "4-D House" (referring to

Einstein's fourth dimension, time), the name was later changed to "The Dymaxion Dwelling Machine." Fuller later developed a future vision of cities covered by geodesic domes, with sunlight and weather easily regulated.

Today's geodesic dome is more likely to be three-eighths or five-eighths of a full sphere of intersecting triangular sections assembled on a conventional foundation. The triangular network design provides for a free-standing, self-supporting structure requiring no internal supports or load-bearing walls, thus opening up the interior of the dome for maximum space and light.

Due to the mathematics involved, the bigger the dome, the more energy-efficient and materials-efficient it becomes. The volume grows by a factor of eight every time the diameter is doubled, and the surface area increases by a factor of four.

Dome Efficiency
Because a dome with the same square footage as a box-type structure has 38 percent less exterior surface area, it takes considerably less energy to heat or cool a spherical structure. The open interiors of domes also encourage better air circulation and uniform temperature.

In a study commissioned by the National Dome Council, geodesic domes were compared with conventional, rectilinear

Geodesic-dome structures are constructed using precision-cut frames and professionally designed dome shells that are noted for their great resiliency and energy efficiency.

homes in five different climate zones. The results conclusively showed domes to be 30 to 40 percent more energy-efficient for heating and cooling than traditional houses. With big skylights bringing sunlight into the open interiors, domes often incur lower lighting costs as well. Domes, it was also discovered, require less insulation to achieve the same energy efficiency as rectilinear homes. For this reason, dome homes qualify for increased financing under the Federal Home Loan Mortgage Corporation regulations.

If you translate the energy savings of a dome home in terms of barrels of oil and use some arithmetic, the results are impressive. According to figures from the Pacific Gas & Electric Company in San Francisco, California, 1 barrel of oil generates approximately 625 kilowatt-hours of electricity. In California the average monthly consumption of electricity is 500 kilowatt-hours per residence. A 40 percent savings represents 200 kilowatt-hours per month, or 32 percent of a barrel of oil, conserved. And for every barrel of oil used to create energy for heating or cooling, additional carbon dioxide is emitted as pollution into the atmosphere, so using less oil via a more energy-efficient domicile means polluting less.

California, which has more than 10 million residences, could cut its annual energy consumption by 38.4 million barrels of oil if all residences were as energy-efficient as geodesic domes. And, on a national level, energy savings would be about nine times that of California, or about 350 million barrels of oil conserved annually.

A Unique Environment
Beyond their energy efficiency, domes are unique, dramatic environments. With no load-bearing walls, the dome environment is open and impressively spacious, with high, vaulted ceilings.

Many artists are drawn to the spaciousness and aesthetics of domes. "There's something very expansive about the dome that inspires creativity," says painter and sculptor Marianne Wyss, who lives and works in a dome in Seneca, Maryland. "The abundance of natural light is also a plus for an artist."

When Jim Hill, an engineer at the National Aeronautics and Space Administration (NASA), isn't designing backup flight-control systems for the space shuttle, he retreats to his Lake Nacimiento, California, geodesic-dome retirement home. "I was going to build a log cabin, but I found domes were very energy-efficient, and I like the open, light interiors." Hill claims his geodesic-dome home has reduced his estimated heating and cooling costs dramatically. "For folks on a fixed income interested in a retirement home or a vacation cabin, I recommend looking into domes," says Hill. "They're spacious and very affordable."

In the United States, dome homes can be found in all 50 states, although the greatest concentrations occur in the Southeast and Northwest. According to some estimates, the U.S. dome industry is currently producing and selling about 1,500 dome kits annually.

Future Applications
Dome designs can vary dramatically in size, materials, and architectural configuration. Their use as permanent structures is unlimited; throughout the world, geodesic domes serve as primary residences, vacation homes, commercial buildings, theaters, and sports arenas. For example, in Japan, plans are under way for entire master-planned communities of domes.

Geodesic Domes, Inc., of Davison, Michigan, sells more domes for use as churches than as family homes and commercial buildings combined. "When you think about it, domes are perfect sanctuaries in the round," says GDI's Tom Ferguson. "There are no walls or supports to block your view, and they're less expensive or at least on a par with other options."

Public familiarity with dome architecture has grown as people have been exposed to domed stadiums, radar stations, conservatories, and theaters. In the Austrian Alps, a large dome covers an all-weather spa, allowing it to stay open throughout the year. In Huaxiang, China, a communal-living complex is made up of a cluster of domes. In Ghana a science-and-technology fair is spread out under a geodesic dome. And in

Home buyers are attracted to the spaciousness and aesthetic appeal of domes. The self-supporting structure is ideal for encompassing a swimming pool (left). The interior of the double-dome abode above is rich with customized features, such as unique window encasements and a spiral staircase.

Africa—countries that are experiencing mass immigration—about the benefits and cost efficiency of building domes to house newcomers. Timberline Geodesics, a dome-manufacturing firm, has made a formal proposal to the city government of Berkeley, California, offering domes at cost to help house the poor and homeless.

Since geodesic-dome shells are available as easily assembled, color-coded kits of precision-cut pieces and patented hardware connectors, they can be shipped all over the world and assembled with simple hand tools, even in the world's remotest places. And science fiction writers and space illustrators have consistently maintained their visions of domed, self-sustaining, extraterrestrial colonies of Earthlings.

Dome enthusiasts see domes in our future—lots of them. They will not only serve our need to conserve energy, but will also change our sense of just what the aesthetics of the buildings we live and work in should be: full of light, full of space, and *round*.

the harsh climate of the Antarctic, a huge dome encloses a research facility and living station for American scientists.

Geodesic domes are being proposed for a variety of applications worldwide. Dome-industry executives are talking to nations in Asia, the Middle East, Eastern Europe, and

Computers

DYNAMIC RANDOM-ACCESS MEMORY

In 1992 computer-chip manufacturers vied to develop memory chips that could hold many millions of bits of information at one time to increase the efficiency of personal computers (PCs). Called DRAMs (for Dynamic Random-Access Memory), these fingernail-sized slivers of silicon contain the data used by a PC's microprocessor to carry out various functions. The largest DRAM chip in production in 1992 contained 16 million bits of information; chips holding 64 million bits should be ready for use within three years. However, a new international venture established in 1992 among IBM Corporation, Siemens AG, and Toshiba Corporation seeks to create chips that hold 256 million bits of information. These chips will take somewhat longer to develop because they will also require new manufacturing technology to produce them. Currently, a precise lithographic machine called an optical "stepper" is used to etch memory cells on silicon chips. To fit 256 million bits of information on a single chip, a stepper must be able to etch lines that are a quarter of a micron in width—400 times thinner than a human hair. This is three times thinner than is commercially possible today. As a result, IBM has also formed a joint venture with Motorola Corporation to develop technology that uses X rays to etch these lines.

▶ **Size and Design.** DRAM chips are measured in megabytes (MBs), which indicate how many millions of bits of information they can hold. For example, the 4MB DRAM chip in widespread use in 1992 holds 4 million bits of information. Within each chip, each bit of information is stored in a single cell, and these cells are arranged in a square grid.

In most chips currently on the market, data can enter and exit from the chip at only one point, and only one bit of data can pass through this "door" at one time. This design works efficiently for small chips that contain 1 million to 2 million bits of information or less. But for the 64MB and 256MB DRAM chips in development, the design creates several problems. Each "super" chip contains many more millions of tiny data cells than its predecessors. As a result, many of these cells are situated far away from the chip's "door" and take a long time to retrieve. A PC's microprocessor must wait for each bit to arrive at the door and then exit before it can complete the task that requires the information. In addition, once retrieved, each bit of information must wait its turn to exit, which creates another delay for microprocessor functions that require multiple bits of information.

Chip developers are currently looking at various ways to change the square-grid design to ensure that the computing potential promised by the new superchips will be fully realized. For example, some manufacturers are redesigning chips to provide more than one access point for passing information along to the microprocessor. Others are simply making these access points wider to allow more bits of data to exit at once. Another design allows the DRAM chip to work synchronously with the microprocessor; that is, instead of waiting while the DRAM chip retrieves data, the microprocessor performs other tasks until the information is available.

ON-LINE BULLETIN BOARDS

Industry analysts estimate that in 1992 nearly 10 million people used their PCs to "log on" to electronic bulletin board services (BBSs), as these on-line information exchanges are commonly known.

An electronic bulletin board is a "space" set up on a central computer where PC users can post ideas and questions about topics ranging from business investments to the best places to sight rare birds. BBS subscribers can also search for and copy information from various on-line information databases, and they can sample "shareware" software programs written by enterprising programmers. (Shareware allows owners of personal computers to try new programs on a trial basis and then pay only for those programs they actually use.)

To participate in BBSs, users need only a PC linked to a phone line via a modem, along with a communications program that can dial BBS phone numbers. Once accessed, bulletin boards display menus of options that list the various topics available. Users then choose the specialty they want to peruse, and read the messages left by other members of the BBS or leave comments or queries of their own.

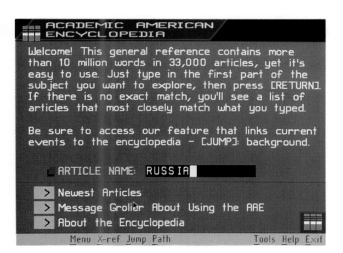

Anyone who has access to a personal computer and a modem can tap into a wealth of information provided by electronic bulletin board services. Subscribers to Prodigy Services can, for example, log onto the Academic American Encyclopedia (above).

▶ **Rapid Growth.** Fifteen years ago, BBSs didn't exist. But by 1987, 4,000 public-access, on-line information exchanges were in operation in the U.S. In 1992 that number grew to over 44,000. Eighty percent of these are nonprofit endeavors maintained by "regular folks" out of their homes. The other 20 percent are commercial operations run by big companies for profit. These services charge an initial fee ranging from $10 to $90 to join, and then monthly fees based on use.

Together, CompuServe Information Services and Prodigy Services account for 74 percent of the commercial on-line BBS industry. In 1992 CompuServe had 1.1 million paid subscribers, while Prodigy boasted 1.75 million members. CompuServe provides over 300 forums that cover 5,000 topics. Prodigy lists over 400 topics, including several forums introduced in 1992: veterans of the armed services; singles; senior citizens; and support groups for AIDS, alcohol abuse, and incest survivors. During the 1992 presidential campaign, Prodigy also offered an "Ask the Candidate" forum, which allowed subscribers to post messages to and about the candidates, and which allowed the candidates to respond.

▶ **Types of BBSs.** The most popular topics on bulletin boards deal with hobbies—in particular, computers, genealogy, and cooking. But many other diverse subject areas are also addressed, including birdwatching, Dante's *Divine Comedy*, herpetology, astronomy, gay issues, and radio/film/television. America Online provides a service called Romance Realtime, which permits singles to find love via their modems. BBSs also provide breaking news and sports information, weather maps, home shopping, encyclopedias, computer games, stock quotes and investment advice, private messaging, banking, and airline information. In addition, many software companies provide BBSs to enhance their customer technical-support system as well as to allow customers to sample software.

One of the fastest-growing electronic information exchanges is CompuServe's Working from Home forum, which has 25,000 subscribers. Small-business owners use it to swap advice on marketing and health insurance, as well as to solve technical computer problems.

Other BBSs cater to persons with long-term illnesses. For example, CompuServe lists forums on diabetes, hypoglycemia, cancer, and AIDS. These on-line information exchanges not only provide medical news to their subscribers, but they also allow housebound patients to stay in touch with the outside world.

Abigail W. Polek

Technology

ROBOTICS

Robots are advancing, if quietly. Transitions Research Corporation in Danbury, Connecticut, is marketing "Helpmate," a robot that can move about a hospital performing such feats as delivering meal trays, documents, and supplies to rooms and offices. The $60,000 robot, about the size and shape of a portable dishwasher, can speak prerecorded messages and operate an elevator. Another service robot developed by Transitions Research, "RoboKent," can clean floors without being supervised.

Integrated Surgical Systems, Sacramento, California, in collaboration with IBM, has developed a surgical robot that cuts hipbones so that hip implants can be inserted. The cuts leave gaps between the bone and replacement part that are only 50 microns wide—much thinner than the results of surgeons.

The doctors first cut through the soft tissue and prepare the area. Then the "Robodoc," working in tandem with a computer, analyzes the geometry of the hip, and uses its blades to cut away at the hipbone while doctors supervise.

On the industrial scene, researchers made progress in giving a robotic arm the ability to identify and grasp irregularly shaped objects that are positioned randomly. Perceptron, Farmington Hills, Michigan, is marketing a robotic system that includes a camera with a laser scanner that can yield a three-dimensional picture of a number of objects. That information is then used to position a robotic arm so that it will successfully grasp one of the objects.

In the lighter vein of robotics, Sembog Corporation, Lansing, Illinois, received a patent for an "animated character system with real-time control." Developed by engineer Victor Lang, Jake, a 2.5-foot (76-centimeter), 68-pound (31-kilogram) robot, can "see" with a color camera in its right eye, can "hear" with

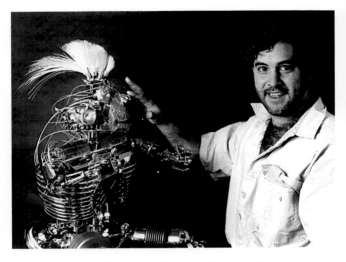

Jake (above left, with creator Victor Lang), a remote-controlled robot, has many human characteristics, including vision, hearing, speech, and mobility.

microphones in its ears, and can respond by moving and speaking—with the aid of a person using a remote receiving and controlling device.

COMMUNICATIONS

Researchers at AT&T have developed a way to increase fourfold the amount of information that can be sent by lasers through fibers. This advance in fiber-optics technology was achieved by sending four different laser beams—that is, four different wavelengths of light—over a fiber in a coordinated fashion. The resulting capacity of transmission is 6.8 billion bits of information per second. Meanwhile, Northern Telecom, a Canadian company, has improved its "surface-emitting circular grating laser," a technology that could lower the cost of fiber-optics systems of the future. This laser emitter has a wider diameter than a typical laser used in fiber optics, so the light is more easily lined up with the small-diameter optical fiber through which it must travel.

Cable-television companies continue to develop technologies for sending telephone messages over existing cable lines —which would compete with traditional telephone communications. Cox Enterprises made a cross-country telephone call using a wireless telephone, a linking system, and its cable-television system operated out of San Diego, California. McCaw Cellular Communications, Kirkland, Washington, began operating such a system on an experimental basis in Ashland, Oregon.

Nimrod International Sales of Clarksburg, New Jersey, is marketing a hand-held device that, when placed by a telephone mouthpiece, automatically dials a number. The device, named "Phone Home," can be programmed with a home number and carried by children who might forget the number. It works by emitting dialing tones that are heard and translated automatically by the telephone system.

ENERGY

The use of windmills to generate power continues to expand. U.S. Windpower of California contracted with Iowa Gas and Electric Company, the Sacramento Municipal Utility District (California), and Niagara Power Corporation (New York) to build power-generating windmill systems. U.S. Windpower's latest-generation windmills have long fiberglass blades operating at changeable pitches, to produce optimal velocities in the midst of varying wind speeds. Computers analyze the flow of air over the blades, and make adjustments that smooth the flow and produce steady currents in electric generators. These new windmills operate in a wide range of winds, from 9 to 60 miles (15 to 96 kilometers) per hour.

A newly developed light bulb uses a radio signal in place of an electric filament. The E-Lamp, or electronic lamp, generates a high-frequency radio signal, which fluctuates and stimulates the gas filling a bulb; the gas then produces an invisible light that stimulates a phosphor coating on the inside of the bulb's glass. The E-Lamp will last for 20,000 hours of use (about 14 years of typical home use) before the phosphor fades. And it can operate in cold temperatures. The bulb's de-

velopment was a venture of American Electric Power and Intersource Technologies, both of Sunnyvale, California.

Solar Kinetics, Dallas, Texas, has produced a solar collector that is easily disassembled and made of less-expensive materials than current collectors. It includes a metal foil membrane stretched between a circular rim and pulled into a parabolic shape by a vacuum. The foil is the same material of which small, helium-filled balloons are made. As in glass systems, the membrane reflects and focuses solar rays on a collector, where the concentrated rays are used to generate electricity.

ENVIRONMENTAL ADVANCES

A Philadelphia, Pennsylvania company has produced an air conditioner based on a desiccating system rather than on the familiar system of compressing chlorofluorocarbon gases (CFCs), which are suspected of harming the Earth's ozone layer. The system created by ICC Technologies is based on an old idea—drying air by using a solid desiccant (drying agent)—yet improves the process thanks to a new desiccant developed by the Englehard Corporation, Iselin, New Jersey. The effective new agent allows the air conditioner to remove nearly all moisture from the air, and then reintroduce some moisture—thereby cooling the air.

Corning Company, San Antonio, Texas, has introduced a new precatalytic converter for automobiles. It is a grill made of powdered metals, which sits in front of a longer ceramic grill. The first grill quickly heats exhaust gases, using the power of a battery. The second grill then converts environmentally harmful emissions to nontoxic substances. Such precatalytic converters are being sought because ordinary catalytic converters do not work in the first few minutes of operating a car—the exhaust is not hot enough. In tests the Corning converter produced emissions that met the particularly strict automobile-emissions standards mandated in the state of California.

Donald Cunningham

Transportation

ON THE ROAD

"Run-flat" tires, which can be used after they have become deflated, are now being marketed by major tire manufacturers. Unlike normal tires, the run-flat models do not actually become flat after losing air. Instead, because of increased support in their sides, these tires maintain integrity, albeit somewhat flatter than normal. Bridgestone's run-flat tire, named the "Expedia," can travel at least 50 miles (80 kilometers) at a speed of 55 miles (88 kilometers) per hour after being deflated. Goodyear's Eagle Extended Mobility Tire will travel 200 miles (320 kilometers) at a speed of 55 miles per hour after being deflated. These tires use a rubber compound that does not heat up as quickly as does the rubber in ordinary flat tires. Also, an electronic sensor is installed to send a signal to the driver when a tire has lost pressure. General Motors offered the Bridgestone run-flat tires on some models of its Corvette sports car.

"Smart" tires for trucks contain an embedded computer chip that sends out radio signals with information that can be read by a driver or trucking company. The information reveals the age, type, and usage of the tires—for example, telling when the tires were recapped. By providing such data in an easy and dependable manner, smart tires can save trucking companies a great deal of money. Goodyear continued to test a version of such tires in the trucks of a number of small freight companies. In the future the chips embedded in smart tires also will be able to monitor other aspects of the tires' histories, such as inflation pressures and cases of overheating.

Electronic toll collection expanded, as authorities in New York joined those in Oklahoma, Texas, and Louisiana in testing systems that monitor passing cars and deduct tolls from prepaid accounts. In these systems a device in a tollbooth sends out a radio wave, which intercepts a sensor mounted on a passing car. The sensor identifies the car, and the reflected radio signal, read by the electronic tollbooth, is used to deduct an amount of money that the driver has placed in an account. The driver need not slow down when passing through the electronic tollbooth. The Amtech Corporation of Dallas, Texas, has been the leader in developing and installing this technology.

Paying Without Stopping

Two electronic toll collectors, one developed by the Amtech Corporation and the other by a cooperative effort by A.T. & T. and Mark IV Industries, are beginning tests at toll plazas on the Garden State Parkway and the Gov. Thomas E. Dewey Thruway.

Similar systems are already being used in Louisiana, Oklahoma and Texas, cutting the costs of toll collection. Officials say such systems can also be used to monitor and regulate traffic more efficiently.

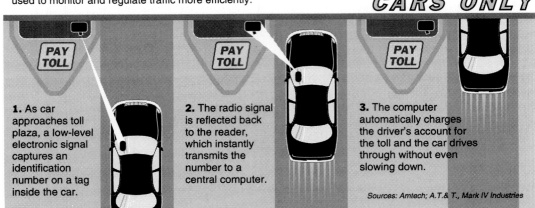

CARS ONLY

1. As car approaches toll plaza, a low-level electronic signal captures an identification number on a tag inside the car.

2. The radio signal is reflected back to the reader, which instantly transmits the number to a central computer.

3. The computer automatically charges the driver's account for the toll and the car drives through without even slowing down.

Sources: Amtech; A.T. & T., Mark IV Industries

IN THE AIR

Rosemount Incorporated of Minneapolis, Minnesota, marketed a device that, when installed on the side of an aircraft's fuselage, will measure the buildup of ice on the plane's surface and automatically turn on heaters to melt the ice. The device consists of a vibrating probe, whose frequency of vibration shortens as ice gradually accumulates on it. The probe's data are stored by a computer chip in an internal canister, which, at a certain point, will electronically alert the plane of a serious buildup of ice. At a certain thickness of ice, a small heating element defrosts the probe itself.

A few airports have introduced a de-icing technology that operates like an automatic car wash. Built by a Swedish firm, the system consists of a large gantry through which a plane is slowly towed while a series of nozzles spray deicing fluid. The spraying, which takes about five minutes, is controlled by a computer program that stores the measurements of planes in computer memory and adjusts the nozzles to accommodate different planes. The systems have been installed at Standiford Airport, Louisville, Kentucky; Stapleton International Airport, Denver, Colorado; and Franz Josef Strauss Airport, Munich, Germany.

Fixed-wing and rotary aircraft have been flying for many decades, but a third flying technology—the flapping-wing aircraft—has remained grounded. That is, until two engineers—Jeremy Harris of the Battelle Institute in Ohio, and James DeLaurier of the University of Toronto—successfully tested a small, flapping-wing craft, or ornithopter, in a field north of Toronto. The plane, which began its flight by being tossed into the air by a researcher, flew up to 100 feet (30 meters) off the ground, made figure eights, and landed safely. It was powered by a model engine and remote-controlled by radio. The plane is 5 feet (1.5 meters) long, with a wingspan of 10 feet (3 meters), and is made of light materials such as Kevlar and carbon fibers, to keep its weight down to 9 pounds (4 kilograms). The basis of flight is a three-part wing structure: Two outer wings are attached to a middle wing segment, which rises and falls three times per second. This produces the flapping of the outer wings, which, in addition, perform a twisting motion at their tips. The downstrokes of the wings provide lift, while the shape and motion of the wings create forward thrust. The plane has a conventional tail, with flap and rudder. The engineers built a large version of the plane, powered by an internal-combustion engine, which they planned to test in 1993. A future goal is to create an ornithopter large enough for a single person to ride and operate.

ON THE WATER

Lightweight materials also showed up in the *Aruba 12*, a new small boat made by Pelican International of Quebec. The hull of the *Aruba 12* comprises a layer of plastic foam sandwiched between two layers of an alloy of polycarbonate and ABS (acrylobutylstyrene). The foam gives the structure rigidity and buoyancy, while the alloy gives it strength. The strong, lightweight alloy, named Triax 2000, was created by the Monsanto Company of St. Louis, Missouri. As a thermoplastic, Triax 2000 can be melted and easily shaped, so that the manufacture of the *Aruba 12* is less labor-intensive and less expensive than if it employed, for example, fiberglass. Without an outboard motor, the *Aruba 12* weighs 210 pounds (92 kilograms), and retails for $1,900.

A sailor in Los Angles developed a new spinnaker, the large, triangular sail fixed to the forward end of a sailboat. Lincoln Baird's design includes a lightweight tube that runs horizontally around the sail and inflates automatically as the sail fills up with wind. Valves in the tube prevent it from deflating quickly. The tube causes the sail to remain spread out, even when the wind shifts direction. The spread sail fills with wind better and propels the boat faster. Ordinarily, spinnakers collapse somewhat as winds shift direction, thereby losing efficiency.

Donald Cunningham

In Memoriam—1992

ALLEN, JOSEPH G. (79), U.S. physician who discovered that the serum-hepatitis virus in plasma could be inactivated by storing it for six months. He also warned about the risk of infection from commercial blood banks. He served as chairman of the department of surgery at the Stanford University School of Medicine. d. Stanford, Calif., Jan. 10.

ANDRONICOS, MANOLIS (73), Turkish-born Greek archaeologist who discovered the ancient tombs of Macedonian royalty, including, in 1977, the tomb of Philip II, the 4th-century-B.C. king who conquered Greece and was the father of Alexander the Great. The tomb also contained the first complete painting from the Hellenistic period, and many ornaments and weapons that helped provide a new perspective on Macedonian culture. d. Salonika, Greece, March 30.

ARTH, MALCOLM J. (61), U.S. anthropologist and educator who served as chairman of the education department of the American Museum of Natural History in New York City from 1970 to 1991. Believing that museums should be lively places that attract a wide range of visitors, he created an overnight "camp-in" for children, the Margaret Mead Film Festival, and the People Center, where visitors could see re-creations of native activities and actually touch and use implements from the country's past. d. New York, N.Y., Jan. 13.

ASIMOV, ISAAC (72), Soviet-born U.S. writer of popular science and science fiction. He wrote nearly 500 books, including the sci-fi classic *I, Robot*, published in 1950. In his nonfiction works, he displayed a gift for explaining complex scientific concepts to laypeople. d. New York, N.Y., April 6.

ATWOOD, KIMBALL C., III (71), U.S. geneticist who helped develop molecular hybridization, a technique for detecting and analyzing individual genes in a chromosome. d. Woods Hole, Mass., Oct. 13.

AUERBACK, ISAAC L. (71), U.S. engineer who helped develop the first Sperry Univac computer, and later worked on the first real-time computer guidance system for the American space program. He also was instrumental in developing communication between different computers, computerizing the airline-ticket-reservation system, and computerizing the country's ballistic-missile early-warning-defense system. d. Lower Merion, Pa., Dec. 24.

AVIGAD, NAHMAN (86), Israeli archaeologist who discovered the Second Temple and remnants of an ancient city in Jerusalem. He discovered the earliest depiction of a menorah in a wall more than 2,200 years old, and found evidence that the city had been destroyed by the Romans. In 1956 he identified the last of the Dead Sea scrolls. d. Jerusalem, Jan. 28.

BALASSA, LESLIE L. (88), Hungarian-born U.S. scientist who invented painkilling treatments for arthritis and tooth extractions, and held the patent on microencapsulation, a process used in time-release medications. d. Pomona, Calif., July 2.

BOGERT, CHARLES M. (84), U.S. herpetologist who researched the temperature-control mechanism of amphibians and reptiles and studied the sounds made by toads and frogs. He chaired the Department of Herpetology at the American Museum of Natural History, retiring in 1968. Twenty-one reptiles and amphibians have been named after him. d. Santa Fe, N.Mex., April 10.

BRONEER, OSCAR T. (97), Swedish-born U.S. archaeologist who discovered the Temple of Poseidon, a Panhellenic shrine dating from the 7th century B.C. The temple had been the site of the Isthmian Games, a national celebration much like the Olympics. d. Ancient Corinth, Greece, Feb. 22.

CAMINOS, RICARDO A. (76), Argentine-born scholar of ancient Egyptian hieroglyphics and writings who taught at Brown University for 28 years and headed its Egyptology department from 1972 until his retirement in 1980. d. London, May 28.

COPLEY, ALFRED L. (81), German-born U.S. physiologist who studied the flowing properties of blood and other body fluids, but who is better known as an abstract artist whose work hangs in the Museum of Modern Art in Manhattan and many international museums. d. New York, N.Y., Jan. 28.

CRILE, GEORGE W., JR. (84), U.S. surgeon who fought against unnecessary surgery, particularly radical mastectomy for breast cancer when a more limited procedure would be just as effective. Although subjected to much criticism by his peers, he was called an "unsung hero" for championing a conservative approach to surgery that now has become the norm in the United States. d. Cleveland, Ohio, Sept. 11.

CRONQUIST, ARTHUR (73), U.S. botanist who developed a means of classifying plant families based on their evolutionary relationships. Known as the "Cronquist System," it is the standard system for studying plant evolution. He also developed a key for identifying flowering plants and ferns that is widely used in field guides. Dr. Cronquist received numerous national and international awards and authored several books, including the *Manual of Vascular Plants of the Northeastern United States and Canada*, known as the "Green Bible." d. Provo, Utah, March 22.

DALES, GEORGE F., JR. (64), U.S. archaeologist who began explorations in Iraq in 1957. More recently he headed the Harappa Project, the only U.S. team authorized by the Pakistani Government to study the Indus ruins where a highly developed civilization existed from about 4000 B.C. to 2000 B.C. d. Berkeley, Calif., April 18.

DONNER, MARTIN W. (71), German-born U.S. radiologist who created a center for the gastrointestinal subspecialty of swallowing disorders. He also founded a medical journal, *Dysphagia*, devoted to the subject. d. Baltimore, Md., April 13.

DRURY, WILLIAM H. (71), U.S. ecologist, former research director of the Massachusetts Audubon Society, and authority on Eastern Seaboard seabirds who helped reintroduce the peregrine falcon in the Northeast and helped return several species of seabirds to the Gulf of Maine. Dr. Drury was an outspoken critic of the indiscriminate use of pesticides. d. Bar Harbor, Maine., March 26.

DUKE, WILLIAM MENG (76), U.S. scientist who, as vice president of Space Technology Laboratories from 1956 through 1962, oversaw the development of the early space program, including the Thor, Atlas, Minuteman, and other projects. d. Santa Monica, Calif., Nov. 8.

EAGLE, HARRY (86), U.S. medical scientist who developed a method of growing cells in a test tube using a mixture of essential compounds that came to be known as "Eagle's Growth Medium." His discovery made possible advanced research on viruses, genes, and cancer. He also discovered the enzyme responsible for blood clotting, and a treatment for arsenic poisoning; discovered a test and treatment for syphilis; described the differences between normal and malignant cells; and was part of a team that employed freeze-drying for long-term storage of serums. He received America's highest scientific honor, the National Medal of Science, in 1987. d. Port Chester, N.Y., June 12.

FEINBERG, GERALD (58), U.S. physicist who in 1958 proposed that there are two kinds of neutrinos, not just one, as was thought at the time. Three of his colleagues confirmed his theory, and were awarded a Nobel Prize in 1962. d. New York, N.Y., April 21.

FLETCHER, GILBERT H. (81), U.S. cancer therapist who developed the cobalt-60 radiotherapy unit and high-voltage linear accelerators, both used to treat cancer. He determined the optimal radiation doses for treating cancer, and was the first to recommend a combination of radiation and surgery. d. Houston, Tex., Jan. 11.

FOX, A. GARDNER (80), U.S. physicist who helped create the first transcontinental microwave system for long-distance telephone calls, and helped develop a basic theory of light behavior that led to the invention of laser devices. d. Harmony, Pa., Nov. 24.

GERSTMAN, LOUIS J. (61), U.S. neuropsychologist who co-invented the talking computer and did extensive research into speech disorders and processes. He was a leading authority on voice tape recordings and "voiceprint" spectrograms. d. Malvern, Pa., March 17.

GIBBS, FREDERIC A. (89), U.S. neurologist who pioneered research in epilepsy and established a laboratory in 1938 to diagnose and treat the disorder. In 1944 he founded the University of Illinois Clinic for Epilepsy, the first of its kind. He also did extensive research with electroencephalographs (EEGs), and was the first to match specific EEG patterns with particular epileptic seizures and other neurological disorders. d. Northbrook, Ill., Oct. 18.

GLASSE, ROBERT M. (63), U.S. anthropologist who in the early 1960s helped determine that *kuru,* a fatal neurological disease found among New Guinea natives, was caused by handling and eating human brain tissue. This project helped win a Nobel Prize for his boss, the chief epidemiologist of the New Guinea Public Health Department. d. New York, N.Y., Jan. 1.

HAMBURGER, JEAN (82), French physician who pioneered and coined the term *nephrology,* founded the International Society of Nephrology, developed one of the first two artificial kidneys, directed the first successful kidney transplant between two non-twins, and successfully transplanted the first kidney from a cadaver. He identified many previously unknown kidney-related disorders and published hundreds of articles and several textbooks. At the time of his death, he was president of the French Academy of Sciences. d. Paris, Feb. 1.

HERRIOTT, ROGER M. (83), U.S. scientist who was the first to suggest that a virus spreads infection by injecting its DNA into host cells. d. Baltimore, Md., March 2.

HILL, GLADWIN (78), U.S. journalist who, as national environmental reporter for the last 10 of his 44 years with *The New York Times,* pioneered the field and helped increase public awareness of pollution and conservation. d. Los Angeles, Calif., Sept. 19.

HOLTFRETER, JOHANNES F. C. (91), German-born U.S. scientist who helped establish embryology as a discipline and invented a medium for growing embryonic cells and tissue in a test tube. d. Rochester, N.Y., Nov. 13.

HOPPER, GRACE M. (85), U.S. mathematician who helped develop modern computer systems and the COBOL language. She coined the term "bug" to describe a mysterious computer problem after a malfunction in an early Mark I computer at Harvard was found to have been caused by a moth. A career naval officer, she had risen to the rank of rear admiral. Admiral Hopper was awarded the National Medal of Technology in September 1991, the first individual woman to receive this honor. d. Arlington, Va., Jan. 1.

HURWITZ, HENRY, JR. (73), U.S. physicist who helped develop the hydrogen bomb, then designed nuclear-power plants and set safety standards that have been adopted internationally. d. Schenectady, N.Y., April 14.

JACOBSON, LEON (81), U.S. physician who, while serving as head of the medical team for the Manhattan Project that developed the atomic bomb, became the first to use chemotherapy to treat cancer. d. Chicago, Ill., Oct. 18.

KESTON, ALBERT S. (80), U.S. chemist who invented Tes-Tape, a glucose-detecting tape that provided an inexpensive and easily used test for diabetes. This led to the identification of millions of previously unknown cases of diabetes. d. New York, N.Y., Feb. 25.

KLERMAN, GERALD L. (63), U.S. psychiatrist who became an expert on depression, schizophrenia, and anxiety disorders. He served as President Carter's chief of the Alcohol, Drug Abuse and Mental Health Administration from 1977 through 1980. He also researched combining drug therapy and psychotherapy, and developed a therapeutic technique known as interpersonal therapy. d. New York, N.Y., April 3.

MAN, EVELYN B. (87), U.S. biochemist who helped develop a test to detect low hormone levels in the thyroid gland, making it possible to treat the disorder and prevent mental retardation. d. West Hartford, Conn., Sept. 3.

MARK, HERMAN F. (96), Austrian-born U.S. chemist who for more than 75 years was a leader in research on polymers, natural or synthetic molecules that are part of plastics, fibers, and many other substances. In 1928 he determined the structure of the natural-polymer molecule, and later helped advance polystyrene and two synthetic-rubber fibers toward commercial use. In 1944 Brooklyn Polytechnic Institute created the Polymer Institute and named Professor Mark director. d. Austin, Tex., April 6.

McCLINTOCK, BARBARA (90), U.S. geneticist considered one of the most influential experts in the field. She discovered that fragments of genetic material move along the chromosome, affecting the genetic control of growth and development. She also identified "crossing over," the breaking and recombining of chromosomes, producing genetic changes. She discovered the chromosomal component that organizes genetic material during cell division. She won the 1983 Nobel Prize in Physiology or Medicine, becoming the first woman to win an unshared prize in that category and only the third woman to win it in science. The Nobel Prize recognized her discovery—made some 40 years earlier—that genetic material is fluid, not fixed, and that DNA fragments move, affecting the activities of the genes. d. Long Island, N.Y., Sept. 2.

NOLAN, THOMAS B. (91), U.S. geologist who directed the course of the U.S. Geological Survey during his 68 years with the agency. As director of the agency from 1956 to 1965, he initiated ambitious projects like mapping the Moon and evaluating the effects of underground nuclear tests. d. Washington, D.C., Aug. 2.

OLDENDORF, WILLIAM H. (67), U.S. physician who helped develop the CAT scan and magnetic-resonance-imaging technology (MRI), but was not included in the 1979 Nobel Prize in Physiology or Medicine awarded for this accomplishment. d. Los Angeles, Calif., Dec. 14.

O'NEILL, GERARD K. (69), U.S. physicist who in 1956 discovered the storage-ring principle for colliding particle beams, the basis for high-energy research. He later advocated colonization of space with self-supporting, solar-powered communities. At the time of his death, he had been working on a high-speed magnetic-transportation system where a car, suspended in a magnetic field, would travel at very high speeds through a special vacuum tube, making it possible to cross the country in less than an hour. d. Redwood, Calif., April 27.

OORT, JAN H. (92), Dutch astronomer who was the first to discover how the Milky Way rotates and the relative velocities of stars in the system. He determined that Earth and the solar system are 30,000 light-years from the center of the galaxy. Known as the "Father of Dutch Astronomy," Oort also proposed that most comets originate from a single area in a remote part of our solar system—an area that has come to be known as the Oort Cloud. d. Leiden, the Netherlands, Nov. 5.

PAGE, ROBERT MORRIS (88), U.S. physicist who developed pulse radar, which sends out pulses of high-frequency electromagnetic radiation to detect or locate the position of a target. He also developed the familiar TV screen with a sweeping beam that indicates both direction and range of an object simultaneously. In the early 1960s, his "Project Madre" radar allowed the United States to "see over the horizon" as part of its early-warning system. d. Edina, Minn., May 14.

PAINE, THOMAS O. (70), U.S. engineer who headed the National Aeronautics and Space Administration (NASA) during the first seven *Apollo* missions, when 20 astronauts orbited the Earth, 14 orbited the Moon, and four walked on its surface. d. Brentwood, Calif., May 5.

PARK, THOMAS (83), U.S. zoologist who helped change ecology from a study of field observations into a science with controlled experiments. He specialized in population ecology, studying how crowding and other factors affect population size and how one species always emerges as dominant to the exclusion—and sometimes extinction—of others. d. Chicago, Ill., March 30.

RAMZY, ISHAK (80), Egyptian-born U.S. psychologist who helped establish the field of child psychology and was one of the founders, and later president, of the International Association of Child Psychoanalysis. d. Topeka, Kans., Feb. 6.

REDISCH, WALTER (94), Czech-born U.S. physiologist who was one of the first to discover the role of microcirculation in vascular disease. He also developed new treatments for gangrene and other vascular disorders, and received numerous awards for his work. d. New York, N.Y., Jan. 11.

ROCKWELL, WILLARD F., JR. (78), U.S. engineer who expanded his father's companies into the modern aerospace conglomerate Rockwell International. His companies manufacture rocket-propulsion and other spaceflight-related components, as well as automotive products and printing presses. d. Pittsburgh, Pa., Sept. 24.

SCHEINER, MARTIN L. (69), U.S. inventor of electronic medical devices, including the first heart monitor for use during surgery, and other monitoring equipment that has since been refined for widespread use in hospital intensive-care units. d. New York, N.Y., Jan. 28.

SELIKOFF, IRVING J. (77), U.S. physician who co-discovered a treatment for tuberculosis and helped discover the health dangers of working with asbestos. Considered a pioneer in environmental and occupational medicine, he established the nation's first hospital division in this specialty. d. Ridgewood, N.J., May 20.

SHEEHAN, JOHN CLARK (76), U.S. chemist who invented a synthetic form of penicillin in 1957 after nine years of research. He also discovered ampicillin, a form of penicillin that can be taken by mouth and is now widely used for infections in children. d. Key Biscayne, Fla., March 21.

SHUGG, CARLETON (92), U.S. engineer who helped invent an underwater-rescue diving bell in 1939, developed a welding technique that replaced riveting in submarine hulls, and in 1955 helped build the first modern nuclear submarine. As deputy general manager of the Atomic Energy Commission in the late 1940s, he appointed then-Captain Hyman Rickover head of the Navy project on nuclear propulsion. d. Mystic, Conn., Jan. 23.

STERN, ARTHUR C. (83), U.S. scientist who in the 1930s was one of the first to identify air pollution as a threat to human health and the global ecology of all living things. He was an international expert on air pollution and led the effort to control and reduce toxic emissions. d. Chapel Hill, N.C., April 17.

STOMMEL, HENRY M. (71), U.S. oceanologist who earned international recognition for his study of Atlantic Ocean circulation. He proposed that the Earth's rotation moves the Gulf Stream along the East Coast of North America, and that there is a balancing southward flow deep beneath it. He theorized that surface water sinks in the north, then flows in a deep current south, while Antarctic water rises to the surface and flows northward. d. Boston, Mass., Jan. 17.

STORMS, HARRISON A. (76), U.S. aeronautical engineer who helped design 48 aircraft and space vehicles, including the B-25 bomber, F-86 fighter, and P-51 Mustang. d. Rancho Palos Verdes, Calif., July 11.

STRONG, JOHN D. (87), U.S. physicist who helped develop the process for coating telescope mirrors with aluminum and applied the technique to the 200-inch-diameter telescope at Mount Palomar in California, then the world's largest. His 1964 claim that he had detected water vapor in the atmosphere of Venus was hotly contested, but his theory was later confirmed. d. Amherst, Mass., March 21.

TOOLAN, HELENE W. (80), U.S. pathologist who in 1960 helped determine that viruses were responsible for eight different varieties of cancer. d. Bennington, Vt., Nov. 29.

WARNER, ROBERT (80), U.S. pediatrician who developed a simple screening test to detect phenylketonuria, a genetic disorder that causes abnormal brain development and mental retardation. Now used on virtually all newborns, the test permits affected babies to be fed a special diet that allows normal development. d. Buffalo, N.Y., May 17.

WEBB, JAMES E. (85), U.S. attorney who headed the National Aeronautics and Space Administration (NASA) during the time of some of its greatest achievements, from the first manned spaceflight to the first walk in space and landing on the Moon. He also oversaw unmanned flights to Venus and Mars and many other technological breakthroughs. d. Washington, D.C., March 27.

WEBER, JOHN JOSEPH (72), U.S. psychiatrist whose early work on schizophrenia was considered a landmark. He received international recognition in the 1950s for his "outcome studies," which tracked the effectiveness of psychoanalysis by following up with the patients. d. Deep River, Conn., April 12.

WESTMAN, JAMES R. (81), U.S. biologist who studied the role of birds and mosquitoes in transmitting a form of encephalitis that can affect human beings. d. Naples, Fla., Feb. 25.

WHITEHEAD, EDWIN C. (72), U.S. industrialist who founded the Whitehead Institute, one of the world's leading biomedical-research institutions. After making a fortune developing scientific and medical equipment, he dedicated himself to financing additional research. d. Greenwich, Conn., Feb. 2.

WICK, GIAN CARLO (82), Italian physicist who in 1951 proposed quantum electrodynamics, a mathematical system for the quantum theory of electromagnetic radiation. He later developed a theory of particle collision that was used by other physicists in the 1960s to analyze particles. d. Turin, Italy, April 20.

WILLETT, HURD C. (89), U.S. meteorologist who helped develop long-range forecasts, and in 1939 established the still-used five-day forecast system used by the U.S. Weather Bureau. He predicted sunspot activity over 20-year cycles and even predicted unusual weather 40 years in advance. d. West Concord, Mass., March 26.

ZYGMUND, ANTONI (91), Polish-born U.S. mathematician who helped describe vibrating objects in mathematical terms—harmonic analysis. In 1986 he was awarded the U.S.'s highest honor, the National Medal of Science, for his work. d. Chicago, Ill., May 30.

Index

R

Radar (electron.)
  satellite imagery 49
Radial steering (mech.) 334
Radiation (phys.) 52
Radio
  electronic lamp 381
  Mars Observer 20
  meteor showers 36
  signal 43–44
  space science 52
Radiotelemetry 41–45, 282
Railroad
  elevated tracks, *illus.* 217
  horsecar 214–20
  technology 332–38
Rainforest Alliance (org.) 97
Ramzy, Ishak (Amer. psych.) 386
Rape (bot.) *see* Canola
Rape (law) 178
Rat (zool.)
  transplant surgery 135, 138–39, *illus.*
    134
Rauvolfia serpentina (bot.) 93
Reaction, Chemical 259–62
Recording *see* Sound recording; Video
    recording
Recycling
  environment 103
  strategic minerals 258
Red-cockaded woodpecker (zool.)
    324
Redisch, Walter (Amer. physiol.)
    386
REM sleep 146
Repetitive stress injury (med.) 365
Reproduction (biol.)
  animal cannibalism 329
  cockroach 319–20
  giraffe 297
  seasonal variation 118
Republican Party (U.S.) 300
Repulsion technology (transportation)
    338
Rescue (firefighting) 343, 346
Reserpine (drug) 93–94
Rhodes (Gr.) 194–95
Rice (grain) 98
Right whale (zool.) 42
Riley, Richard (Amer. pub. offi.), *illus.*
    99
Ring of Fire (volc., Pac.O.) 110
Rio Declaration on Environment and
    Development 60
Rio de Janeiro (Brazil)
  Earth Summit 58–65
Robinson, Doane (Amer. hist.) 210
Robotics 110, 380
Robot probe (spacecraft) 52
Rock (geol.)
  mantle of the earth 105
  meteorite 36, 38–39
  volcano 106
Rockwell, Willard F., Jr (Amer. eng.)
    386
Roentgen Satellite (ROSAT) 51–52
Rome, Ancient
  helmet, *illus.* 253
  topiary 287
ROSAT *see* Roentgen Satellite
Rosy periwinkle (bot.) 92
Rotation of the Earth 26
Roundworm (zool.) 170
Rowing shell (sport) 362
Rubbia, Carlo (It. phys.) 250
Rushmore, Mount *see* Mount Rushmore
    National Memorial
Russia *see also* Union of Soviet Socialist
    Republics
  space program 21, 53
Rutherford, Ernest (Eng. phys.) 248

S

Sacrifice, Human *see* Human sacrifice
SAD *see* Seasonal affective disorder
Safety
  building codes 71, 340–41
  sports equipment 357–58
SAGE (experiment) 277
Sail (naut.) 383
Salam, Abdus (Eng. phys.) 250
SAMPEX *see* Solar Anomalous and
    Magnetospheric Particle Explorer
San Andreas fault (Calif.) 108–9
Sand dollar (zool.) 232
Sandhill crane (zool.) 325
San Diego Zoo (Calif.)
  topiary 286–87, *illus.* 289
Satellite, Artificial
  animal tracking 41–45
  environmental surveillance 102–3
  space photography 47–49
  space shuttle repair and deployment
    54–55
  wave tracking 90
Saudi Arabia, *illus.* 48
Scheiner, Martin L. (Amer. inv.) 386
Scientific fraud 263–69
Scientific Freedom and Responsibility
    Award 269
Scientific method 265
Screw (mech.) 230
Scuba diving 341
Sculpture
  Mount Rushmore 209–13
  prehistoric 222
  wonders of the ancient world 191–
    96
Sea (oceanog.), *illus.* 86
Sea cow (zool.) 43–44
Seal (emblem)
  Great Seal of the United States, *illus.*
    302
Seal (zool.) 45
Sealant (chem.) 212
Sea lion (zool.) 327
Seasonal affective disorder (SAD)
    (psych.) 117–19
Sea turtle (zool.) 44–45
Sea urchin (zool.), *illus.* 229
Security, Computer 353–54
Seiche (oceanog.) 86
Seismic-cycle theory (geol.) 109
Seismology *see* Earthquake
Selikoff, Irving J. (Amer. phy.) 386
Sensor (electron.)
  electronic toll collection 382
  flat tire 382
Sequence (genetics) 124, 126
Serotonin (biochem.) 119
Seven Wonders of the Ancient World
    188–96
Shareware (computer) 378–79
Shark (fish) 329
Sheehan, John Clark (Amer. chem.)
    386
Shift workers 120
Ship, Magnet-driven 241–47
Shipwrecks 89–90, 223
Shugg, Carleton (Amer. eng.) 386
Siamese twins 128–32
Sickle-cell anemia (med.) 169–70
Sierra Club 103
Sign language *see* American Sign
    Language
Silicon chip (computer) 378
Silicone (chem.) 212
Similaun man (anthro.) *see* Iceman
Simpson Paper Company 89
Sirenia (zool.) 281
Skin cancer 169
Skin transplant 139, *illus.* 134
Skyscraper (arch.), *illus.* 341

Sleep
  disorders 120–21
  dreams 141–46
Smokey the Bear 301
Smoking
  disease risks 179
  health-care costs 179
  memory loss 167
  teenage suicide 166
Snakeroot (bot.) 93–94
Snow (meteorol.) 112
Software (computer) 378–79
  virus 354
Soil 312, 315
Solar Anomalous and Magnetospheric
    Particle Explorer 52
Solar eclipse (astron.) 26, *illus.* 27
Solar energy 381
Solar neutrino (phys.) 277
Solar system (astron.) 23, 50 *see also*
    names of planets
Solution (chem.) 275
Sorghum (bot.) 324
Sound, Underwater 86, 107
Sound recording
  ocean sounds 86
Soybean (bot.) 98
Space and Rocket Center, U.S. 29
Space Arc (time capsule) 208
Space camp (Huntsville, Ala.) 28–32
Spacelab-J 53, 55
Space science and technology 52–53
  manned spaceflight 54–55
  Mars spacecraft missions 14–21
  meteoroids 40
  photography 47–49
  planetary colonization 27
  satellite tracking 41–45
  space camp 28–32
  time capsules 208
Space shuttle 53, *illus.* 118
  U.S. manned spaceflights in 1992,
    table 54–55
Space station 53
Spectrometer (instru.)
  Mars Observer 19
  X ray 55
Spectroscope (instru.)
  femtochemistry 260–61
  squeezed light 276
Spider (zool.)
  carnivorous plants 315
  reproduction 329
  venom 96
  web 231, *illus.* 232
Spinnaker (sail) 383
Sports
  high-tech equipment 356–64
Sprague, Robert L. (Amer. psych.) 267–
    68
Spray (oceanog.) 88
Sprinkler system (firefighting) 341
Spurr, Mount (volc., Alas.) 110
Squeezed light (phys.) 276
Stanford Linear Accelerator (Calif.) 251
Star (astron.) 26, 50–51
Star Trek IV: The Voyage Home (film),
    illus. 77
Statement of Forest Principles
    (internat. agreement) 63
Statue of Liberty (N.Y.C.) 195
Stealth technology 243, 246–47
Steel (alloy)
  sports equipment 357–59
  strategic minerals 254
Steller's sea cow (zool.) 281
Stellwagen Bank Sanctuary (Mass.)
    102
Sterling, Bruce (Amer. writ.) 355
Stern, Arthur C. (Amer. sci.) 386
Stillwater Complex
  strategic minerals 254

Acknowledgments

Sources of articles appear below, including those reprinted with the kind permission of publications and organizations.

LIFE ON AN OLDER EARTH, Page 22: Copyright © 1993 by Neil F. Comins. Reprinted by permission of Maria Carvainis Agency, Inc. All rights reserved.

WELCOME TO SPACE CAMP, Page 28: Reprinted with permission; article originally appeared in the June/July 1992 issue of *Air & Space/Smithsonian.*

GOING INTO ORBIT, Page 41: Reprinted from WILDLIFE CONSERVATION Magazine, published by NYZS/The Wildlife Conservation Society.

LOOKING AT EARTH, Page 46: Reprinted with permission; article originally appeared in the August/September 1992 issue of *Air & Space/Smithsonian.*

THE GREENING OF HOLLYWOOD, Page 72: Reprinted by permission of the author; article originally appeared in the March/April 1992 issue of *Audubon,* the magazine of the National Audubon Society.

A WEATHER EYE ON THE WEATHER CHANNEL, Page 79: Reprinted from the April/May 1992 issue of *Weatherwise* magazine by permission of the Helen Dwight Reid Educational Foundation. Copyright © 1992.

THE ECOLOGY OF WAVES, Page 85: Copyright 1992 by the National Wildlife Federation. Reprinted from the August/September 1992 issue of *National Wildlife.*

JUNGLE POTIONS, Page 91: *American Health* © 1992 by Cathy Sears.

THE POWER OF LIGHT, Page 116: Copyright © 1992 by The New York Times Company. Reprinted by permission.

THE TRANSPLANTED SELF, Page 133: Mark Caldwell/ © 1992 *Discover* Magazine.

DIRECTING YOUR DREAMS, Page 141: Excerpted with permission from *Crisis Dreaming: Using Your Dreams to Solve Your Problems* (HarperCollins, 1992).

THE BODY ELECTRIC, Page 147: Carl Zimmer/ © 1993 *Discover* Magazine.

LSD MAKES A COMEBACK, Page 152: Reprinted by permission of Associated Press.

AMERICAN SIGN LANGUAGE, Page 157: Reprinted by permission of the author; article originally appeared in the July 1992 issue of *Smithsonian.*

WHO WAS THE ICEMAN?, Page 197: Reprinted by permission of the author; article originally appeared in the February 1993 issue of *Popular Science* Magazine, copyright © 1993 Times Mirror Magazines, Inc.

HORSE POWER, Page 214: Reprinted by permission of AMERICAN HERITAGE Magazine, a division of Forbes Inc., © Forbes Inc., 1992.

DESIGN BY NATURE, Page 228: Reprinted by permission of the author; article originally appeared in the May/June 1992 issue of *Audubon,* the magazine of the National Audubon Society.

NUCLEAR DETECTIVES, Page 234: Deborah Blum/ © 1993 *Discover* Magazine.

SUPERCONDUCTIVITY GOES TO SEA, Page 241: Reprinted by permission of the author; article originally appeared in the November 1992 issue of *Popular Science* Magazine, copyright © 1992 Times Mirror Magazines, Inc.

THE STRATEGY OF STRATEGIC MINERALS, Page 252: Adapted with permission from "Strategic Minerals" by Peter Harben (**Earth** Magazine, July 1992). Copyright 1992 by Kalmbach Publishing Company.

SCIENTIFIC FRAUD, Page 263: Reprinted by permission of OMNI, © 1992, Omni Publications International, Ltd.

THE MAGIC OF TOPIARY, Page 285: Reprinted by permission of the author; article originally appeared in the March 1993 issue of *Smithsonian.*

FAMILIAR STRANGERS, Page 292: Copyright 1993 by the National Wildlife Federation. Reprinted from the January/February 1993 issue of *International Wildlife.*

PARTY ANIMALS, Page 299: Reprinted by permission of the author; article originally appeared in the November/December 1992 issue of WILDLIFE CONSERVATION Magazine, published by NYZS/The Wildlife Conservation Society.

NO HONKING MATTER, Page 304: Copyright 1993 by the National Wildlife Federation. Reprinted from the December/January 1993 issue of *National Wildlife.*

PLANTS THAT EAT MEAT, Page 309: Reprinted by permission of the author; article originally appeared in the December 1992 issue of *Smithsonian.*

THE CONTEMPTIBLE COCKROACH, Page 317: Reprinted by permission of the author; article originally appeared in the December/January 1993 issue of *National Wildlife*.

FIGHTING TODAY'S FIRES, Page 339: Reprinted by permission of the author; article originally appeared in the May 1992 issue of *Smithsonian*.

TEACHING COMPUTERS TO TELL A "G" FROM A "C", Page 347: Reprinted from December 7, 1992, issue of *Business Week* by special permission, copyright © 1992 by McGraw-Hill, Inc.

HIGH-TECH OLYMPIANS, Page 356: Reprinted with permission from *Technology Review*, copyright 1993.

QUIRKY KEYBOARDS, Page 365: Copyright © 1992 by The New York Times Company. Reprinted by permission.

BIONIC BOTANY, Page 368: Reprinted with permission from *Popular Science* Magazine, copyright © 1992 Times Mirror Magazines, Inc. Distributed by Los Angeles Times Syndicate.

THE RETURN OF THE GEODESIC DOME, Page 374: Reproduced, with permission, from THE FUTURIST, published by the World Future Society, 7910 Woodmont Avenue, Suite 450, Bethesda, Maryland 20814.

Manufacturing Acknowledgments

We wish to thank the following for their services:
Typesetting, Dix Type Inc.; Color Separations, Colotone, Inc.;
Text Stock, printed on S.D. Warren's 60# Somerset Matte;
Cover Materials provided by Holliston Mills, Inc. and Decorative Specialties International, Inc.;
Printing and Binding, R.R. Donnelley & Sons Co.

ILLUSTRATION CREDITS

The following list acknowledges, according to page, the sources of illustrations used in this volume. The credits are listed illustration by illustration — top to bottom, left to right. Where necessary, the name of the photographer or artist has been listed with the source, the two separated by a slash. If two or more illustrations appear on the same page, their credits are separated by semicolons.

399

233 © Claudia Parks/The Stock Market; © Oxford Scientific Films/Animals Animals
234- All photos: Jeffrey Newbury/© 1993 *Discover*
239 *Magazine*
241 © Danilo Ducat/Hankins & Teigenborg
242 Art: © Steve Karp
244 © Dennis Gray; © Mitsuhiro Wada/Gamma-Liaison
245 © Mitsuhiro Wada/Gamma-Liaison
247 © Steve Karp
248 © Photo courtesy of Fermilab
250 © Copyright 1993 Time Inc. Reprinted by permission.
252- © Dan Simonsen/Check Six; helmet: © Erich
253 Lessing/Art Resource
255 Photo: © Cameramann International, Ltd.; map by George Stewart
256 Photo: © Nance S. Trueworthy/Gamma-Liaison; map by George Stewart
257 Photo: © Kenneth Garrett/Woodfin Camp & Assoc.; map by George Stewart
258 Photo: © Cameramann International, Ltd.; map by George Stewart
259 Photos courtesy of Dr. Ahmed H. Zewail/California Institute of Technology
260- Photo: Eadweard Muybridge, Courtesy, George
261 Eastman House; art by George Stewart
263 © Brad Holland
264 © Professor Luc Montagnier/Institut Pasteur/CNRI/Science Photo Library/Photo Researchers; © Shepard Sherbell; © J. Andanson/Sygma
265 © The Bettmann Archive
266 © The Bettmann Archive
267 © The Granger Collection
268 © The Bettmann Archive
269 © The Granger Collection
271 The Bettmann Archive; UPI/Bettmann
273 © 1992 by The New York Times Company. Reprinted by permission.
274 Art by Laurie Grace
275 © Armel Brucelle/Sygma; © Kevin Argue/Gamma-Liaison
276 Photo courtesy of California Institute of Technology
278- © Mitch Reardon/Tony Stone Images; inset:
279 © Color Box/FPG International
280 © Douglas Faulkner/Photo Researchers; © Marilyn Kazmers/SharkSong
281 © Jeff Foott
282 © Marilyn Kazmers/SharkSong; © Douglas Faulkner/Photo Researchers

283 © Pat Canova/Southern Stock Photos
284 © Timothy O'Keefe/Southern Stock Photos
285 © Chad Slattery
286 © Chad Slattery; © Lee Snider/Photo Images
287- © Chad Slattery
290
291 © Lee Snider/Photo Images
292 © Mark N. Boulton/Photo Researchers
293 Both photos: © Stephen J. Krasemann
294 © Gregory G. Dimijian/Photo Researchers
295 © Frans Lanting/Minden Pictures
296 © Renee Lynn; © Rita Summers
297 © Tim Davis
298 © Renee Lynn/Photo Researchers
299 © Glenn Wolff
300 Both illustrations: The Granger Collection
301 © Glenn Wolff
303 The Granger Collection; © 1982, The Christian Science Publishing Society
304 © Stephen J. Krasemann/DRK Photo
305 © Scott Camazine
306 © J.M. Mejuto; © Joe McDonald/Animals Animals
307 © Marilyn Wood/Photo/NATS; © Stephen J. Krasemann/DRK Photo
308 © Link/Visuals Unlimited
309 All photos: © Robert & Linda Mitchell
310 Both photos: © Patti Murray/Animals Animals
311 Both photos: © Nuridsany et Perennou/Photo Researchers
313 © Nuridsany et Perennou/Photo Researchers; © Robert & Linda Mitchell
314- From left: © Kathie Atkinson/Oxford Scientific
315 Films/Animals Animals/Earth Scenes; © Robert Sisson; © Robert & Linda Mitchell; © Robert & Linda Mitchell
317 © Oxford Scientific Films/Animals Animals
318 © Holt Studios/Animals Animals
319 Top two photos: © James L. Castner; bottom: © Scott Camazine
320 © Ken Kerbs/Dot
321 © James L. Castner
322 Courtesy Calgene Fresh, Inc.
323 © Ray F. Evert, University of Wisconsin, Madison
325 © Pat & Tom Leeson/Photo Researchers
326 © D. Clendenen/U.S. Fish & Wildlife Service
327- All art: © 1992 by The New York Times
329 Company. Reprinted by permission.

330- © Ron Sandford/Allstock; inset: © David
331 Parker/SPL/Photo Researchers
332- Crossover art: © Marc Ericksen; inset: © Ian
333 Worpole
334 © Ian Worpole
335 Photo: Courtesy, AMTRAK; art: © Ian Worpole
337 Courtesy, Metro North Commuter Railroad; courtesy, Los Angeles DOT
339 © Jack Van Antwerp/The Stock Market; inset: © J. Barry O'Rourke/The Stock Market
340 © John Mantel/Sipa
341 © AP/Wide World
342 © P. Howell/Gamma-Liaison
343 © Bob Daemmrich/Stock Boston
345 Both photos: © Layne Kennedy
347 Adapted from the December 7, 1992, issue of *Business Week* by special permission, copyright © 1992 by McGraw-Hill, Inc.
348 Both photos: © Dennis Howland/Ogden IRS Center, Utah
349 © U.S. Postal Service
350- Art by Joseph Van Severen
354
356- © Ronald C. Modra/*Sports Illustrated;* © UPI/
357 Bettmann
358 © David Leah/Allsport; © Yann Guichaqua/Vandystadt/Allsport; © UPI/Bettmann
359 © Yann Guichaqua/Vandystadt/Allsport
360 © Peter Read Miller/*Sports Illustrated;* © UPI/Bettmann
361 © David Gannon/Allsport; © AP/Wide World
363 © AP/Wide World; © Richard Martin/Vandystadt/Allsport
365 © Kathryn J. MacDonald
366 © NYT Pictures
367 © The BackCare Corporation; © NYT Pictures
368 © Andrew Washnik/The Stock Market
370 Both photos courtesy of the Monsanto Company
371 Both photos: © Scott Bauer/Agricultural Research Service
373 Art by Laurie Grace
374 © Melinda Holden/Oregon Dome
375 Oregon Dome
377 Oregon Dome; Domes America
379 Photo courtesy of Prodigy Services Company
380 © William A. Cotton/Colorado State University
382 © 1992 by The New York Times Company. Reprinted by permission.

400